T0336930

Autowave Processes in Kinetic Systems

Mathematics and Its Applications *(Soviet Series)*

Autowave Processes in Kinetic Systems

Spatial and Temporal Self-Organization in Physics, Chemistry, Biology, and Medicine

V. A. Vasiliev, Yu. M. Romanovskii,
D. S. Chernavskii, V. G. Yakhno

with 113 figures

D. Reidel Publishing Company

A MEMBER OF THE KLUWER ACADEMIC PUBLISHERS GROUP 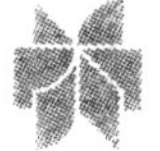

Dordrecht / Boston / Lancaster / Tokyo

Library of Congress Cataloging-in-Publication Data

Autowave processes in kinetic systems.

(Mathematics and its applications. Soviet series)
Translated from the Russian.
"Published in 1987 by VEB Deutscher Verlag der Wissenschaften, Berlin in co-edition with D. Reidel Publishing Company, Dordrecht, Holland" — T.p. verso.
Bibliography: p.
Includes index.
1. Self-organizing systems — Congresses.
I. Vasil'ev, V. A. II. Series: Mathematics and its applications (D. Reidel Publishing Company). Soviet series.
Q325.A89 1987 003 86-24879
ISBN 90-277-2379-6

Distributors for the Socialist Countries
VEB Deutscher Verlag der Wissenschaften, Berlin

Distributors for the U.S.A. and Canada
Kluwer Academic Publishers,
101 Philip Drive, Norwell, MA 02061, U.S.A.

Distributors for all remaining countries
Kluwer Academic Publishers Group,
P.O. Box 322, 3300 AH Dordrecht, Holland

Published in 1987 by VEB Deutscher Verlag der Wissenschaften, Berlin in co-edition with D. Reidel Publishing Company, Dordrecht, Holland.

Printed in the German Democratic Republic.

"... In the complex field of nonlinear oscillations, some specific general concepts, ideas, and methods will crystallize and be adopted for general use by the physicist to an even greater extent than is the case today. These concepts, ideas, and methods will become obvious and natural, and they will enable the physicist to analyze complicated sets of phenomena. They will become a powerful heuristic tool for new research.

The physicist interested in the modern problems of oscillations must, in my opinion, be taking this approach right now. He must understand the existing mathematical methods and approaches which are at the basis of these problems, and he must learn how to use them."

L. I. Mandelshtam

(Preface to the book "Theory of Oscillations", by A. A. Andronov, A. A. Witt, and S. E. Khaikin, published in 1935.)

Preface

Probably, we are obliged to Science,
more than to any other field
of the human activity,
for the origin of our sense
that collective efforts are necessary indeed.

F. Joliot-Curie

The study of autowave processes is a young science. Its basic concepts and methods are still in the process of formation, and the field of its applications to various domains of natural sciences is expanding continuously.

Spectacular examples of various autowave processes are observed experimentally in numerous laboratories of quite different orientations, dealing with investigations in physics, chemistry and biology. It is our opinion, however, that if a history of the discovery of autowaves will be written some day its author should surely mention three fundamental phenomena which were the sources of the domain in view. We mean combustion and phase transition waves, waves in chemical reactors where oxidation–reduction processes take place, and propagation of excitations in nerve fibres. The main tools of the theory of autowave processes are various methods used for investigating nonlinear discrete or distributed oscillating systems, the mathematical theory of nonlinear parabolic differential equations, and methods of the theory of finite automata. It is noteworthy that the theory of autowaves has been greatly contributed to be work of brilliant mathematicians who anticipated the experimental discoveries in their abstract studies. One should mention R. Fisher (1937), A. N. Kolmogorov, G. I. Petrovskii, and N. S. Piskunov (1937), N. Wiener and A. Rosenbluth (1946), A. Turing (1952).

Remarkably, these venerable mathematicians turned to problems in biology; the contemplation of mechanisms of life promited them to state new ideas in mathematics. Their success in this field was due, in essence, to their understanding of two fundamental factors acting simultaneously in distributed active systems: autocatalysis and diffusion.

An important milestone in the progress of the theory of autowave processes was the invention of models for nervous conductivity, by A. Hodgkin and A. Huxley, and a model for the propagation of excitations in conducting system of heart muscle by D. Noble. These models — systems of partial differential equations — described not mental, but real experiments.

We have a right to claim that the Soviet school incorporating mathematicians, physicists, chemists and biologists is highly experienced in the study of autowave processes, its contribution to the subject is an essential one. So it seems fairly natural that the term itself, "autowave process" or "autowave", was proposed by an outstanding Soviet physicist R. V. Khokhlov who possessed a deep understanding of the role of the theory of nonlinear oscillations, not only in electronics and mechanics, but also in other sciences. Soviet scientists have contributed substantially, in particular, to the theories of combustion, formation of the universe structures, wave propagation and formation of strata in plasmas, chemical reactors, living tissues, and to the theory of signal transport in active lines.

During recent years, the role of the concept of autowave processes has become clearer in the development of synergetics, the study of open systems remote from thermodynamical equilibrium. It would be reasonable to assert that autowaves, besides the processes creating "new" information in such systems, are the mainsprings of self-organization. An understanding of this profound relation has been inspired to an appreciable extent by the work of I. Prigogine and H. Haken. Probably the most exhaustive exposition of the problems relevant to the history and recent developments of the study of autowave processes was presented at two International Symposia which took place at the Humboldt University, GDR (January, 1982) and in Pushchino, USSR (June, 1983). Nevertheless, a presentation of fundamental experimental facts and the main concepts of the theory of autowave processes has embarassed the present authors in a number of points. Among the difficulties we met were the absence of a commonly adopted terminology, an uncertainty in determining the boundaries for the new field of science (as it is the case with any new trend originating at frontiers of different sciences), and the multitude of publications which are scattered in numerous books and a great number of various periodicals. Above all, the authors are physicists, so the exposition of mathematical aspects is presented at a "physical" level of rigor.

The authors emphasize that their book is essentially a collective work; each author is responsible for all chapter. The presentation of the Material would be impossible without preliminary discussions with many colleagues.

The authors want to express their profound gratitude for goodwill and encouragement and for discussion of a number of problems to S. A. Akhmanov, M. T. Grekhova, Yu. L. Klimontovich, E. V. Astashkina, H. Bartsch, S. I. Beylina, B. N. Belintsev, Yu. M. Galperin, Yu. A. Danilov, S. D. Drendel, G. G. Elenin, Yu. D. Kalafati, B. N. Klochkov, I. K. Kostin, V. I. Krinskii, S. P. Kurdyumov, K. K. Likharev, A. S. Logginov, I. I. Minakova, A. S. Mikhailov, G. A. Nepokoichitskii, M. S. Polyakova, A. V. Priezzhev, M. I. Rabinovich, S. A. Regirer, N. V. Stepanova, V. A. Teplov, I. V. Uporov, R. Feistel and others. The authors are extremely thankful to A. V. Kraev and L. Schimansky-Geier for their kind assistance in the translation of the book. We also greatly appreciate help rendered by L. M. Andrianova and N. I. Vasileva in arranging the manuscript.

Finally, we are very grateful to Professor Werner Ebeling of the Humboldt University Berlin who initiated the composition of the present book. Without his constant attention and support the text would never have been completed to the extent of the authors' ability.

Editor's Preface

Approach your problems from the right end and begin with the answers. Then one day, perhaps you will find the final question.

'The Hermit Clad in Crane Feathers' in R. van Gulik's *The Chinese Maze Murders*.

It isn't that they can't see the solution. It is that they can't see the problem.

G. K. Chesterton. *The Scandal of Father Brown* 'The point of a Pin'.

Growing specialization and diversification have brought a host of monographs and textbooks on increasingly specialized topics. However, the "tree" of knowledge of mathematics and related fields does not grow only by putting forth new branches. It also happens, quite often in fact, that branches which were thought to be completely disparate are suddenly seen to be related.

Further, the kind and level of sophistication of mathematics applied in various sciences has changed drastically in recent years: measure theory is used (non-trivially) in regional and theoretical economics; algebraic geometry interacts with physics; the Minkowsky lemma, coding theory and the structure of water meet one another in packing and covering theory; quantum fields, crystal defects and mathematical programming profit from homotopy theory; Lie algebras are relevant to filtering; and prediction and electrical engineering can use Stein spaces. And in addition to this there are such new emerging subdisciplines as "experimental mathematics", "CFD", "completely integrable systems", "chaos, synergetics and large-scale order", which are almost impossible to fit into the existing classification schemes. They draw upon widely different sections of mathematics. This programme, Mathematics and Its Applications, is devoted to new emerging (sub)disciplines and to such (new)interrelations as exempla gratia:

- a central concept which plays an important role in several different mathematical and/or scientific specialized areas;
- new applications of the results and ideas from one area of scientific endeavour into another;
- influences which the results, problems and concepts of one field of enquiry have and have had on the development of another.

The Mathematics and Its Applications programme tries to make available a careful selection of books which fit the philosophy outlined above. With such books, which are stimulating rather than definitive, intriguing rather than encyclopaedic, we hope to contribute something towards better communication among the practitioners in diversified fields.

Because of the wealth of scholarly research being undertaken in the Soviet Union, Eastern Europe, and Japan, it was decided to devote special attention to the work emanating from these particular regions. Thus it was decided to start three regional series under the umbrella of the main MIA programme.

The phrase "Autowave processes" is perhaps not yet well and universally known. The subtitle of the book "Spatial and temporal self-organisation in physics, biology and medicine" is perhaps an easier understandable description of the subject of this book. Mathematically speaking we are dealing with evolution equations which have both autocatalytic and diffusion terms, as follows

$$\frac{\partial u_i}{\partial t} = f_i(u_1, \ldots, u_m) - \sum_{l=1}^{n} \frac{\partial}{\partial x_l} (v_{il} u_i) + \sum_{l,k} \frac{\partial}{\partial x_l} \left(D_{ik} \frac{\partial u_k}{\partial x_l} \right)$$

They are also called (generalized) reaction-diffusion equations. Here $u_1, \ldots, u_m$ are e.g. m substances, $f_i(u_1, \ldots, u_m)$ is a "reaction term" which may embody autocatalytic phenomena (think of an equation like $y^{\cdot} = y^2$, $y(0) = c^{-1}$, $c \neq 0$, with the finite escape time solution $y(t) = (c - t)^{-1}$), v_{il} is the l-th component of the velocity of substance i, and the D_{ik} are diffusion coefficients which are positive functions of x, t and which often are assumed to be constant. The reaction terms $f_i(u)$ may give rise to steep gradients, shock waves, etc., a tendency which is balanced by the diffusion terms. As a result there may arise phenomena of pattern generation such as stationary spatially inhomogeneous stable solutions. In addition, such equations can exhibit localized wave sources, reverberation phenomena, travelling shape preserving solitary waves of a ompletely different kind than solitons, and dissipative structures. The equations often arise as mass balance equations

$$\frac{\partial u_i}{\partial t} = f_i(u) - \operatorname{div} I_i$$

where I_i is the flux of substance i, $I_i = v_i u_i \Sigma D_{ik} \operatorname{grad} u_k$, and they serve e.g. as equations describing nerve impulse transmission (Hodgkin-Huxley model), a phase transition wave, a chemical reaction process (e.g. the oscillating Belousov-Zhabotinskii reaction) combustion processes, and the evolution of a multispecies population with random migration. Quite generally they occur in connection with all kinds of processes in excitable media.

The term "autowave processes" for all these (and other) phenomena was coined by the USSR physicist R.V. Khokhlov. There are definite and important relations between these autowaves and ideas from synergetics and self-organisation. As indicated above, autowaves occur in medicine, physics, biology, chemistry; they also occur in mechanics and electronics. Thus the field of autowaves is not so easy to delimit and describe. Mathematically, we just have to study the equation type given above. Thus one way to view the field is a part of applied mathematics with lots of applications in lots of different fields and with lots of experimental facts to guide intuition and to help one's thinking about these equations. A topic therefore which fits the underlying philosophy of the book series perfectly. I hope that — like me — many readers will find it an enormously stimulating introduction to the fascinating world of self-organisation and pattern formation.

The unreasonable effectiveness of mathematics in science ...

Eugene Wigner

Well, if you know of a better 'ole, go to it.

Bruce Bairnsfather

What is now proved was once only imagined.

William Blake

As long as algebra and geometry proceeded along separate paths, their advance was slow and their applications limited.

But when these sciences joined company they drew from each other fresh vitality and thenceforward marched on at a rapid pace towards perfection.

Joseph Louis Lagrange.

Bussum, October 1986

Michiel Hazewinkel

Contents

Chapter 1
Autowave processes and their role in natural sciences

Even if there is no direct relation between sciences, they illuminate each other with analogies.

H. Poincaré

1.1 Autowaves in non-equilibrium systems

In the progress of natural sciences there are stages when different domains, seemingly quite separate, are united by common ideas and methods. Such a stimulating co-operation of sciences has always been fruitful for the domains involved. A remarkable feature of modern developments is an extensive penetration of mathematics, as well as various methods of experimental and theoretical physics, into chemistry and biology. The process of contact between different sciences has always led to the consolidation of common concepts and regularities.

One of the common problems arising in the study of nature, both animate and inanimate, is creating a language appropriate for the description of the dynamics of distributed systems consisting of many interacting elements in states far from thermodynamical equilibrium. Such systems are quite manifold. The relevant examples are semiconductor and magnetic films, chemical reactors where processes may occur with autocatalysis at some stages, extended communities of living organisms, nerve fibres, aggregates of neurons constituting nerve tissues, and many others.

Needless to say, all these complicated systems have been under investigation for a long time, and there are interesting approaches to their space–time dynamics. However, at present a possibility has appeared of proposing a classification of the dynamical processes in terms of so-called specific characteristic structures engendered in non-linear media. The concept of specific nonlinear structures is now commonly adopted in physics. Regular and stochastic self-excited oscillations, and solitons are examples of the structures in view. A new contribution to the list of nonlinear structures is the multitude of autowave processes. By the definition adopted, an autowave process (AWP) is a self-sustained wave process (including stationary structures) in an active nonlinear medium maintaining its characteristics at a constant level at the expense of an energy source distributed in the medium. These characteristics (period, wave length or momentum, propagation velocity, amplitude and shape) of a steady state depend only on local properties of the medium and are independent of initial data and boundary conditions, provided that the boundary of the medium is sufficiently remote. It is assumed that space regions communicate by means of transport processes of the diffusion type. At present, with numerous works devoted to the study of autowaves, it is clear that AWPs constitute an adequate basic concept providing a general mathematical language enabling one to describe similar types of space–time phenomena in active media quite different in their nature. Properties of nonlinear structures, known as autowaves, are the main subject of the present book. The purpose of Chapters 1 and 2 is to present a classification of various AWPs, compile the most important experimental facts from physics, chemistry and biology which are relevant to AWPs, and to expound physical and mathematical methods used in constructing "basic models" of the phenomena considered. Chapter 3 deals with mathematical methods of the analysis of the basic models. The most important types of autowave phenomena are considered in Chapters 4 to 7. Stochastic phenomena related to AWPs are considered in Chapter 8 and Chapter 6 (Section 6.3). Finally, the last Chapter 9 describes AWPs in living channels.

As was mentioned in the preface, in the present exposition we consider some problems akin to the domain which is now called self-organization theory or "synergetics". In our opinion, an AWP is one of the effects specific for self-organization which occurs in distributed open systems far from thermodynamical equilibrium. Without the creation of such structures and the appearance of inhomogeneities (gradients) in non-equilibrium media, a competitive selection is impossible in pre-animate and living systems. It is owing to the competitive selection that the creation of new "useful" information is possible, which is a necessary condition of biological evolution in general (Eigen and Schuster 1979; Romanovskii, Stepanova, Chernavskii 1975, 1984, Shnol 1979, Ebeling and Feistel 1982; Ebeling and Klimontovich 1984).

There is no doubt that the principles of synergetics including the theory of AWP was founded on remarkable achievements of modern statistical physics of non-equilibrium processes. It is this field of physics that secures the understanding of synergetics processes at a level of interactions between atoms or molecules and grounds the next kinetic level of descriptions that involve mean values (for example, concentrations of reacting molecules) as variables. To look through a modern statistical theory of non-equilibrium processes readers can exploit books by Bogolyubov (1946), Prigogine (1964), Lifshits and Pitaevsky (1978), Zubarev (1971), Balescu (1975), Klimontovich (1982), Stratonovich (1985), as well as Springer-Verlag-books "Synergetics", printed since 1976.

We must admit from the very beginning that the qualitative description of AWPs and methods of their study presented in this book does not pretend to be an exhaustive exposition of the problems arising in the investigation of manifold self-organization phenomena. Nevertheless, it is the authors' hope that the book will be useful to those who study autowaves in various specific media, and help investigators in the construction of adequate mathematical models.

1.2 Mathematical model of an autowave system

We shall formulate a general mathematical model describing processes in an active distributed kinetic system.

Let x_i be a set of interacting components. For example, in physical problems the quantities may be temperature, an ion density, or concentration of a substance in superconducting state and in chemical and biological problems concentrations of reagents and/or numbers of some living objects per unit volume, area or length. One should have in mind that sometimes physical, chemical and biological factors are coupled together in a unified model which may be rather complicated. The kinetic equations accounting for interactions of components x_i and describing the transport phenomena are quasi-linear parabolic equations.

The models are usually based upon the substance balance equations

$$\frac{\partial x_i}{\partial t} = F_i(x_1, x_2, \ldots, x_n) - \operatorname{div} \boldsymbol{I}_i. \tag{1.1}$$

Here $\boldsymbol{I}_i$ is the flux of the i-th component,

$$\boldsymbol{I}_i = \boldsymbol{v} x_i - \sum_{k=1}^{n} D_{ik} \operatorname{grad} x_k, \tag{1.2}$$

where $\boldsymbol{v}$ is a directed velocity of the component, and D_{ik} is the diffusion coefficient matrix. Its off-diagonal elements are called coefficients of mutual diffusion, as they describe fluxes of the i-th "substance" induced by fluxes of the k-th substances. It was found (Haase 1967) that mutual diffusion is absent only if all the proper diffusion coefficients D_{ii} are equal. Sometimes mutual diffusion can be neglected, but there are cases in chemistry (Haase 1967) and biology (Lightfoot 1977) where induced fluxes are comparable with proper diffusion.[1])

Equations in form (1.1) describe a very wide range of phenomena: turbulence in the making, chemical kinetics under the condition of convection and so on. The purpose raised in the present volume is to show properties of "a simplified model", namely, the model (1.1) under $v_i \equiv 0$. However, as it will be shown in next chapters, even the simplified model gives possibilities to describe and to understand a wide range of objects and processes.

In the simplest case of one-dimensional space, Eqs. (1.1) and (1.2) become

$$\frac{\partial x_i}{\partial t} = F_i(x_1, x_2, \ldots, x_n) + \frac{\partial}{\partial r}\left(\sum_{k=1}^{n} D_{ik}(x_1, x_2, \ldots, x_n)\, \frac{\partial x_i}{\partial r}\right). \tag{1.3}$$

Here, as in (1.1), F_i are nonlinear functions describing local interactions of the components. In some cases we deal with more general models taking into account active fluctuations, or other space coordinates. However, we will concentrate mainly on systems described by Eq. (1.3) where we put $D_{ii} = \text{const}$, $D_{ik} = 0$ for $i \neq k$. The boundary conditions for Eqs. (1.1) to (1.3) are determined by the problem we are concerned with, but the problem arising more frequently is that of a finite segment $[0, L]$ with impermeable boundaries,

$$\left.\frac{\partial x_i}{\partial t}\right|_{\substack{r=0\\ r=L}} = 0. \tag{1.4}$$

---

[1]) Properties of the diffusion matrix D_{ik}, and the reduction of the AWP models of physical, chemical and biological systems to Eqs. (1.1), (1.2) are considered in more detail in Chapter 2.

The system is most autonomous under these conditions, and an AWP is less sensitive to boundary effects. And above all, these conditions are specific for an ample class of systems.

If mixing within the region $[0, L]$ is fast enough, processes are synchronous at every point, and the system is described by the so-called local equations

$$\frac{dx_i}{dt} = F_i(x_1, x_2, \ldots, x_n). \tag{1.5}$$

Eqs. (1.3) are reduced to (1.5) if the diffusion is very fast, i.e. $D_{ik} \to \infty$, while the system size L and other parameters are fixed.

As a matter of physics, it signifies that the transition to Eq. (1.5) corresponds to a zero approximation with respect to the ratio of characteristic diffusion time and characteristic time of chemical order kinetic processes.

One of the main mathematical indications that the distributed kinetic system is active, and that its state is far from thermodynamical equilibrium, is the existence of unstable singularities in the point model (1.5). Note that the concept of "activity" is extended for a distributed system if $D_{ik} \neq 0$ for $i \neq k$ (more details are given in Chapter 3).

In the next Chapter we present and justify basic models for a number of important AWPs which are particular cases of the general model of Eqs. (1.3). In Section 1.3 we give a classification of essential AWPs; a table is presented containing a compilation of experimental data on AWPs in physical, chemical and biological systems. In conclusion, in Section 1.4, we discuss the most interesting experiments with autowaves.

1.3 Classification of autowave processes

Experience gained by the theory of AWPs enables one to indicate specific types of the phenomena observed. That is to say, one can determine the basic set of nonlinear structures for a non-equilibrium dissipative distributed system. Using these structures, or corresponding images, it is possible to describe complicated motions and to formulate assumptions on the conditions leading to a certain regime.

The following specific structures appear in homogeneous active media.

1. Propagation of a solitary excitation front (motion of a phase transition interface, flip-over front) or travelling front (TF).
2. Propagation of a pulse having a stable shape (travelling pulse — TP).
3. Autonomous localized wave sources: the "echo" regime and a stable leading centre (LC).
4. Standing waves.
5. Reverberation.

6. Synchronized self- or auto-oscillations in space (SO).
7. Quasistochastic waves.
8. Steady-state inhomogeneous spatial distributions — dissipative structures (DS).[1]

This classification is based upon evident physical ideas; its mathematical justification stems from the existence of special auto-model variables in terms of which the variable space dimension of the original model is decreased substantially with the auto-model variables, the structures in view are represented by special trajectories connecting singular points, or limit cycles in the phase space or configuration space. The possibility to indicate such trajectories is related to the fact that they are either attracting, or separating. This property is reflected in the equivalent term "intermediate asymptotics" (Barenblatt and Zeldovich 1971).

The structures in nonlinear dissipative media (or AWPs) may have some features in common with the structures occurring in nonlinear conservative media which are usually called solitons. However, the physical nature and the behaviour of these structures are quite different. For instance, colliding solitons penetrate through each other (cf. e.g. "Solitons in Action" 1978, Scott, Chew and McLaughlin 1973), whereas collisions of travelling pulses or travelling fronts result in their annihilation. The shape and velocity of a steady-state travelling pulse are independent of the initial conditions within a wide range, while the shape and velocity of a soliton are determined completely by the initial excitation. In general, the above features of AWPs are specific for their phenomenology.

1.4 Basic experimental data

At present we know a lot of AWPs taking place in active media of various natures. An attempt to perform a satisfactory classification of the processes starting from the type of the active medium (biological, physical etc.) would be rather unsuccessful. Therefore we try the following composition in this Section. First, we point to basic types of media having different character or nature of the local positive feedback which is responsible for their autocatalytic properties. Second, we describe the most spectacular examples of AWPs in these media. Third, we present summary tables of the experimentally observed AWPs with the reference to their relevance to chemistry, biology etc.

1. A large class of AW media is described by the following scheme. A substance enriched in energy is injected into an open distributed system. The input flux is controlled by local properties of the system surface or, to be more precise, of a thin boundary layer. In turn, the local properties of the surface depend on temperature, potential and other waves propagating in a thin boundary layer, as well as on processes

[1]) Let us note that dissipative structures of term are understood in the more broad sense as compared with our book. For example, processes of auto-oscillations and so on are related to dissipative structures by G. Nicolis and I. Prigogine (1977). In our consideration this term is used only in narrow sense as stated above.

which occur in the substrate. This is the case for autowaves in bio-membranes (see Chapter 2, Section 2.1), ammonia and carbon monoxide oxidation waves on platinum surface, propagation of "normal" zones in homogeneous superconductor wire etc.

2. In another class of autowave media there are no manifest surface effects. The local positive feedback is due to an N-shaped characteristic of the medium, having a segment of a "negative" resistance for any elementary volume. Space–time structures in such media are, for instance, Belousov-Zhabotinskii auto-oscillating reactions, strata in gas discharges, strata or domains in semiconductor electron–hole plasmas etc.

3. In nature and technology one sometimes encounters complicated multi-phase media where a non-equilibrium state and AWPs are maintained by an energy flux from laser radiation, thermochemical reactions and other external sources. Phenomena of this type are generated and controlled not only by diffusion and heat conduction, but also by hydrodynamical currents (in particular, convection), evaporation, boiling, surface tension. Here we mention some work along this line. Dynamics of processes and generation of structures induced by laser radiations have been considered by Bunkin, Kirichenko and Lukyanchuk (1982). Generation of structures related to surface phenomena has been described by Kahrig and Besserdich (1977). (See also work by Linde 1983.) Interaction of convective currents with chemical reactions was studied by Ibanez and Velarde (1977). Another relevant phenomenon is boiling of a liquid on a solid heat exchange surface. There may be two different physical processes, nucleate boiling and film boiling. A study of sufficiently long fuel elements (FEs) have shown that the transitions from one regime to the other do not happen simultaneously over the whole surface; a new regime appears first on a local area of FE surface and then replaces the preceding regime by means of wave propagation. It is known that film boiling results in a drastic overheating of FEs, power output from FEs is decreased substantially in order to prevent this regime. There are some data indicating that if one manages to control nucleate–film transitions the power output from FE may be raised considerably (Zhukov, Barelko, and Bokova 1981).

Of course, mathematical models of all the phenomena mentioned are not given exactly by Eqs. (1.1), (1.2). However, one can make their analysis more simple at certain stages if it is possible to use physical concepts and methods developed in the theory of AWPs in homogeneous media with "pure" diffusion. The same is true also for autowave arising in living tissues at morphogenesis.

Now we review concrete experimental data on AWPs.

A. Propagation of phase transition fronts and creation of structures in solids

Propagation of fronts and creation of structures in solids where phase transitions are possible have not been investigated sufficiently. Nevertheless, numerous AWPs have been observed in many media: superconductors, semiconductor and single-crystal films etc. Relevant information can be found in a review by Balkarey, Nikulin and Elinson (1981). We present below only some more recent examples.

Materials in which semiconductor–metal phase transitions occur are very interesting. This phase transition is associated with a drastic change in electrical and optical properties, such material are thus promising for use in optoelectronics. Phase transition wave patterns in VO_2 films are shown in Fig. 1.1 (see Gulyaev, Kalafati et al. 1981).

AWPs are also observable in optically transparent single-crystal magnetic films. Such films are used in the construction of memory units of the magnetic-domain type. In the films an AWP appears in a travelling front; it is a propagation of the domain walls. We should mention the following interesting experimental results. Logginov and Nepokoichitskii (1982) have discovered a propagation of magnetic moment flipping wave in pulse gradient magnetic fields. Velocities of such waves are considerably higher than the minimum phase velocity of spin waves and amount to 6×10^4 m/s. In work by Ivanov, Logginov and Nepokoichitskii (1983) the authors reported that if the pulse magnetic field strength is above a critical level

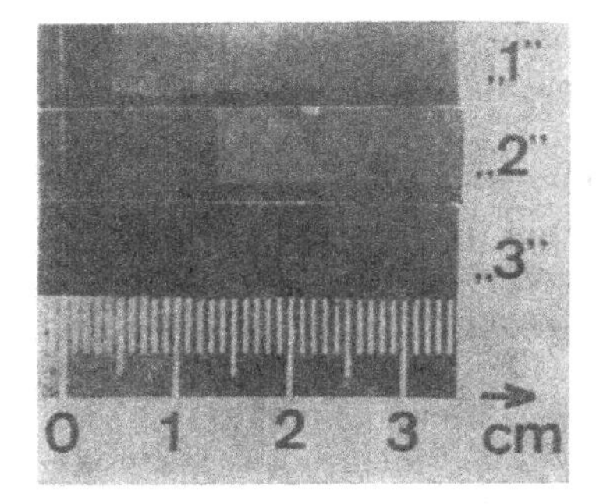

Fig. 1.1
The propagation of TF of the metal—semiconductor phase transition in VO_2 film which coincides with TF of a heat wave. The interval between successive frames is 2 s. (Kalafati, Serbinov, Ryabova 1979)

magnetic excitations appear ahead of a travelling domain boundary, and their velocity is higher than that of the domain boundary. The excitations fuse and microdomains are created; then the latter fuse with the initial domains. The theory of AWPs in magnetic films is at present in its initial stage (cf. e.g. Baryakhtar, Ivanov and Sukstanskii 1978, Eleonskii, Kirova and Kulagina 1978).

The next example of an AWP is relevant to applications of superconductivity. It is the propagation of normal metal zones in a homogeneous superconductor wire (Altov et al. 1975, Lvovsky and Lutsev 1979). In the first approximation, the state of a cooled superconductor is described by the one-dimensional heat excitation equation where the source function is the difference between the heat production per unit length and the heat release from the surface. The temperature dependence of the source function is N-shaped, so two stable stationary states are possible: superconductor and the normal metal. The heat transfer in the wire is responsible for autowave transitions between these states. We are not in a position to dwell on details of the phenomena observed or their analysis; it is sufficient to emphasize a deep analogy with autowaves appearing in other areas: combustion, heat production on surfaces of boiling liquids, heterogeneous catalysis etc. Unfortunately, until now these analogies were beyond the scope of applied investigations of superconduction.

A few experiments are known in which dissipative structures were observed in active solid-state media. Interesting experiments were described by Kerner and Sinkevich (1982). The authors have observed multi-pinch and multi-domain states in the electron–hole plasma in a homogeneous semiconductor n-GaAs specimen, 0.25×10^{-4} cm in thickness and 20×10^{-4} cm in length. First, as the electric field is applied, the electron–hole plasma is stratified normally to the field direction. As the field strength is increased, longitudinal pinches are splitted and glowing drops (domains or strata) are formed. The voltage–current curve of the specimen has corresponding steps, hysteresis is observed at every step. According to the authors, the phenomenon considered can be explained on the basis of a theory of contrast dissipative structures developed by Kerner and Osipov (see in Chapter 7).

Finally, we mention a review by Rozanov (1982) where the author considered hysteresis phenomena and flipping waves in optically active semiconductor media and "nonlinear" interferometers. It should be noted that heat conduction is an essential factor in the generation of autowaves for some of the effects mentioned above. Heat conduction is also described by a nonlinear equation of the type (1.1) and takes part in the formation of AWPs with which we are concerned here and which we consider in Chapters 4 and 7.

In conclusion we point to a "classical" AWP that is a propagation of thermal waves or structures which has been observed in iron wire heated by electric current in an atmosphere of hydrogen and/or helium. Being in series connection with a load resistor, such a wire is able to stabilize the current (the barretter effect). This phenomenon has been used in radio engineering for a long time, but it was found recently that it is due to stable "temperature" dissipative structures (Barelko, Beibutyan et al. 1981, Voloddin, Beibutyan et al. 1982). This example is a bridge between active solid-state media and chemical non-equilibrium media where autowaves can arise.

B. *AWPs in chemical reactions*

As we have mentioned in the foreword, combustion autowaves were one of the first examples of the phenomena considered which was observed and analysed. A review is given in a monograph by Zeldovich et al. (1980). Experimental data on the relevant space–time structures are presented with some completeness in a monograph by Polak and Mikhailov (1983).

A class of similar processes which are known as the Belousov-Zhabotinskii reactions is of special interest; AWPs of practically all known types have been observed in these processes. It is relatively easy to observe oscillations taking place in such reactions. On the other hand, auto-oscillations and AWPs in biological systems have many features analoguous to these model processes. In thin tubes and layers where self-sustained oscillations of concentrations take place sources of autowaves were discovered, leading centres generated only because of an inhomogeneous initial excitation (Fig. 1.2). The existence of two-dimensional wave processes, "reverberators", (Fig. 1.3), was also observed for the first time here, and the regime of creation of dissipative structures was revealed. A complicated regime may arise because of interaction of autowaves and breakages of their fronts. Sometimes this state is called

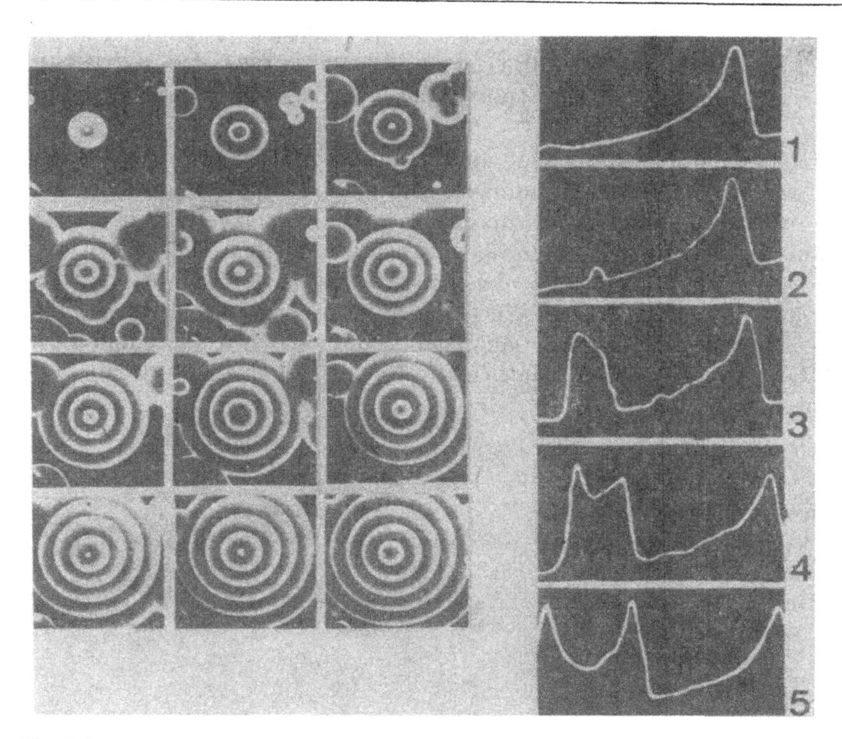

Fig. 1.2

Leading centres (LC) in the Belousov-Zhabotinsky auto-oscillating reaction proceeding in a fine layer of an aqueous solution of reagents. a) LC frames, the interval between the frames is 30 s, the auto-oscillation period is 56 s., the wavelength is 0.55 cm (Zhabotinsky 1974). b) Oscillograms of profiles of concentration AW (coloured by dark blue in the reaction) at the same point of the reactor: t(s) = 0 ("1"); 0.8 ("2"); 5.4 ("3"); 9.2 ("4"); 23.6 ("5"). (Vasilev and Zaikin 1976)

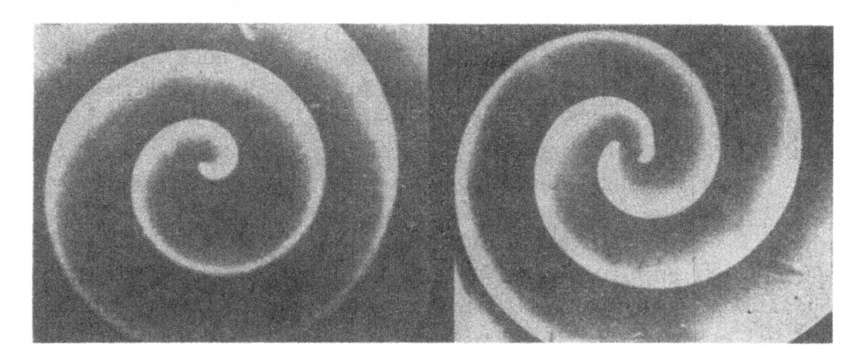

Fig. 1.3

Reverberators of 1st order (a) and 2th order in the Belousov-Zhabotinsky reaction (Agladze 1982)

"chemical turbulence" (Fig. 1.4). More details on this subject may be found in the books by Zhabotinskii (1974), Ivanitskii, Krinskii and Selkov (1978), Winfree (1980), and in the pioneer work by Belousov (1981). We note other important examples in the following.

Barelko (1977) and Barelko, Merzhanov and Shkadinskii (1978) presented experimental data on waves of oxidation of ammonia and carbon monoxide on platinum. In this system the propagation of the autocatalytic reaction waves may be caused by three mechanisms: gas-diffusion, migration and heat conduction. The combination of all three processes results in a number of new interesting phenomena, in particular near halting points of the autowaves. We mean wave propagations with periodic changes in velocity, pulsations in the wave front coordinate near the equilibrium positions, and dissipative structures. Mathematical models of these effects are considered in Chapter 4.

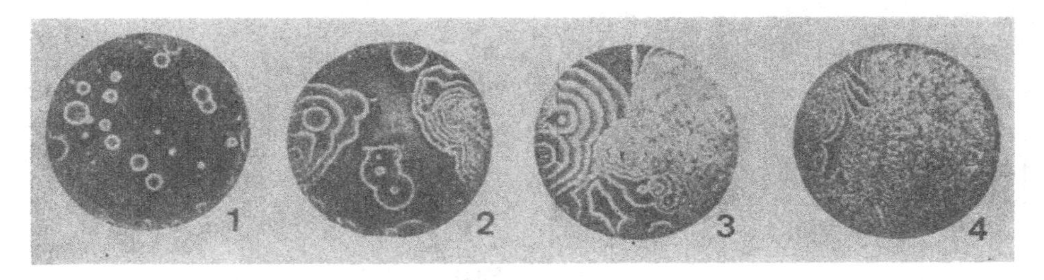

Fig. 1.4
Formation of quasi-stochastic AWPs of the Belousov-Zhabotinsky reaction. The interval between frames is 4 min. (Zaikin and Zhabotinsky 1970)

Autowaves known as self-propagating high-temperature synthesis (SPHTS) have been under investigation since 1967. For example, titanium, zirconium, niobium, tantalum can react with carbon, boron, silicon, nitrogen, producing carbides, borides etc. The maximum temperatures in the waves amount to 3000 °C, and the wave velocities are from 0.2 up to 12 cm/s. Waves of this type have been generated in about 150 different systems. The SPHTS method has been introduced in technology for the production of refractory materials. Investigations of self-exciting and spin regimes of AWPs in SPHTS are very interesting. Both regimes are of the same nature: they result from a thermal instability of the process with respect to small perturbations which are longitudinal for auto-oscillations, and transverse for spin combustion. Photographs of spin combustion are given in Fig. 1.5 (see Section 4).

A new phenomenon has been discovered recently in radiational cryochemistry: a stratified propagation of chemical transformation autowaves in frozen mixture of reagents irradiated previously by γ-rays. Halogenation and hydrohalogenation reactions have been induced in hydrocarbons. The phenomenological basis of the reaction is a positive feedback to the chemical transformation from the brittle failure of solid reagent specimens. Cracking increases the active surface and involves a new

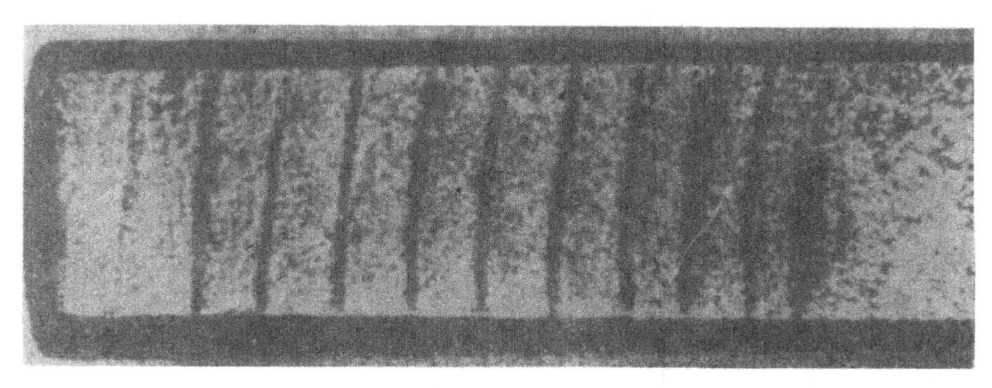

Fig. 1.5
The frame of a longitudinal ground end of a specimen obtained after spin burning in titanium-ferroboron system. Spiral lines represent the tracks of spin burning. (Merzhanov 1980)

reaction process, and the chemical reaction, in turn, stimulates failure of the material. The resulting autowave of the brittle failure and chemical transformation travels in the solid specimen (Barelko, Barkalov, Vaganov et al. 1982). It has been suggested (Zanin, Kiryukhin et al. 1981) that one of the reasons of the brittle failure is a difference between the densities of the initial reagents and the reaction products, another reason is thermal stresses produced by heat released in the chemical processes.

Other interesting AWPs associated with chemical reactions and heat release are now known. Some of them will be considered in subsequent chapters. The last process to be mentioned here is the travelling polymerization front (see e.g. work by Arutyunyan and Davtyan 1975).

C. AWPs in living systems

AWPs taking place in biological systems can be divided into four categories: autowaves in excitable cells and tissues, in biochemical reactions, in communities of living organisms and AWPs at morphogenesis.

1. In the first category the active distributed element (i.e. the excitable medium) is the nerve fibre membrane of neuron or other cells. AWPs in excitable media are investigated thoroughly, this subject has been considered in numerous reviews and monographs. In Chapter 2 we show why the propagation of excitations in living electric cables must be described by parabolic equations, like (1.3), and not by the hyperbolic telegrapher's equation. The mathematical models describing AWPs in nerve tissue (brain) can not be reduced in all the cases to Eqs. (1.1)—(1.3); integro-differential equations may be necessary (see e.g. Markin, Pastushenko and Chizmadzhiev 1981). Nevertheless, we will mention AWPs occurring in some remarkable systems.

Waves of a collective pulsed activity may propagate in networks of checking and exciting neurons. For example in one of the brain structures — the hippocampus — a wave has been detected (the so-called summary spike potential) which is a sequence of discharges of pyramidal cells (exciting neurons). The spike potential propagation velocity is 20 to 60 cm/s, the period in the sequence of enhanced pulsation regions is about 10^2 ms by the order of magnitude (Shaban 1976). Propagation of the activity wave was observed also in an isolated stria of the cerebral cortex (Burns 1968); the wave velocity is 20 cm/s, the duration of the activity "volley" is up to 300 ms.

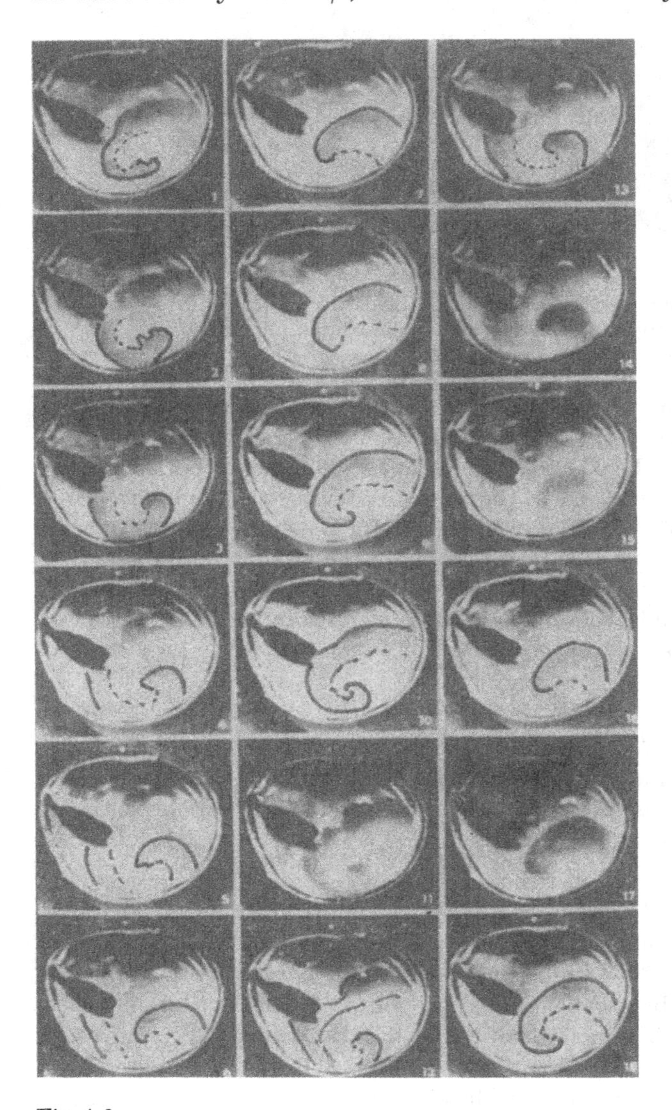

Fig. 1.6

Frames of two revolutions of reverberator which occurs when a depression propagation (DP) proceeds in the retina of a chicken eye. DP manifests itself by the dark region with sharp contrasts of TP edges. (To make the picture clear the leading edge is marked with the dark line and the trailing edge is marked with the dotted line.) Interval between two frames is 20 s. (Gorelova and Bureš 1983)

If the cerebral cortex is stimulated locally by a chemical, mechanical or electrical factor, a depression appears in the electrocorticogram; it propagates as a travelling front from the stimulated point and has a small velocity (30 to 90 m/s), as compared with the spike potential velocity. The propagating depression is always associated with a change in the level of the stationary cortex potential (Leão 1944, Koroleva and Kuznetsova 1971, Shibata and Bureš 1974). An enhanced neuron pulsation is observed at the travelling front of the propagating depression. For this wave the propagation mechanism is determined by changes of inter- and intra-cellular concentrations of K^+, Na^+ and C^{2+} ions (see e.g. Reshodko and Bureŝ 1975, Tuckwell and Miura 1981). A remarkable fact is that the propagating depression wave is observed simultaneously in the retina. This wave is associated with a change in the optical properties of the retina, so it can be registered in photographs. An example of such a wave which is of the reverberator type is shown in Fig. 1.6. The pictures were obtained from a chicken retina (Gorelova and Bureš 1983).

The conducting system of heart controlling contractions of the cordiac muscles consists of a large number of various cells. The activity of every cell is maintained by mechanisms analogous to those which operate in the axons of nerve cells. In the intercellular medium the ion currents and special substances ("mediators") circulate and mediate the coupling between cells. Under normal conditions an electric potential wave having the form of a travelling pulse can propagate along the fibres of cordiac muscle. The appearance of autonomous wave sources (reverberators) results in arhythmia and fibrillations (unsynchronized activity of the heart), which are pathological for the organism. Figure 1.7a shows a set of "motion pictures" representing two reverberators rotating in the auricle of rabbit. The scheme has been reconstructed by means of the analysis of electrical signals from many electrodes injected into a stria of the auricle. The electrical stimulus initiating the reverberators is indicated with a delay of 60 ms with respect to the original stimulus. The size of the reverberator core is about 2.5 mm, the rotation frequency is 7.2 Hz for each reverberator. As two reverberators are operating together, the autowave source frequency is 14.2 Hz. For comparison, the normal wave propagation in the right auricle is shown in Fig. 1.7b. It is seen that a single leading centre exists under normal conditions; it is the sinoatrial node setting the main rhythm for contraction stimuli in the cordial muscle. The experiment described here (Allessie, Bonke and Schopman 1973) is unique though it was performed more than a decade ago. More details on the propagation of autowaves in the heart can be found in a monograph by Ivanitskii, Krinskii and Selkov (1978) and in the book "Autowave Processes in Systems with Diffusion" (1981) (see also Chapter 5).

2. Probably there are also autowaves involving biochemical transformations. Processes like the travelling front of flipping and the creation of dissipative structures manifest themselves most clearly at the differentiation of living tissues. Genetic control systems and the active transport of substances through cellular membranes take part in the generation of AWPs in this case. In the process of morphogenesis such phenomena are associated with mechanical autowaves of various types (see Table 1.1). There is an extensive literature on this subject (monographs by Romanovskii, Stepanova and Chernavskii 1975 and 1984, and Murrey 1983, review articles by Vasilev, Romanovskii and Chernavskii 1982 and Belintsev 1983).

3. Various autowaves have been observed in distributed communities of living objects. Activities of such systems may be of different types: proliferation (equivalent of autocatalysis), "predator-prey" interaction and some others. In such systems mutual diffusion may exist together with proper diffusion; terms with $D_{ij} \neq 0$ $(i \neq j)$ may be present in the system models. Besides, such systems can be inhomogeneous.

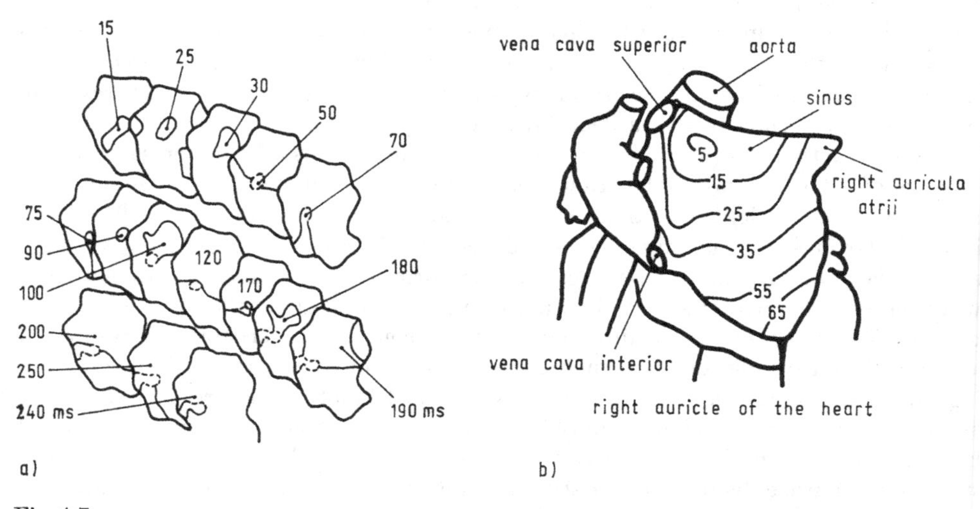

Fig. 1.7

a) Rotation of two excitation waves in tissue of a rabbit auricle. The figures in consecutive order show positions of excitation TF (dark line). The region which is not excited is restricted by dotted lines. Numbers indicate observation moments (ms) (Allessie, Bonke, Schopman 1973).

b) TF position during the normal heart activity (Ivanitskii 1982)

The character of AWPs in communities is more complicated because of chemiotaxis, phototaxis, and the generation of convective currents in aqueous suspensions of microorganisms (Levandovsky et al. 1975). Communities or ecosystems are subject to the influence of various external geophysical factors. Let us mention two relevant examples. At a certain stage in the existence of amebiform cells Dictyostelium discoideum, an aggregation is observed which is associated with a wavelike motion of these cells. The aggregation proceedes as follows: One of the cells begins to periodically secrete a special substance, an attractant (cAMP). In response, the neighbouring cells secrete cAMP pulses about 15 s after receiving the signal from the first cell. Then the cells move toward the signal source. While the cell is moving, it is apparently not susceptible to further stimulation. Such a mechanism of signal and response guarantees the propagation of the signal wave and the convergence of cells toward the central source (Cerish 1978). (See also Beylina et al. 1984.)

A similar wave motion has been detected recently in the development of the chicken embryo. This process is also governed by intercellular cAMP. An external addition of this substance perturbs the normal propagation of the cell motion waves and leads sometimes to the creation of a reverberator (Stern and Goodwin 1977).

Recall that the history of AWP theory dates from the description of gene wave propagation in distributed populations (Kolmogorov, Petrovskii and Piskunov 1937, Fischer 1937). At present autowaves like the spreading of epidemics or pests are again under extensive investigation in genetics and ecology (Svirezhev, Giguri and Razzhevaikin 1983).

4. The autowave mechanisms essentially determine the development of embryonal structures; one of the examples has been mentioned above. They also determine peristaltic motions in the walls of living channels. This problem is discussed in some detail in Chapter 9 for the examples of intestine, blood vessels, and strands of plasmodium myxomycete Physarum. So we restrict ourselves to a few examples of autowaves of activities for isolated living cells.

Autowaves have been observed at the flattening of fibroplast (Smolyaninov and Bliokh 1976). It was shown there that two-fold waves circulate along the edge of the fibroplast lamella (the flattened part of the cell, less than 1 μm in thickness). A more frequent process observed involves quasi-stochastic thickness oscillations in domains of 5 to 10 μm, the oscillation period being a few minutes. Such processes determine the mechanisms providing displacements of a cell on a flat surface.

AWPs have been observed at various stages of embryogenesis and morphogenesis. The data have been compiled by L. V. Belousov, they are presented in Table 1.1 that is taken from his work (Belousov 1984). For example, pulsation autowaves have been observed practically in all investigated fertilized roe-corns (Reznichenko 1982). A study of roe-corns of freshwater fishes has also revealed that a wave of Ca^{2+} ions propagates in the ovum (about 1 mm in diameter) during two minutes after fertilization; the wave is excited in the point where the spermium has penetrated and travels along the diameter. The generation of this concentration wave is, evidently, caused by release of bound Ca^{2+} ions from intra-cellular compartments (Lea 1977).

We conclude this Chapter with a summary, Table 1.2, presenting some data on the AWPs mentioned above and other effects given previously in other work: Vasilev, Romanovskii and Yakhno (1979), Yakhno (1981), Romanovskii, Stepanova and Chernavskii (1984).

We note particularly that Table 1.2 can be expanded essentially by means of interaction of various phenomena such as formation of turbulence structures, waves in plasma and so on. However, we must consider equations (1.1) with "non-zero transfer velocity" as generalized AWP model that drop out range of our book.

Table 1.1

AWPs in developing biological systems (Belousov 1984)

Object	Wave type	Velocity (μm/min)	Period (min)	Reference
Hydroide polyps				
a) Hydrozoa (Thecaphora)	Proximal-distal waves of cell rearrangement	30–70	5–15	Belousov 1973, Zaraysky et al. 1983
b) Hydra attenuta	Peristaltic ejections	10–100	5	Labas et al. 1981, Campbell 1980
Ova				
a) Round worms Pontonema vulgaris (Hematoda)	Cortical-peristaltic waves	15–20	12–16	Cherdantsev
b) Pollicipes (Crustacea)	Cortical-peristaltic waves	14–20	2	Vacquier 1981
c) Gastropoda (Lymnaea stagnalis)	Subcortical motions of yolk granules	12	30	Meshcheryakov 1976
d) Teleostei	Cortical-peristaltic waves			Reznichenko 1982
	Calcium ion (Ca^{2+}) concentration wave (TP)	8		Lea 1977
e) Amphibians	Cortical-peristaltic waves	60	30	Vacquier 1981, Kirschner et al. 1980, Yoneda et al. 1982
Blastoderm of chicken embryo	Cell contraction waves	200	2–3	Stern and Goodwin 1977
Amphibian explants				
a) Neuroectodermal explants at gastrula stage (mid-late phase)	Polarization wave in contacting cells (TF)	5–10		Belousov and Petrov 1983
b) Explants of ectoderm and mesoderm (tailbud phase)	Polarization wave in contacting cells (TF)	10–60		Litchinskaya and Belousov 1977
Polymerization of actin (in vitro)	TF	6		Pollard and Mooseker 1981

Note: Dispersion in velocities is due to temperatures under which the experiments were performed.

Table 1.2

AWPs observed in experiments

Object	Specific velocity	AWP type
Physics		
Polymer film	10^{-2}–10 cm/min	TF
Boiling film	0–1 cm/s	DS, TF
VO_2 film on a substrate	1 cm/s	TF, TP
Magneto-crystal film	10^4 m/s	TF
Electron–hole plasma (n-GaAs)	0	DS
Technology		
Networks of coupled self-sustained oscillators		SO
Active RC-line		TF
Distributed electronic luminiscence image converters		DS
Optically active media in interferometers		TF
Laboratory gas discharge plasma		DS, TP
Superconducting wire		TF
Chemistry		
Burning substances		TF
Barretor (conducting wire in H_2 and/or He atmosphere)	0–2 cm/s	DS, TF
Belousov-Zhabotinsky reactions	10^{-2}–10 cm/s	DS, SO, TP, AWS
Iron wire in nitric acid	2 m/s	TP
Oxidation of ammonia on platinum	0.5 cm/s	TF
Oxidation of carbon monoxide	5 m/s	TF
High-temperature synthesis (titanium–carbon and other similar chemical reactions)	1–15 cm/s	TF
Halogenation and hydrohalogenation of solid hydrocarbons at low temperatures	0.1–2 cm/s	TF
Amine-induced polymerization reaction in epoxidial oligomer	10^{-2} cm/s	TF
Biology		
Squid axon	21 m/s	TP
Conducting system of the heart	25–300 cm/s	TP, AWS
Myocardium muscle	30 cm/s	TP
Nonstriated muscles	5–10 cm/s	TP, AWS
Neuron networks: a) fast waves	10–50 cm/s	TP
b) slow waves	2–5 mm/min	TP, AWS
Eye retina	2–5 mm/min	TP, AWS
Active filaments in algae	50 μm/s	TP
Coral polyps	50 cm/s	TP
Myxomycetes plasmodium	10–50 μm/s	SO, TP, AWS
Populations of ameboid cells	1–5 μm/s	TP, AWS
Populations of numble lizards	0	DS
AWP in morphogenesis — see Table 1.1		

Notes:

(i) Notation AWS stands for autonomous wave source (see Chapter 5).

(ii) For AW media in technology and for burning substances the order of magnitude of specific velocities depends on particular conditions and special features of the devices, so we do not present these data here.

Chapter 2

Physical premises for the construction of basic models

The multitude of objects in which AWP have been observed is far too great to dwell on the construction and analysis of mathematical models for each of them. Their common features, with respect to AW phenomena, are revealed in the structure of simplified basic models which are analysed in the subsequent chapters of this book. Evidently, a model must be a sort of a "sketch" of the system concerned, so that it can be studied with a fair degree of completeness and the results obtained would be relevant to a wide class of objects. The basic models considered below meet this requirement. On the other hand, one should bear in mind the limits of applicability of the results obtained in this way to concrete objects. To this end, one should have, of course, a sufficiently complete mathematical description of the object in view, to be able to compare it with the corresponding basic model. Meanwhile, some essential limitations can be considered from a general point of view. These aspects are the subject of the present Chapter where we deal with mathematical models for systems with finite transport velocities, direct interactions between remote elements, AWPs in media with anisotropy in the transport phenomena and with mutual diffusion of the system components.

In the last Section we present some examples of the basic models, paying special attention to the choice of the source functions and the number of components in the models which are sufficient to describe the specific AWP.

2.1 Finite interaction velocity. Reduction of telegrapher's equations

Dynamics of components of the fundamental basic model is described in terms of quasilinear parabolic equations which are of first order in the time derivative and second order in the space derivatives (cf. Eqs. (1.1) to (1.3)). Equations of this type lead to an infinite velocity of transfer of interactions between elements of the medium. Actually, if there are no distributed sources ($F_i \equiv 0$), and the initial data are localized ($x_i(0, r) = \delta(r)$), the solution is

$$x_i(t, r) = \frac{1}{2\sqrt{\pi^2 D_{ii}^2 t}} \exp\left\{-\frac{r^2}{4D_{ii}t}\right\}. \tag{2.1}$$

It is seen that at any time ($t > 0$) the concentration x_i is *nonzero* at *all* points of space, so the substance transport proceeds with infinite velocity. Of course, the amount of the substance transported decreases rapidly with distance, and this is just the reason why the diffusion approximation is valid for the transport phenomena. However, in the presence of distributed sources with a low excitation threshold the use of equations of this type in the models concerned leads to physically incorrect results; for instance, one gets overestimated (and sometimes even inadmissible) velocities of wave front propagation. On the other hand, in many cases the diffusion approximation is quite reliable as a model of an AWP. Let us discuss the reasons for reduction to this approximation taking a few simple examples of models.

1. The mass transfer is due to particle motion. The time evolution of the velocity and coordinate distribution function $f(\boldsymbol{v}_0, r)$ is governed by the equation

$$\frac{\partial W}{\partial t} + (\boldsymbol{v}_0 \operatorname{grad} W) = \int\limits_{\boldsymbol{v}_0 \in \Omega} k(t, r, \boldsymbol{v}_0', \boldsymbol{v}_0)\, W(t, r, \boldsymbol{v}_0')\, \mathrm{d}\boldsymbol{v}_0', \tag{2.2}$$

the right-hand side of this equation is called the collision integral. Under the assumption of local balance in velocities which are distributed according to the Boltzmann–Maxwell law, one gets the usual diffusion equation, like Eqs. (1.1) to (1.3), for the description of the particle density distribution (Falkenhagen and Ebeling 1965; Ebeling 1965, 1967).

Let us consider now the situation where the particle velocities are limited. For the sake of simplicity we suppose that the velocity acquires only two values, $\boldsymbol{v}_0 = \pm a$, and the system is uniform. Then Eq. (2.2) reduces to linear equations for the densities of particles moving with the velocities of $+a$ and $-a$, respectively,

$$\begin{aligned} \frac{\partial n_+}{\partial t} + a\,\frac{\partial n_+}{\partial r} &= \frac{\alpha}{2}\,(n_- - n_+), \\ \frac{\partial n_-}{\partial t} - a\,\frac{\partial n_-}{\partial r} &= \frac{\alpha}{2}\,(n_+ - n_-). \end{aligned} \tag{2.3}$$

Instead of the partial densities n_+ and n_-, we introduce the full total density, $n = n_+ + n_-$, and the flux, $I = n_+ - n_-$,

$$\frac{\partial n}{\partial t} + a\,\frac{\partial I}{\partial r} = 0, \tag{2.4a}$$

$$\frac{\partial I}{\partial t} + a\,\frac{\partial n}{\partial r} = -\alpha I. \tag{2.4b}$$

The continuity equation (2.4a) and the diffusion law (2.4b) describe the mass transfer in the model considered. The diffusion law reflects a finite current relaxation time. An account of this aspect in the framework of linear thermodynamics of irreversible processes was given by A. V. Lykov (1965, 1978).

Putting Eq. (2.4b) into the continuity equation, (2.4a), we get

$$a\,\frac{\partial^2 n}{\partial r^2} - \frac{\alpha}{a}\,\frac{\partial n}{\partial t} - \frac{1}{a}\,\frac{\partial^2 n}{\partial t^2} = 0 \tag{2.5}$$

or, introducing the diffusion coefficient, $D = v_0^2/\alpha$,

$$\frac{\partial n}{\partial t} + \tau \frac{\partial^2 n}{\partial t^2} = D \frac{\partial^2 n}{\partial r^2}, \tag{2.6}$$

where $\tau = D/v_0^2$ is the relaxation time for the mass transfer. The statistical-mechanical foundation of equations of type (2.6) has been studied by Kowalenko and Ebeling (1968).

Evidently, for $v_0 \to \infty$ Eq. (2.6) reduces to the usual diffusion equation. Before the discussion of its solution, we shall consider another spectacular example of a physical system, an active cable, which has as model Eq. (2.6).[1])

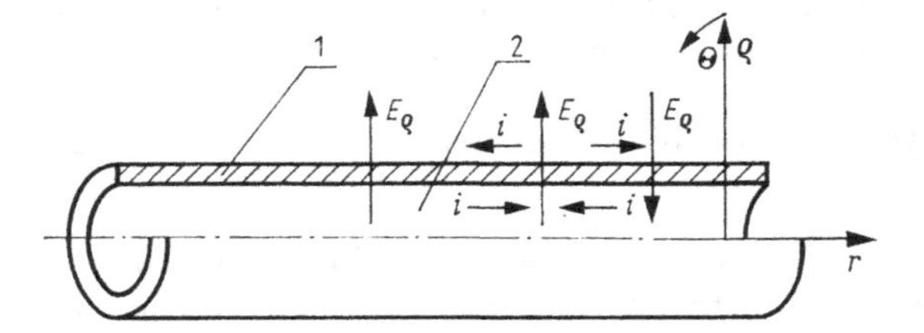

Fig. 2.1
Electric field (E) and currents (i) in a nerve fibre: (1) membrane, (2) intracellular protoplasm

2. Propagation of electric excitations along nerve fibres and other similar systems (Fig. 2.1) must be described, strictly speaking, by means of the Maxwell equations (cf. e.g. Scott 1977). As the specific resistance of protoplasm is high ($\varrho \approx 0.3\ \Omega$m) and the fibre radius is small (less than 1 mm), the damping factor is high for passive propagation of travelling pulses. Active properties are due to membranes. Evidently, these properties can manifest theirselves only in presence of transverse magnetic waves ($E_R \neq 0$, $H_\varrho = H_r = E_\theta = 0$) which are shown in Fig. 2.1. Transverse electric modes do not interact with active elements of membranes and therefore can be discarded. Based on these arguments, we represent the equivalent scheme of the fibre as shown in Fig. 2.2 (it is assumed that the internal and external longitudinal active resistance of the fibre can be given by the same quantity, $\approx R_1$). For this line we have the following equations,

$$\begin{aligned} &\frac{\partial i}{\partial t} + C \frac{\partial U}{\partial t} + i_M(U) = 0, \\ &\frac{\partial U}{\partial t} + L \frac{\partial i}{\partial t} + R_1 i = 0, \end{aligned} \tag{2.7}$$

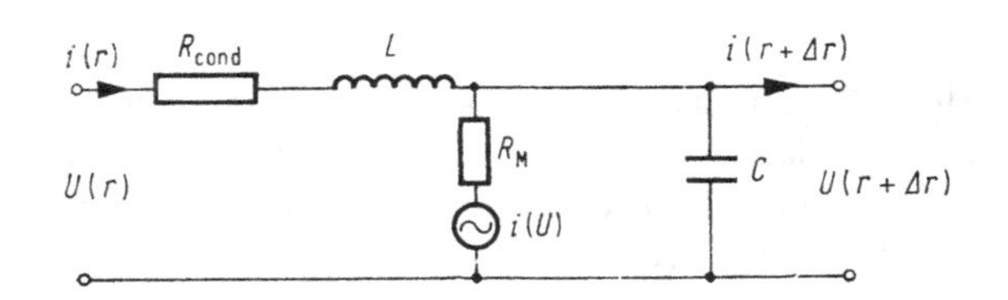

Fig. 2.2
Equivalent electric scheme of a nerve fibre

[1]) It should be mentioned that there are numerous examples of mass transfer, among ecological systems, porous bodies etc., where τ cannot be considered as small *a priori*, and the parabolic diffusion equation is inadequate.

where C and L are the linear capacitance and inductance of the fibre, R_1 the linear resistance of its protoplasm, $i_M(U)$ the voltage dependence of the current through the membrane.

Combining the two equations in (2.7) one gets

$$\frac{\partial^2 U}{\partial r^2} - LC \frac{\partial^2 U}{\partial t^2} + \left(L \frac{\partial}{\partial t} - R\right) i_M = R_1 C \frac{\partial U}{\partial t}. \tag{2.8}$$

It is seen that this telegrapher's equation is similar to (2.5), (2.6), where $v_0^2 = (LC)^{-1/2}$, $\tau = L/R_1$.

Let us consider passive properties of these equations. The complete solution of the Cauchy problem for (2.6) on the infinite coordinate axis can be obtained by means of the method of characteristic Riemann's functions (Tikhonov and Samarsky 1977). In the case where $u(0, t) \equiv n(0, t) = 1$, and $u \equiv n(r > 0, 0) = 0$, it is

$$n(r, t) = \frac{r}{2} \sqrt{\frac{1}{D\tau}} \int\limits_{t=r\sqrt{\tau/D}}^{t} \exp\left\{-\frac{t'}{2\tau}\right\} \frac{I_1\left(\sqrt{t'^2 - r\tau/D'}/2\tau\right)}{\sqrt{t'^2 - r^2\tau/D}}\, dt'$$
$$+ \exp\left\{-\frac{r}{2}\sqrt{\frac{1}{D\tau}}\right\} \tag{2.9}$$

for $t \geqq r(\tau/D)^{1/2} = rv_0$, and $n(r, t) = 0$ for $t < r(\tau/D)^{1/2} = rv_0$, where $I_1(x)$ is a modified Bessel function of the first kind.

Solution (2.9), the isochrome for semi-space, is shown in Fig. 2.3. As should be anticipated, the penetration velocity of perturbations (mass) cannot exceed v_0. For telegrapher's equations the space damping factor is $\delta = 2(L/C)^{1/2}/R$. The specific capacitance of membrane in nerve fibre is about 1 pF/cm², the linear inductance is about 10^{-8} H/cm, the specific resistance of protoplasm is 0.3 Ωm, and the fibre radius is less than 0.1 mm. Hence one gets an estimate for the space damping factor, $\delta \leqq 0.1$ cm. Evidently, the nerve fibre cannot transmit information as a passive conductor.

For $t \gg \tau$ one can use the asymptotic expansion of the modified Bessel function, $I_1(x) \sim \exp(x)/(2\pi x)^{1/2}$. Putting it into Eq. (2.9) one gets for $t \gg r(\tau/D)^{1/2}$,

$$n(r,t) = 1 - \frac{2}{\pi} \int\limits_0^q \exp\{-q^2\}\, dq, \qquad \text{where } q = \frac{1}{4}\frac{r^2}{Dt}. \tag{2.10}$$

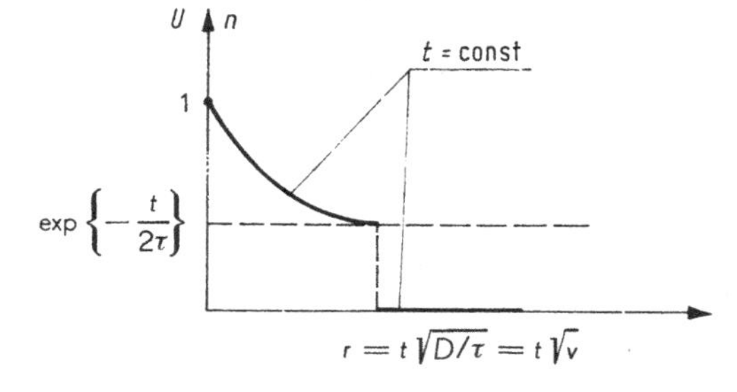

Fig. 2.3
Isochrome in half-space: solution (2.9) of the linear hyperbolic equation, (2.6) or (2.8)

This expression coincides with the solution of the corresponding problem for the classical diffusion equation with an instant relaxation of the flux.

Thus one can conclude that the basic models (1.1)—(1.3) are satisfactory if the specific times of local kinetics $\left(F_i(x_1, \ldots, x_n)\right)$ are much greater than $r_s(\tau/D)^{1/2} = r_s/v_0$, where r_s is a parameter characterizing the system size.

Look at the nerve fibre from this standpoint. The pulse propagation velocity in the fibre is 2×10^3 cm/s, and the front duration is $\approx 10^{-3}$ s, so the front length is about 2 cm. Meanwhile the space damping factor (wave-to-active resistance ratio, $2(L/C)^{1/2}/R_1$) is about 0.1 cm. Therefore one can neglect the wave properties of Eq. (2.8), reducing the hyperbolic equation to the form specific for Eqs. (1.1) to (1.3). Recall that the damping factor is low because of the high resistance of protoplasm; if its conductance were comparable with that of metals, the damping length would amount to sizes of living organisms, and equations of the type (2.9) would be used in models.

3. Active lines with high linear conductances, like nerve fibres, may be investigated in the framework of a basic model which has been obtained by reduction from Eqs. (2.6) to (2.8) (or from an analoguous equation). Propagation of excitation waves is due to active (but not wave) properties of such media. We shall elucidate this statement, as well as relations between parameters of the medium for which the reduction is possible, taking the following model as an example,

$$\tau \frac{\partial^2 x}{\partial t^2} + \frac{\partial x}{\partial t} - D \frac{\partial^2 x}{\partial r^2} = -2\gamma x(1 - x)(x_0 - x), \tag{2.11}$$

where $r \in (-\infty, \infty)$, $0 < x_0 < 1$, and all parameters of the model are positive. The left-hand side is similar to that in Eqs. (2.6) and (2.8), while the right-hand side describes distributed sources responsible for the activity of the medium to be simulated. The equation has three stationary solutions: $x(r, t) = x_0$ (instable), and two stable solutions, $x(r, t) = 0$, $x(r, t) = 1$. Let us evaluate the propagation velocity of excitations (travelling fronts) in the medium, which are "flip-over" waves transforming the 0-state into the 1-state.

With the auto-model variable $\eta = r - vt$, Eq. (2.11) is written as an ordinary differential equation,

$$\left(\tau v^2 \frac{d^2}{d\eta^2} - v \frac{d}{d\eta} - D \frac{d^2}{d\eta^2}\right) x = -2\gamma x(1 - x)(x_0 - x). \tag{2.12}$$

Its solution must satisfy the following conditions: (i) $x(\eta) \to 0$ for $t \to -\infty$, (ii) $x(\eta) = 1$ for $t \to +\infty$. The parameter $\boldsymbol{v}$ (the TF propagation velocity), as yet unterminated is an eigenvalue of the problem in (2.11) with the boundary conditions (i) and (ii).

By means of a transformation, $w = dx/d\eta$; $w(dw/dx) = d^2x/d\eta^2$ one gets the following eigenvalue problem,

$$(\tau v^2 - D)\, w(dw/dx) - vw = -2\gamma x(1 - x)(x_0 - x),$$
$$w(0) = w(1) = 0. \tag{2.13}$$

Setting here $w = \overline{w}(1 - x)\,x$ one gets a system of equations for $\overline{w}$ and v from the requirements that the coefficients of x, x^2, and x^3 must vanish. A solution of these equations does always exist. Introduce the notations

$$v_0^2 = D/\tau\,, \qquad v_{\rm d}^2 = \gamma D(2x_0 - 1)^2\,, \tag{2.14}$$

where v_0 is the propagation velocity for waves due to inertial-elastic properties of the medium in (2.8), or the maximum velocity of the particle motion in (2.6), and $v_{\rm d}$ is the velocity of the excitation waves described by the basic model (1.1) to (1.3), i.e. that obtained from (2.11) for $\tau = 0$, which are determined by active properties only. The eigenvalue of the problem in (2.12), the autowave propagation velocity is written in terms of these parameters,

$$v^2 = v_{\rm d}^2/[1 + (v_{\rm d}/v_0)^2] = V_0^2/[1 + (v_0/v_{\rm d})^2]\,. \tag{2.15}$$

The velocity $v_{\rm d}$ of waves of the medium excitation described by basic model (1.3) can take infinite values that answers the existence of tail for Maxwell's distribution of medium particles and small values of the medium excitation threshold (see also(2.1)). It is obvious that the velocity v must be always less than the permissible maximum of the information transfer velocity in this medium — v_0. It determines a physical sense the relation $v_{\rm d}/v_0 \ll 1$ as the condition of the basic model applicability. As one should expect, the excitation propagation velocity lies between $v_{\rm d}$ and v_0, and the necessary condition under which the basic model is adequate is $(v_{\rm d}/v_0) \ll 1$. A somewhat longer derivation leads to similar relations for any nonlinearity satisfying the conditions $F(0) = F(x_0) = F(1) = 0$, $F_x'(0) < 0$, $F_x'(1) < 1$, so the reduction is possible under the above condition for a wide class of systems. Since $v_{\rm d}$ is determined also by parameters characterizing nonlinearity of the medium (γ and x_0 in Eq. (2.14)), the applicability of the reduced basic model (1.1)—(1.3) must be evaluated more accurately at the stage when the AW solutions are analysed. One should have in mind that this question is not as simple as it may seem at first glance. For instance, for an electric excitation pulse in nerve fibre one has $v_{\rm d} \approx 10$ m/s and $v_0 \approx 10^8$ m/s, but for thermal waves or for systems with reactions and diffusion in porous bodies we encounter situations where the simplest basic models are inadequate and one has to use either a set of hyperbolic equations or introduce a variable diffusion coefficient (see the next Section), i.e. to take into account a finite transport velocity.

It is instructive to compare the automodel solution of Eq. (2.11) at large values of R_1 (see Chapter 4) with propagation of waves in active lines with $R_1 \to 0$, say, in lines with tunnel diodes (Ilinova and Khokhlov 1963). In the first case $v \approx v_{\rm d}$, so it is determined by $i(u)$ and is practically independent of LC. In the second case $v \simeq (LC)^{-1/2}$ and is independent of $i(u)$, but the nonlinearity of $i(u)$ determines the stationary form of the travelling pulse which is independent of the initial conditions. It is reasonable to assert that the reduced model of Eqs. (1.1)—(1.3) describes a limiting case, a nonlinear propagation of excitations in the medium.

2.2 Nonlinear diffusion equation. Finite diffusion velocity

Diffusion equation (1.3) can be used also for the description of mass and heat transport with a finite velocity, even in the absence of sources $\big(F(x) = 0\big)$, if the diffusion coefficient depends on the variable x, vanishing at some value (Zeldovich and Kompaneets 1950, Barenblatt 1952, Barenblatt and Vishik 1956). We are going to elucidate this point with an example and then to discuss briefly some media with such a dependence of D on x.

Let us find the simplest particular solutions of the equation

$$\frac{\partial x}{\partial t} = \frac{\partial}{\partial r}\left(\mathrm{D}(x)\,\frac{\partial x}{\partial r}\right), \tag{2.16}$$

$D = \tilde{D}x^\sigma$, $\sigma > 0$. Equation (2.16) will be considered for the semi-axis, $r > 0$, with the boundary condition

$$x(0, t) = x_0 t^p . \tag{2.17}$$

The initial condition will be

$$x(r, 0) = 0 , \qquad r > 0 . \tag{2.18}$$

Introduce the variable $\eta = vt - r$ and look for solutions of the form

$$x(vt - r) = \varphi(\eta) , \tag{2.19}$$

where φ is an unknown function. Putting (2.19) into Eq. (2.16), we get an ordinary differential equation for the function $\varphi(\eta)$,

$$v\,\frac{\mathrm{d}}{\mathrm{d}\eta}\,\varphi = \frac{\mathrm{d}}{\mathrm{d}\eta}\left(\tilde{D}\varphi^\sigma\,\frac{\mathrm{d}}{\mathrm{d}\eta}\,\varphi\right). \tag{2.20}$$

After integration over η, setting the integration constant to zero, we get

$$\tilde{D}\varphi^\sigma\,\frac{\mathrm{d}}{\mathrm{d}\eta}\,\varphi = v\varphi \quad \text{or} \quad \frac{\tilde{D}}{v^\sigma}\,\frac{\mathrm{d}}{\mathrm{d}\eta}\,\varphi^\sigma = 1 . \tag{2.21}$$

A second integration leads to

$$\frac{\tilde{D}}{v\sigma}\,\varphi^\sigma = \eta + C_0 . \tag{2.22}$$

The constant C_0 is determined by the condition $\varphi = 0$ for $t = r = \eta = 0$. So $C_0 = 0$, and (2.22) determines the function $\varphi(\eta)$.

$$\varphi(\eta) = \left(\frac{v\sigma}{\tilde{D}}\,\eta\right)^{1/\sigma} \quad \text{or} \quad x(r, t) = \left(\frac{v^2\sigma}{\tilde{D}}\right)^{1/\sigma} t^{1/\sigma}\left(1 - \frac{r}{vt}\right). \tag{2.23}$$

Comparing this expression for the solution $x(r, t)$ with the boundary condition (2.17), we find relations between the parameters,

$$p = 1/\sigma \quad \text{and} \quad v^2\sigma/\tilde{D} = x_0{}^\sigma . \tag{2.24}$$

Putting (2.24) into (2.23) we obtain finally the desired solution,

$$x(r, t) = \begin{cases} \dfrac{x_0}{v^{1/\sigma}} (vt - r)^{1/\sigma}, & 0 \leqq r \leqq vt, \\ x(r, t) = 0, & r \geqq vt. \end{cases} \tag{2.25}$$

Solutions of this type are called temperature waves. Their propagation velocity is determined by the parameters ($\tilde{D}$ and σ) and the boundary conditions (x_0),

$$v = \sqrt{\tilde{D} x_0{}^\sigma / \sigma}. \tag{2.26}$$

For simplicity, we have considered just a simple example. In the literature mentioned above the existence of temperature waves like (2.25) was shown for quite general classes of boundary and initial conditions. Thus the existence of temperature waves (with a finite TF velocity) is due to the dependence of D on x in the model presented in Eq. (2.16). The expression for the velocity v, Eq. (2.24), shows that the case of linear diffusion equation ($\sigma = 0$) corresponds to $v \to \infty$, an infinite velocity of mass (heat) transport, according to solution (2.1) discussed above. Thus a nonlinearity of the diffusion coefficient is responsible for new physical effects. In Chapter 7 we discuss this subject and show that even an autonomous problem (neutral boundary conditions) for a nonlinear diffusion equation with sources, which corresponds to autocatalysis, has automodel solutions describing localized structures.

Note a fundamental difference between the definitions of wave velocities in (2.16) and (2.11). In problems of type (2.11) (cf. also Chapter 4) the velocity of the flip-over wave transforming one stationary state into the other is an eigenvalue of a nonlinear boundary problem in the configuration space. In problems of type (2.16) the velocity is determined from Eq. (2.24) which matches the automodel solution to the boundary condition (2.17). Respectively, solutions of nonlinear differential equations are usually divided in two classes: intermediate asymptotics (automodel regimes) of the first (2.16) and second (2.11) kinds (Barenblatt and Zeldovich 1971).

This book deals with automodel solutions of the second kind. The most important qualitative effects can be analysed in the linear transport approximation. So we mainly consider models with constant diffusion coefficients. If the x-dependence of D is low at scales specific for the AWP under investigation, this assumption is quite satisfactory for many real systems. Nevertheless, it is appropriate to mention some physical reasons of why such a dependence may appear.

Thermal conductivity is responsible for the transport of energy. In liquid and gaseous media energy is transferred by means of particle motion in a diffusion process (in the absence of convection). A useful representation of the temperature dependence of the diffusion coefficient D is $D = \tilde{D} \exp(-A/T)$ ($\tilde{D}$ and A are constants). As the diffusion coefficient and the thermal conductivity are proportional to each other in such media, we conclude that $D_q(T) \sim \exp(-A/T)$. Note that a temperature dependence such as $D \sim \exp(-A/T)$ or the Einstein-Fokker law ($D \sim bkT$, b — mobility, k — the Boltzmann constant) cannot be universal, as they do not reflect the presence of critical points in the system.

Actually, in the linear thermodynamics of irreversible processes the flux is given by

$$\boldsymbol{I}_i = -a(\text{grad } \mu_i)_{T,p}, \qquad a = \text{const}, \tag{2.26a}$$

where μ_i is the chemical potential of the i-th component, T the temperature, p the pressure. Using the fundamental equations (1.1)—(1.2), it is easy to get an expression for the diffusion coefficient: $D_i(\partial\mu_i/\partial x_i)_{T,p}$, hence we conclude that $D = \text{const}$ only for ideal gases (solutions). At critical points one has $\partial\mu_i/\partial x_i = 0$. So no simple representation of $D(x)$ can work near the critical points.

In the studies of plasmas, electron gas in solids etc., a dependence on a power of the temperature is assumed as a rule for the thermal conductivity, $D \sim T^2$ (see Section 7.7). If the variation of temperature in autowave phenomena is not too large $(T_{\max} - T_{\min})/T_{\min} \ll 1$ results of AWP theory obtained in the linear transport approximation must hold qualitatively.

In principle, mass transfer processes must be described by a nonlinear equation. Deviations from the linear approximation are especially large for solutions of strong electrolytes, plasmas etc. However, in this case the x-dependence of D (x is a concentration) is weaker than the temperature dependence of D. Thus far from critical points, the use of transport equations with constant coefficients does not lead to appreciable inaccuracies. One should bear in mind, nevertheless, that there exist systems in which the x-dependence of D must be taken into account, for instance, porous bodies (Barenblatt and Vishik 1956). In the consideration of mutual diffusion of components one must also account for an x-dependence of D, otherwise absurd results may arise.

2.3 Diffusion in multicomponent homogeneous systems

The AWP theory has not given much attention to crossing effects in the transport of system components. As a rule, one assumes when analysing basic models that the "diffusion matrix" in the generalized Fick law, (1.2), is diagonal. In many cases this assumption (reduction) must be considered as quite reasonable. There are systems, however, for which it is inadmissible in principle. First, in some systems off-diagonal elements D_{ij} ($i \neq j$) have magnitude close to those of D_{ii}. We mean, for example, aqueous solution of strong electrolytes, $NaSO_4$ and H_2SO_4, where $D_{ij} \approx 0.5D_{ii}$ (Haase 1967, Falkenhagen 1971). Similar effects are known for diffusion of ions in plasmas. Second, though in passive systems the ratio D_{ij}/D_{ii} is small enough to neglect crossing fluxes, in an active system a "local" autocatalysis results in qualitatively new AW properties even for small, but nonzero, values of D_{ij}/D_{ii}. In other words, in such cases one can decide whether D_{ij} ($i \neq j$) can be neglected only with a due account for local kinetics (cf. Section 3.3). Let us dwell on some problems of the theory of crossing effects and consider the example of mass transfer in multicomponent systems. Similar effects appear in combined mass and heat transfer processes, mass and electric charge transfer etc.

Phenomenological theory of mass transfer uses the generalized Stefan–Maxwell equations (Lightfoot 1977),

$$f_i = \sum_{j=1,\, j\neq k}^{n} \frac{x_i x_j}{D'_{ij}} (\boldsymbol{v}_j - \boldsymbol{v}_k) + \beta_{i0}\, \mathrm{grad}\, T\,, \tag{2.27}$$

$$\sum_{i=1}^{n} \boldsymbol{f}_i = 0\,, \quad D'_{ij} = D'_{ij}\,, \quad \sum_{i=1}^{n} \frac{x_i}{D'_{ij}} = 0\,, \tag{2.28--30}$$

where x_i is the molar fraction of the i-th component, $\boldsymbol{v}_i$ its velocity in the laboratory coordinate frame, T temperature, and D'_{ij} and β_{i0} are kinetic coefficients. Other notations used in the present Section are: c_i — molar concentration of i-th component, c — total molar concentration, V_{0i} — partial molar volume of i-th component, ω_i — mass fraction of i-th component. Note that the basic equations (1.1)—(1.2) are written for c_i, not for the x_i which are involved in Eqs. (2.27)—(2.30); the above notations are those used in literature of physical chemistry. Finally, $\boldsymbol{f}_i \cdot (cRT)$ is the force applied to a unit volume and acting on the i-th component.

Methods of thermodynamics of irreversible processes enable one to find a relation of these forces to macroscopic characteristics (Lightfoot 1977),

$$cRT\boldsymbol{f}_i = c_i\, \mathrm{grad}_{T,p}\, \mu_i + (c_i \boldsymbol{V}_{0i} - \omega_i)\, \mathrm{grad}_T\, p\,, \tag{2.31}$$

where μ_i is the chemical potential of i-th component. For simplicity we restrict ourselves to isobaric and isothermal systems, grad p = grad $T = 0$.

Eq. (2.37) expresses a linear relation between driving forces and fluxes, the equality $D'_{ij} = D'_{ji}$ is a consequence of the reversibility of equations of mechanics, and is the Onsager principle. The equality holds in systems which are in local thermodynamical equilibrium. On the other hand, the symmetry of kinetic coefficients is not relevant to any coordinate frame. For instance, for the Fick currents, (1.2), which are measured with respect to the volume-averaged (convective) particle velocity, the matrix is not symmetrical, in general. Let us derive an expression for the Fick fluxes from Eqs. (2.27)—(2.30).

First of all, replace the fluxes measured in the coordinate frame referred to the k-th component by the fluxes measured with respect to the convection velocity. The expression for the flux is

$$\boldsymbol{I}_i = c_i(\boldsymbol{v}_i - \boldsymbol{v})\,, \quad \boldsymbol{v} = \sum_{i=1}^{n} b_i \boldsymbol{v}_i\,, \quad \sum_{i=1}^{n} b_i = 1\,, \tag{2.32}$$

where $\boldsymbol{v}$ is a characteristic velocity. Summing over all the fluxes, we get $\sum_{i=1}^{n} (b_i/c_i)\, \boldsymbol{I}_i = 0$. A similar identity, $\sum_{i=1}^{n} (b_i'/c_i)\, \boldsymbol{I}_i' = 0$, holds for fluxes in another characteristic frame. Comparing these identities we get the desired formula for the transformation of fluxes corresponding to substitution of one characteristic velocity by another,

$$\boldsymbol{I}_i' = \boldsymbol{I}_i - c_i \sum_{j=1}^{n} (b_j'/c_j)\, \boldsymbol{I}_j\,. \tag{2.33}$$

The following characteristic velocities are most widely used:

(i) the Fick system (the average volume velocity), $\boldsymbol{v} = \sum_{i=1}^{n} c_i V_{0i} \boldsymbol{v}_i$ $(b_i = c_i V_{0i}, \sum_{i=1}^{n} c_i \boldsymbol{v}_i = 1)$,

(ii) the Hittof system, $\boldsymbol{v} = \boldsymbol{v}_k$ $(b_k = 1, b_i = 0$ for $i \neq k)$,

(iii) the particle system (average molecular velocity), $\boldsymbol{v} = \sum_{i=1}^{n} x_i \boldsymbol{v}_i$ $(b_i = x_i)$,

(iv) centre-of-mass system, $\boldsymbol{v} = \sum_{i=1}^{n} (\varrho_i/\varrho) \boldsymbol{v}_i$.

Other systems are also possible, for instance, those referring to deformation velocities in crystal lattices. The choice of characteristic velocity is determined by special features of the problem to be solved in the model.

The Stefan–Maxwell equations are in the Hittof system: $\boldsymbol{I}_i' = c_i(\boldsymbol{v}_i - \boldsymbol{v}_k)$, $\boldsymbol{v} = \boldsymbol{v}_k$. Eqs. (2.33) enable one to perform a transformation to any other characteristic system. For example, the Fick system is most appropriate for the investigation of AWPs (1.1) in liquid media (solutions), as it makes it possible to separate the convective motions explicitly. To illustrate the use of Eqs. (2.26)—(2.33), let us consider the construction of equations for Fick's fluxes. For the sake of clarity, we shall restrict ourselves to a system with $n = 3$, where the first component $(k = 1)$ is a "solvent". In terms of the basic model (1.1), this is a two-component system, so one has to relate the matrix D_{ij} in Eq. (1.2) to D'_{ij} in Eq. (2.27). The reduction of the system dimensionality is due to a relation between the fluxes $\sum_{i=n}^{n} V_{0i} \boldsymbol{I}_i = 0$.

Using (2.33), one gets $(b_1' = 1, b_2 = b_3 = 0)$

$$\boldsymbol{I}_i' = \boldsymbol{I}_i - \frac{c_i}{c_1} \boldsymbol{I}_1 = \boldsymbol{I}_i + \sum_{j=2}^{n} \frac{c_i V_{0j}}{c_1 V_{01}} \boldsymbol{I}_j = \sum_{j=2}^{n} \alpha_{ij} \boldsymbol{I}_j, \tag{2.34}$$

where $\alpha_{ij} = \delta_{ij} + (c_i V_{0j}/c_1 V_{01})$ and δ_{ij} is the Kronecker symbol. Substituting the Fick fluxes, (2.34), for $\boldsymbol{I}_j' = c_j(\boldsymbol{v}_j - \boldsymbol{v}_1)$ in (2.26), one gets the following equations (if grad $T = 0$),

$$\boldsymbol{f}_i = \sum_{j=2}^{n} \tilde{D}_{ij} \boldsymbol{I}_j. \tag{2.35}$$

In contrast to D'_{ij}, the matrix D_{ij} is not symmetrical. In the construction of relations between D_{ij} and D'_{ij} one should take into account Eqs. (2.28) to (2.37), retaining only D'_{21}, D'_{31}, D'_{32} in the final expressions, i.e. the mutual diffusion coefficients between the components and between each component and the solvent. After simple algebra one has

$$\tilde{D}_{22} = -[(x_3/D'_{32}) + (x_2/D'_{21})(1 + c_2 V_{02}/c_1 V_{01})]/c, \tag{2.36}$$

$$\tilde{D}_{23} = (x_2/c)(1/D'_{32} - V_{03}/D'_{21} V_{01}). \tag{2.37}$$

The expressions for D_{33} and D_{32} are obtained from Eqs. (2.36), (2.37) by interchanging the subscripts "2" and "3".

Solving the set of linear equations (2.35) with respect to the fluxes and using (2.31) (T and p are fixed), one gets

$$\boldsymbol{I}_i = \sum_{j=2}^{n} \tilde{a}_{ij} \left(\frac{c_j}{cRT}\right) \operatorname{grad}_{p,T} \mu_j = \sum_{j=2}^{n} a_{ij} \operatorname{grad}_{p,T} \mu_j . \tag{2.38}$$

The chemical potential for ideal mixtures or weak solutions is $\mu_i = RT \ln (x_i) + \text{const}$ (Landau and Lifshits 1976), thus Eq. (2.38) becomes

$$\boldsymbol{I}_i = \sum_{j=2}^{n} (\tilde{a}_{ij}/c) \operatorname{grad}_{p,T} c_j = \sum_{j=2}^{n} D_{ij} \operatorname{grad}_{p,T} c_j . \tag{2.39}$$

Transforming this expression to the reference frame with zero characteristic velocity, one gets the flux of the i-th component as given in Eq. (1.2), where $\boldsymbol{v}$ coincides with the characteristic velocity in the Fick system, being the average volume velocity of the medium motion.

Below we present the diffusion coefficients D_{ij} for the case of ideal mixtures. If $x_2/x_1 \ll 1$ and $x_3/x_1 \ll 1$, the expression for the determinant of D_{ij} is simplified, $\det \{\tilde{D}_{ij}\} \approx \alpha_{22}\alpha_{33}/(c^2 D'_{21} D'_{31})$, so

$$\begin{aligned} D_{22} &= -[x_1 D'_{21}/\alpha_{22} + x_2 D''_{21} D'_{31}/(D'_{23}\alpha_{22}\alpha_{33}], \\ D_{23} &= x_2[D'_{21}(V_{02}/V_{01}) - D''_{21} D'_{31}/D'_{23}]/(\alpha_{22}\alpha_{33})] \end{aligned} \tag{2.40}$$

and D_{33} and D_{21} are obtained from (2.40) by interchanging the subscripts "2" and "3".

The coefficients D'_{ij} in the Stefan-Maxwell equation have a weak concentration dependence, in contrast to the Fick coefficients D_{ij}; the dependence is totally absent for ideal dilute mixtures. Therefore the expressions in (2.40) provide a correct representation of the concentration dependence for the Fick diffusion coefficients. These expressions suggest a number of important conclusions.

The mutual diffusion coefficients for the components, D_{32} and D_{23}, vanish for $c_3 = 0$ and $c_2 = 0$, respectively. The matrix D_{ij} in Eqs. (1.2) and (2.39) is not symmetrical. Moreover, D_{32} and D_{23} may have opposite signs. It should be emphasized, however, that the determinant of D_{ij} is positive if $\det \{D'_{ij}\} > 0$; the latter inequality must hold for locally stable (metastable) states of the medium. Hence one concludes that $D_{ii} > 0$ (see also (2.40)).

Subtracting D_{33} from D_{22} and using (2.40) one can show that the mutual diffusion coefficients in Eq. (1.2) may vanish only if the proper diffusion coefficients D_{22} and D_{33} are identical,

$$D_{22} - D_{33} = (1 - x_3)(D_{23}/x_2) - (1 - x_2)(D_{32}/x_3). \tag{2.41}$$

An analogue of Eq. (2.41) can be obtained in the general case, it is sufficient to use the Onsager symmetry for the diffusion coefficients of fluxes in the Hittof system and the expression (2.38) for the Fick fluxes, together with the transformation (2.34) (Haase 1967):

$$\sum_{i=2}^{n} \sum_{k=2}^{n} \alpha_{ik} \frac{\partial \mu_i}{\partial c_m} D_{kl} = \sum_{i=2}^{n} \sum_{k=2}^{n} \alpha_{ik} \frac{\partial \mu_i}{\partial c_l} D_{km}, \tag{2.42}$$

where $l, m = 2, 3, \ldots, n$. Eq. (2.41) is also a consequence of (2.42) for ideal ternary mixtures: $(\partial\mu_2/\partial c_3) = (\partial\mu_3/\partial c_2) = 0$, $(c_2\,\partial\mu_2/\partial c_2) = (c_3\,\partial\mu_3/\partial c_3) = RT$, and $V_{01} = V_{02} = V_{03} = RT/p$. Thus one can see that the diffusion matrix in Eq. (1.2) may be diagonal only in the case that the diagonal coefficients are all equal. For ideal mixtures their order of magnitude is $x_i D_{ii}$; in particular, they are usually small, but nonzero, for weak solutions.

The present thermodynamical approach is easily extended to electrolyte solutions, plasmas and similar systems with additional relations between fluxes. For instance, for electrolyte solutions such a relation is a consequence of the condition that the mixture is electrically neutral (Falkenhagen, 1971; Ebeling, 1967; Ebeling et. al., 1983, 1984),

$$\sum_{i=2}^{n} (z_{i+}c_{i+} + z_{i-}c_{i-}) = 0, \tag{2.43}$$

where z_{i+}, z_{i-} are the electrochemical valencies of positive and negative ions of the i-th substance. For example, for $n = 2$ and in the absence of an external field one has $\boldsymbol{v}_+c_+z_+ + \boldsymbol{v}_-c_-z_- = 0$, so that $\boldsymbol{v}_+ = \boldsymbol{v}_-$, or $\boldsymbol{I}_+ = \boldsymbol{I}_-$, and the number of independent moving components is reduced by one. In general, the number of independent fluxes is the sum of the molecule and ion types minus the number of reactions of the local chemical equilibrium, and minus one. Therefore for a single substance dissociating into two ions with the dissociation constant α, one has a single independent flux, $\boldsymbol{I}_2 = -D \operatorname{grad} c_2$, $\boldsymbol{I}_+ = \nu_+\alpha\boldsymbol{I}_2$ and $\boldsymbol{I}_- = \nu_-\alpha\boldsymbol{I}_2$ (ν_+, ν_- are the ion decay coefficients). Let us find an expression for D using the Stefan-Maxwell equation (2.27) to (2.30).

We shall consider only the case of complete dissociation. The system is three-component: x_w (the solvent), x_+ and x_- (the ions). The expressions presented above, (2.36)—(2.37), are valid for the "forces" $\boldsymbol{f}_i$ (cf. Eq. (2.35), and one should take into account the relation between the fluxes, $c_+z_+\boldsymbol{v}_+ = c_-z_-\boldsymbol{v}_-$ and $x_+z_+ = -x_-z_-$. Substitute these relations into Eqs. (2.35) to (2.37) and sum over the expressions obtained for $\boldsymbol{f}_+$ and $\boldsymbol{f}_-$. The chemical potentials are written as follows: $\mu_i = RT\ln(a_i) = RT \times \ln(\gamma_i x_i)$ (a_i and γ_i are activity and the activity factor for the i-th component). After a simple calculation we get the usual representation for the Fick flux,

$$\boldsymbol{I}_s = -D_{sw}c^2 V_{0w} \operatorname{grad} x_s = -D_{3w} \operatorname{grad} c_3, \tag{2.44}$$

where we have taken into account the following relations $\mathrm{d}(V) = \mathrm{d}(1/c) = \sum V_i\,\mathrm{d}x_i + (\partial V/\partial T)\,\mathrm{d}T + (\partial V/\partial p)\,\mathrm{d}p$. The result for isobaric isothermal processes is $c^2 V_{0w}\,\mathrm{d}(x_i) = \mathrm{d}(c_i)$. The single diffusion coefficient is (Ebeling 1967; Falkenhagen 1971; Ebeling and Feudel 1983)

$$D_{sw} = \left(\frac{1}{D'_{+w}} + \frac{1}{D'_{-w}}\right)^{-1} \left(1 + \frac{\partial \ln \sqrt{\gamma_+\gamma_-}}{\partial \ln x_s}\right). \tag{2.45}$$

The "cation-solvent" and "anion-solvent" diffusion coefficients (D'_{+w} and D'_{-w}) can be found from the experimental data on electrical conductivity of different solutions. It should be noted that the activity coefficients for ions in solutions have a complicated dependence on the concentrations, so as a rule D_{sw} is a nonlinear function of x_s. However, the dependence is not too strong in regions remote from the phase transitions, and the function can be treated as invariable in basic models of AW phenomena.

Thus, the representation (1.2) for fluxes in the basic model is sufficiently universal. Recall that it would be incorrect to put $D_{ij} = 0$ $(i \neq j)$ in Eq. (1.2) from the very beginning. For example, the transport in systems like $(A_- \cdot B_+)_2$, $(A_- \cdot C_+)_3$ and $(w\text{-solvent})_1$ must be described by four coefficients, D_{22}, D_{33} and D_{23}, D_{32}, and two independent fluxes I_2 and I_3. An analysis, such as that in Eqs. (2.36), (2.37), (2.45) but much more cumbersome, shows that under the condition of complete dissociation one should expect $D_{23} < 0$ and $D_{32} < 0$, while in order of magnitude they are comparable with D_{22}, D_{33}. Therefore such a system can be considered as a basic model for investigating effects of mutual diffusion in AW phenomena. Note that in the following we shall call a system n-component according to the number of equations n in (1.2) (in other words, according to the number of independent fluxes).

2.4 Integro-differential equations and their reduction to the basic model

Evidently, the mathematical models analysed in the present book starting from the basic equations cannot be considered as a unique way to describe AWPs. Actually, in the general case any set of partial differential equations can be transformed to a set of integral or integro-differential equations. To this end one can use, for example, the Green function of the linearized boundary problem. The inverse statement is not true, in general, since integral equations provide a more comprehensive class of descriptions of AW systems. Below we give some examples where systems of this type are reduced to the basic models, (1.1) to (1.2). This subject is important, as a lot of basic structure types are already known for Eqs. (1.1) to (1.2), and qualitative investigation methods have been developed in this field. On the other hand, an understanding of the conditions under which the reduction is possible must be useful as an indication of the observable phenomena, the adequate models of which must be based on more general scheme than those given by (1.1) to (1.2).

It is noteworthy that the basic models (1.1) to (1.2) bear the same relation to the integro-differential models, as the axiomatic models (Section 3.5) to the basic models. From this viewpoint, the basic models are intermediate, as to the complication level of the proposed description of the objects and phenomena themselves. As compared with the axiomatic models (finite automates) they provide a possibility to investigate the mutual stability of AWP under study. It is difficult, and sometimes impossible in principle, however, to account in their framework for specific features of space interactions which take place in some media. For instance the starting point of the theory of mass transport must be the integro-differential equations (2.2), which can be reduced with sufficient accuracy to the basic form (1.1), (1.2) only in separate, though very important, cases. For ecological systems, finiteness of the particle velocity should be, seemingly, always taken into account. In this case Eq. (2.6) containing a flux relaxation time is, probably, more realistic than Eq. (1.1), but no accurate reduction to differential equations can be performed in the case of limited, not delta-shaped, velocity distributions. In heat conduction one also encounters examples of problems where integro-differential equations must be used (Tolubinskii 1967). Recall (cf. Section 2.1) that one of the necessary conditions under which the reduction is possible is $v_d \ll v_0$, (2.15), where v_d is the velocity of a nonlinear AWP, and v_0 is the maximal velocity of the "particle motion" in the medium. Further we note that integro-differential equations appear also in the theory of AWPs in ionic systems (Ebeling and Feistel 1983). Now we turn to another class of problems which can be reduced to the basic model. Namely, we consider media where one has direct couplings not only between adjacent "elements", but also between remote ones. Integro-differential equations are involved in the description of such media. The relevant examples are neuron ensembles (Markin, Pastushenko and Chizmadzhiev 1981, Smolyaninov 1980), electronic luminescence amplifier-converters of images (ELA) (Dubinin 1973, 1974). Let us consider a model of an ELA.

A typical scheme of an ELA is shown in Fig. 2.4. The ELA is a glass plate "6" covered with a transparent layer of SnO_2 (or InO_3, or CdO), which is the rear contact of the amplifier. An electroluminophor layer "4" is deposited on the contact. A screen "3" is placed between the electroluminophor and a photosensitive layer "2"; the screen is necessary to produce (or, inversely to break) positive feedback between the electrodes. The input optical image which is formed in the photosensitive layer "2" is converted into an image at the surface of electroluminophor "4", with given resolution, contrast range, and brightness. The second electrode is either a transparent conductive film, or a grid of metallic conductors positioned in the photosensitive

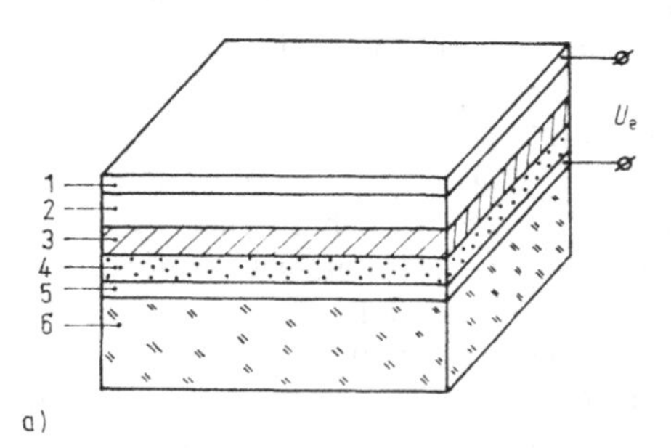

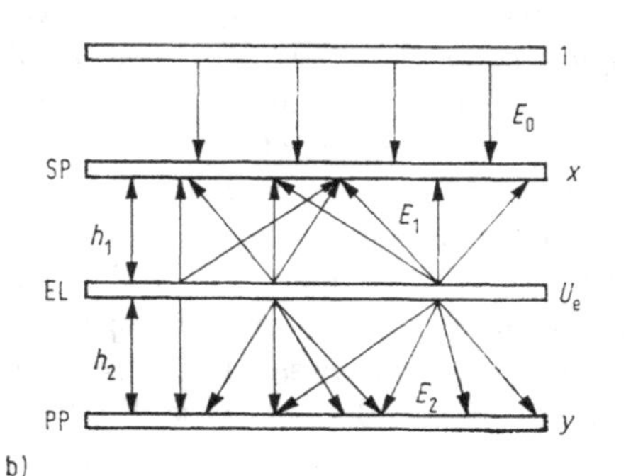

Fig. 2.4

a) Schematic of the electroluminescent image converter (the legend is given in the text);

b) Schematic of a homogeneous model for a biological system constructed of electroluminescent light amplifiers (Dubinin 1974):

1 — transparency; SP — photoresistors connected in series; EL — electroluminophor; PP — photoresistors connected in parallel; E_0 — exciting extrinsic luminosity; E_1 and E_2 — exciting and stopping intrinsic luminosities; h_1 and h_2 — distances between the plates; U_e — voltage applied to the mosaic electroluminescent capacitors

layer. An alternating exciting voltage is applied between the electrodes. ELAs are manufactured in mosaic and continual variants.

Under certain simplifying assumptions, the mathematical model for the scheme considered can be written as (Dubinin 1973, 1974) (see Fig. 2.4b)

$$\tau_1 \frac{\partial x}{\partial t} = x_0 - x + \left\{ E_{01} + \int_{-\infty}^{\infty} S_1(\xi - r)\, B(\xi, t)\, d\xi \right\},$$
$$\tau_2 \frac{\partial y}{\partial t} = y_0 - y + \left\{ \int_{-\infty}^{\infty} S_2(\xi - r)\, B(\xi, t)\, d\xi \right\}, \tag{2.46}$$

where x, $y(x_0, y_0)$ are operating (dark) conductions of the photoresistors, τ_1 and τ_2 are their time constants, $S(\xi - r)$ is a bell-shaped coupling function, and $B(\xi, t)$ depends on x and y,

$$B(\xi, t) = bq^{\alpha} \left\{ (x + y)/\sqrt{(x + y)^2 + \omega^2 C^2} \right\}, \tag{2.47}$$

$b =$ const, q is the exciting voltage amplitude, ω is its frequency, c is the system capacitance, $\alpha = 2 \cdots 6$.

Let us substitute $\xi = \eta + r$, and write the expansion of $B(\eta + r, t)$ around the point r,

$$\int_{-\infty}^{\infty} S(\xi - r)\, B(\xi, t)\, d\xi = \int_{-\infty}^{\infty} S(\eta)\, B(\eta + r, t)\, d\eta = \sum_{i=0}^{\infty} b_i \frac{\partial^i B}{\partial r^i};$$
$$b_i = \int_{-\infty}^{\infty} \frac{1}{i!} S(\eta)\, \eta^i\, d\eta. \tag{2.48}$$

As $S(\eta)$ is symmetrical with respect to the origin, $b_i = 0$ for odd i. The even coefficients in (2.48), b_{2l}, are determined mainly by the shape of the coupling function $S(\eta)$ which is the kernel of the integral transformation. Suppose, for definiteness, that $S(\eta)$ is a Gaussian distribution, $S(\eta) = \exp(-\eta^2/r_0^2)$ where r_0 is the distribution width. Then we get $b_i = b_{2l} = b_i = b_{2l} = \left[1 \cdot 3 \cdot 5 \cdots (2l - 1) \sqrt{\pi}\, (r_0)^{2i+1}\right]/(2i)!\, 2^i = \beta_i r_0^{i+1}$ ($\beta_i < 1$ for $i > 2$). The scale of the sharpest variations of x and $y(\lambda)$ can be used to evaluate derivatives of the function B; e.g. for quasiharmonical processes the scale is given by wavelengths. The order of magnitude of the terms in the expansion is $\beta_i (r_0/\lambda)^i$. Therefore in the long-wave approximation one may retain only the first terms in the expansion (2.48),

$$\tau_1 \frac{\partial x}{\partial t} = x_0 - x + (b_{10} B + E_{01}) + A_b b_{b2} \frac{\partial}{\partial r} \left[D_{xx} \frac{\partial x}{\partial r} + D_{xy} \frac{\partial y}{\partial r} \right],$$
$$\tau_2 \frac{\partial y}{\partial t} = y_0 - y + b_{20} B + A_T b_{20} \frac{\partial}{\partial r} \left[D_{yy} \frac{\partial y}{\partial r} + D_{yx} \frac{\partial x}{\partial r} \right], \tag{2.49}$$

where

$$D_{xx} = D_{yy} = D_{xy} = D_{yx} = \frac{\partial B}{\partial x} = \frac{\partial B}{\partial y} = bq^{\alpha}\omega^2C^2/[(x + y)^2 + \omega^2C^2].$$

Thus, in the long-wave approximation the model (2.49) can be reduced to the basic model describing a number of AWPs including the localized stationary dissipative structures, travelling fronts. It should be emphasized that the question of whether the long-wave approximation is valid can be answered reliably only after an analysis of the reduced model (2.49).

Note that the "mutual diffusion" coefficients in the model discussed are equal to the "proper diffusion" coefficients, and $D_{xy} \neq 0$ at $x = 0$, $D_{yx} \neq 0$ at $y = 0$. The "generalized" basic models are different in this point from the transport models considered in the preceding Section. Models of this type include those with $D_{ii} < 0$ (or det $\{D_{ij}\} < 0$). For example, if $B(x, y)$ in Eq. (2.46) is a function such as $x/\left((x+y)^2 + \omega^2C^2\right)$, one has $D_{yy} < 0$ and $D_{xx} > 0$. Of course, the generalized models are interesting, but they deserve a special approach. So at present we restrict ourselves to the models with $D_{ii} > 0$ and det $\{D_{ij}\} > 0$.

Concluding this illustrative discussion of integro-differential equations, we should note that their reduction to Eqs. (1.1) and (1.2) is rather useful for a qualitative analysis of possible solutions; e.g. it is simpler to classify the basic models (1.1), (1.2) according to the solution types. As to an accurate investigation, the original models may have certain advantages; in particular, numerical experiments are simpler for integral equations.

2.5 Anisotropic and dispersive media

One of the most specific attributes of AWPs and their models is a spontaneous breaking of the system symmetry, for example, transitions from space-homogeneous states to some structures. The admissible variety of forms of the corresponding automodel solutions is determined both by the geometry of the system boundaries and characteristics of the media. However, the spatial form of the solutions depends not only on the configuration of boundaries but also on the anisotropy of the medium. The fact that the symmetry of the structures is reduced, is remarkable in our opinion. The anisotropy of the medium may produce an essential effect on the conditions under which basic structures arise, and on their mutual transformations. Therefore, though AWPs in anisotropic media are in general beyond the framework of the present book, we feel it would be expedient to present here a suggestion concerning a modification of the basic model in the presence of an anisotropy.

The starting point of the deduction is again the Stefan-Maxwell equations (Section 2.3). The diffusion coefficients of the components must be replaced in these equations by matrices $\{D'_{ij}\}_{pq}$ $(p, q = 1, 2, 3)$. The principle of microscopical reversibility leads to Onsager symmetry: $\{D'_{ij}\}_{pq} = \{D'_{ij}\}_{qp}$. Introducing the expression for independent fluxes, as in (1.2), we get $\{D_{ij}\}_{pq} = \{D_{ij}\}\, a_{pq}$, and this factorized representation holds for an arbitrary choice of characteristic velocity (Section 2.3). The tensor $\mathbf{a}_{pq}$ is always symmetrical. There are five groups of such tensors, according to the number of nonzero elements: $\mathbf{A}_{(1)}$ (6 nonzero elements), $\mathbf{A}_{(2)}$ (4 elements), $\mathbf{A}_{(3)}$ (3), $\mathbf{A}_{(4)}$ (2), $\mathbf{A}_{(5)}$ (1 element, isotropic medium).

In solids the type of the transport tensor is determined by the crystal pattern: $\mathbf{A}_{(1)}$ holds for a triclinic lattice, $\mathbf{A}_{(2)}$ for a monoclinic lattice, $\mathbf{A}_{(3)}$ for rhombic and triagonal lattices (e.g. in quartz), $\mathbf{A}_{(4)}$ for triagonal, tetragonal and hexagonal lattices, and $\mathbf{A}_{(5)}$ for a regular (cubic) lattice. Of course, $\mathbf{A}_{(5)}$ is specific also for amorphous bodies. It is noteworthy that the symmetry of the tensors $\mathbf{A}_{(i)}$ reflects the microscopic symmetry of the medium, so the space symmetry of nonequilibrium structures cannot be higher than the internal symmetry of the medium.

The basic representation of flux (cf. Eq. (1.2)) can now be written as follows

$$I_{ip} = -\sum_{j=1}^{n} D_{ij} \left(\sum_{q=1}^{3} a_{pq} \frac{\partial x_j}{\partial r_q} \right), \quad i = 1, 2, \ldots, n, \quad p = 1, 2, 3. \tag{2.50}$$

It is a particular consequence of (2.50) that for media with $\mathbf{A}_{(1)}$ and $\mathbf{A}_{(2)}$ plane waves cannot propagate in a direction orthogonal to the wave fronts. In all the cases, except for $\mathbf{A}_{(5)}$, rotational symmetries are absent. This fact must result in a great variety of structures (leading centres, dissipative structures, reverberators etc.) in different media.

AWs are observed in solids, mainly as thermal waves. The expression in (2.50) holds for the heat flux. A considerable amount of anisotropic media can be found among dispersive media, in particular, among biological tissues. As a rule, they have a transport anisotropy of the $\mathbf{A}_{(4)}$ type. Even intracellular protoplasm can be anisotropic because of intracellular microstructures; syncytial meshs and other similar objects, as a rule, have an anisotropy in the number of intercellular contacts. The cell geometry itself is the reason of an anisotropy in tissues (media). The nature of this phenomenon can be illustrated as follows.

In living tissues, cells are ellipsoidal, and the orientation of their cardinal axes is ordered. Distances between the cell membranes is considerably less than the cell sizes. The membranes and the protoplasm (intercellular liquid) have substantially different diffusion resistivities. The biochemical transformations responsible for the substance activity (in the AWP sense) take place inside the cells. Strictly speaking, such a medium must be simulated in terms of a system of integro-differential equations. At the qualitative level, however, cells can be represented as rectangular bodies and fluxes can be considered as being quasistationary; in other words, one can suppose that the relaxation times for fluxes between adjacent cells is substantially less than the specific times of intracellular reactions.

Let the distance between the cell centres along the $0r_1$ axis be R_1, the membrane thickness be l, the diffusion coefficient for a component x with respect to water be D_x [cm²/s], and the membrane penetrability be P (cm/s). An expression for the effective diffusion coefficient is obtained from the requirement that the fluxes must be invariable at the length of R_1,

$$\frac{1}{D_{x_1}} = \frac{(1 - h/R_1)}{D_x} + \frac{1}{PR_1}. \tag{2.51}$$

Analogous expressions are obtained also for D_{x2} and D_{x3}. It is seen that D_{xp} depends on the structure of the tissue because the parameter R_p is involved, and such a tissue is an anisotropic medium. One may consider, for example, nonstriated muscle tissues (Section 9.1); in this case the anisotropy is quite essential: actually, travelling autowaves can propagate only along the direction with the maximal value of R_p. In other directions the coupling coefficients are below the threshold, and AWs are not excited.

The presence of threshold properties in media, combined with an anisotropy, require special care when general results of the theory are compared with the behaviour of real objects. Evidently, qualitatively new AW phenomena are possible in strongly anisotropic media, but they have not yet been studied at present.

2.6 Examples of basic models for autowave systems

An accurate mathematical description of most of real AWPs can be performed only in terms of multicomponent models (1.1) to (1.3). However, as often noted above, some qualitative properties of AWPs, and sometimes their important quantitative characteristics, like the velocity, approximate shape of TF or TP, and others, can be obtained by means of first-, second-, or third-order basic models.

The basic models are constructed either by means of a reduction of systems of higher-order equations, or by starting from known physico-chemical mechanisms, or identifying solutions of the model with the observed kinetic space–time dependence. A model may be considered as "basic" if it can be used for a simple and spectacular elucidation of the conditions necessary for the creation and existence of certain AWPs. The theory cannot be built with a single basic model, for either the model would be too complicated and awkward, or it would only be able to describe the AWP properties in part. On the other hand, different basic models are not at all independent, their solutions may correspond to the same types of AWP. Below we present a list of the most important basic models are used in this book, as well as in most of the papers dealing with the theory of AWP.

1. The simplest class of AWP described with a mathematical model initially is the travelling front (TF) or the flip-over wave. As was mentioned in the Foreword, two first studies along this line were performed independently and treated the same problem in biology, propagation of dominant genes in space (Kolmogorov, Petrovskii and Piskunov 1937, Fisher 1937). The subject of the work was the solutions of a one-component model of the type (1.1)—(1.3), for $n = 1, D_1 =$ const. The same model was used in the theory of spreading flame (Zeldovich and Frank-Kamenetsky, 1938) and propagation of waves of phase transitions, both in volumes and on surfaces (see Chapter 1). Of course, TF solutions appear in more complicated models; these aspects are considered in Chapter 4. Note also another class of solutions specific for the simplest models, time-independent and spatially periodical solutions. However, as was shown in a number of works (see e.g. Chafee 1974) such stationary solutions are unstable for any $F_1(x_1)$ and constant D_1. Besides, spontaneous creation of metastable localized structures is possible in one-component systems, but only if D_1 has a special dependence on x_1 (see in Chapter 7).

2. The asymptotic behaviour of a planar TF may be not only a uniform translational motion, but also more complicated motions, say, propagation with pulsating velocity. No less than two variables are necessary to describe such situations within the basic model. The corresponding AWP models appeared first in connection with combustion problems (see Sections 1.4 and 4.3). The following system of equations is used as the basic model,

$$\frac{\partial x}{\partial t} = \gamma Q(x, y) + D_x \frac{\partial^2 x}{\partial r^2},$$
$$\frac{\partial y}{\partial t} = Q(x, y) + D_y \frac{\partial^2 y}{\partial r^2}. \tag{2.52}$$

The variables involved here are usually: x — temperature of the medium, y — concentration of a reagent limiting the combustion reaction. The function $Q(x, y)$ satisfies Arrhenius' Law, i.e.

$$Q(x, y) = y^q \exp\{-E/kx\},$$

where E is the activation energy, k Boltzmann's constant, q a constant. A simplified representation of the source function is used in qualitative investigations, $Q \sim (x - x_1)(x - x_2)\, y$. If $D_x = D_y$, after a change of variables, $x' = x - \gamma y$, $y' = x + \gamma y$, we get the single equation discussed above. Thus, the simplest model of a TF is a particular case of the model represented as Eqs. (2.52).

3. The next class of AWPs are travelling pulses (TPs). In this case the basic model is given by equations of the form,

$$\varepsilon \frac{\partial x}{\partial t} = F(x, y) + D_x \frac{\partial^2 x}{\partial r^2},$$
$$\frac{\partial y}{\partial t} = \varphi(x, y). \tag{2.53}$$

The zero isoclinic curve for variable x ($P(x, y) = 0$ in the (x, y) phase plane) is N-shaped, and ε is a small parameter. Such a form of the isoclinic line results in a local positive feedback: $\partial P/\partial x|_\Omega > 0$, where Ω is a domain in the (x, y) plane. Thus in this domain the variable x has autocatalytic properties. In the following x stands for the autocatalytic variable, unless specified otherwise.

A particular, but vast, family of models has a special name: the Fitz-Hew-Nagumo-model (Scott 1975). In this case

$$P(x, y) = -y + x - x^3, \qquad Q(x, y) = b - y + \gamma x.$$

This model enables one to investigate stationary propagation and formation of TPs (Chapter 4), fission of TP fronts, AW "echo" regimes (Chapter 5) and other phenomena. If b and γ acquire values such that the point system has three singular points, only TF can propagate in the distributed system. Putting $y =$ const, one can see quite clearly that the model in (2.53) is an extension of the pioneer model of Kolmogorov, Petrovskii, Piskunov, and Fisher, as the first equation in (2.53) is transformed to that describing the TF propagation. If a third variable is introduced into Eqs. (2.53) (one may consider, say, γ as a variable), we get a model for stable leading centres (Section 5.3). From this point of view, the model considered is basic for the investigation of AWPs incorporating a TP as the main element.

The study of the propagation of a nerve pulse is an example giving reliable evidence that some very complicated systems can be reduced to a basic model of TPs. The analysis of experimental data, which was performed in the first studies in this field (Hodgkin and Huxley 1952, Noble 1962), was the basis for four-component models solved by means of computers. V. I. Krinskii and his collaborators, in a series of studies (Krinskii and Kokoz 1973, Ivanitskii, Krinskii and Selkov 1978), have found not only the conditions under which complicated models can be reduced to the basic form (2.53); they also proposed an experimental method enabling one to construct zero-isoclinic lines for the basic model.

Thus, the Fitz-Hugh-Nagumo-type equations may be considered as the simplest model describing TPs. Other models use either more complicated functions $P(x, y)$ and $Q(x, y)$, or more general diffusion couplings, or both. For instance, functions $P(x, y)$ have been considered for which the equation $P(x, y)|_{y=\text{const}} = 0$ has five and more zeroes. One should bear in mind that there are variations of the form of zero-isoclinic curves which do not substantially affect the behaviour of the local system, but manifest themselves quite clearly in the character of AWPs in a distributed system. Therefore any single basic model is not sufficient for the theory of TPs. The inhomogeneous Fitz-Hugh-Nagumo model with coefficients depending on the coordinate can be exploited in solving some synchronization problems. If the diffusion of the y component is also introduced, with $D_y \gg D_x$, the extended model describes stationary periodical distributions, or DSs. Other models are usually used, however.

4. The "Brussellator" model plays a unique role in the theory of DSs and, in particular, in theoretical biology; the number of studies devoted to this object is very large. The model is named after I. Prigogine's Brussel School, members of which contributed greatly to its investigation (see monographs by Glansdorff and Prigogine 1971 and Nicolis and Prigogine 1977; other references are given in Chapter 7).

The scheme of chemical reactions which is the basis for the local kinetics of a Brussellator originated in the pioneering work of A. Turing 1952; it is

$$\text{(i) } A \underset{k_{-1}}{\overset{k_1}{\rightleftarrows}} X, \qquad \text{(ii) } 2X + Y \underset{k_{-2}}{\overset{k_2}{\rightleftarrows}} 3X,$$

$$\text{(iii) } B + X \underset{k_{-3}}{\overset{k_3}{\rightleftarrows}} Y + D, \qquad \text{(iv) } X \underset{k_{-4}}{\overset{k_4}{\rightleftarrows}} E.$$

The resulting reaction corresponding to this scheme is $A + B \rightleftarrows E + D$, while the ingredients X and Y are intermediate. The second reaction is autocatalytic. Assuming for simplicity that all the direct reaction rate constants are equal to unity, and the inverse reaction rate constants are all k_-, one gets the following kinetic equations for the concentrations $[X] = x$, and $[Y] = y$,

$$\frac{\mathrm{d}x}{\mathrm{d}t} = A + x^2y - Bx - x + k_-(yD + E - x - x^3),$$

$$\frac{\mathrm{d}y}{\mathrm{d}t} = Bx - x^2y + k_-(x^3 - yD).$$

As a rule, the distributed version of the model is analysed under an additional assumption; namely, that the inverse reactions are neglected ($k_- = 0$). This simplified model is called a Brussellator.

If $B > A^2 + 1$, the Brussellator has a limit cycle. After the substitution $x = x'$ and $x + y = y'$, one gets a system with an N-shaped x'-zero isoclinic line. At this point, the Brussellator is like the Fitz-Hew-Nagumo fundamental basic model. However, unlike the latter, here $F(x', y')$ contains a quadratic term which is responsible for some special features of the model relevant to DS (Chapter 7).

The equations of the basic model, the distributed Brussellator, which will be used in subsequent Chapters, are as follows,

$$\frac{\partial x}{\partial t} = A + x^2y - (B + 1)\,x + D_x \frac{\partial^2 x}{\partial r^2},$$
$$\frac{\partial y}{\partial t} = Bx - x^2y + D_y \frac{\partial^2 y}{\partial r^2}, \tag{2.54}$$

where A and B are constants.

5. It was found that duely accounting for mutual diffusion of the components enables one to construct AW models in some cases where the local systems have no active properties. An interesting example is provided by the following scheme of reactions,

(i) $A + Y \to X$, (ii) $B + X \to Y + D$,

(iii) $X + X \to C$, (iv) $Y + P \to E$.

This system tends to a stationary state from arbitrary initial conditions.

According to the results of Section 2.3, the distributed model is given by the equations,

$$\frac{\partial x}{\partial t} = Ay - Bx - x^2 + \frac{\partial}{\partial r}\left(D_{xx}\frac{\partial x}{\partial r} + D_{xy}\frac{\partial y}{\partial r}\right),$$
$$\frac{\partial y}{\partial t} = Bx - P_{yy} + \frac{\partial}{\partial r}\left(D_{yy}\frac{\partial y}{\partial r} + D_{yx}\frac{\partial x}{\partial r}\right). \tag{2.55}$$

It can be shown (see Chapters 3 and 7) that for negative values of D_{xy} and D_{yx} one can always find concentrations of the substance A leading to DSs in the system. It will be shown in Chapters 3 and 5 that the system (2.55) with a third variable introduced is a model for AWPs such as leading centres, standing waves etc. in the cases where no active kinetics appear in the local systems. Probably, models of the type (2.55) should be considered as a separate class of basic models. Such models describe various AWPs under minimal requirements on the local kinetics. The main requirement is that there must be a sufficiently large "influx" from outside; this is the influx of the substance A in the case of Eqs. (2.55). In contrast to the basic models presented above, in this case the solutions can hardly be analysed qualitatively.

6. The so-called $\lambda - \omega$ systems are sometimes considered in the theory of AWPs,

$$\frac{\partial W}{\partial t} = [F(R) + jP(R)]\, W + \operatorname{div} \operatorname{grad} W, \tag{2.56}$$

where $W = x + jy$ ($j = (-1)^{1/2}$). The complex equation is equivalent to a pair of real equations,

$$\frac{\partial x}{\partial t} = F(R)\, x - P(R)\, y + \operatorname{div} \operatorname{grad} x,$$

$$\frac{\partial y}{\partial t} = F(R)\, y + P(R)\, x + \operatorname{div} \operatorname{grad} y.$$

Here F and P are real smooth functions, $R = W = (x^2 + y^2)^{1/2}$. It is also assumed that $F(R) > 0$ for $R < 0$, $F(R) < 0$ for $R > 0$, and $F(0) = 0$; $P(R) > 0$ for $R > 0$. Because of these functions, the local system has a stable circular limit cycle. Iterative methods based upon the Floquet theory are appropriate for analysis of AWPs in $\lambda - \omega$ systems, as the exact zero (homogeneous) approximation is always known. These methods were used to show the variety of AWPs in reactive systems with diffusion (Ortoleva and Ross 1974); one- and two-dimensional waves, and in particular rotating waves (reverberators) were detected.

7. The AWP theory was substantially contributed to by investigations of the models for the Belousov-Zhabotinskii distributed chemical reaction. The gross scheme of this reaction is (Zhabotinskii 1974)

(i) $Ce^{3+} \xrightarrow[H^+]{BrO_3^-} Ce^{4+}$, (ii) $Ce^{4+} \xrightarrow[Br^-]{BMA} Ce^{3+}$.

Here BMA is brommalon acid, and cerium ions can be replaced by ions of iron F^{2+} (Fe^{3+}), manganese etc. Thus there is a class of different reactions proceeding according to the same scheme. The experimental data indicate that autocatalysis takes place in some intermediate reactions, even though it is not clear yet which substances are responsible for the autocatalysis. Actually, the system is extremely complicated, although its gross scheme is simple, and several dozen of components are present at the intermediate stages. A number of models were proposed for the reaction; the first of these was given by Zaikin, Korzukhin and Kreytser (1971):

$$\frac{dx}{dt} = \beta y(1 - x) - \gamma x,$$

$$\varepsilon \frac{dy}{dt} = \beta y[1 - x(1 + \alpha_1 + (y - \alpha_1)^2)] + \varepsilon\beta,$$

where x is the concentration of Ce^{3+}(Fe^{2+}) ions, and y is the concentration of the autocatalysing substance. The coefficients in the model were chosen phenomenologically by comparison of the auto-oscillation parameters with the experimental data.

Field and Noyes proposed a scheme for the reaction and used it to construct a model known under the name "Oregonator". The equations of the model are

$$\frac{dx}{dt} = py - \gamma x, \quad \frac{dy}{dt} = py - qyz - \alpha y^2 + \tau z,$$
$$\frac{dz}{dt} = \gamma f x - qyz - \tau z, \tag{2.57}$$

where z is the concentration of Br^- ions, x and y are the same as above. Assuming that z is a rapid variable in the model (2.57) and reducing the system one gets a two-component model with the structure of the (x, y) phase plane completely coinciding with that of the Zaikin-Korzukhin-Kreytser model. The most essential (and, perhaps, the only) feature distinguishing these models from the Fitz-Hew-Nagumo system is the presence of a quadratic nonlinearity. We consider the Oregonator as one of the basic models; the complete system (2.57) is analysed as a model for standing waves (Section 5.5), in other cases we investigate the system with diffusion reduced in the z-component (Sections 5.4 and 7.4).

In conclusion, we remind the reader once more that a complete list of basic models for AWPs is not presented here; this is hardly possible at present. The examples of the investigation of models given above, as well as some other models, are just an illustration of the general concepts of AWP theory. They provide a fairly exhaustive description of characteristic structures in the AW media which are known at present.

Chapter 3
Ways of investigation of autowave systems

We do not use the word "methods" in the title of this section since we wish to emphasize that a qualitative theory of AW processes is still to be constructed. This can be illustrated by comparing the undeveloped approaches used for studying AW systems with the state-of-the-art investigation of the nonlinear systems of the second order whose behaviour in the parameter space is known in ample detail. So far, only some particular, though rather important, cases have been considered. However, qualitative and computer methods for studying AW processes simulated by parabolic quasilinear equations seem to be promising. The point is that it is sufficient to take into account a finite (usually small) number of degrees of freedom. Therefore the main problem which arises in analyzing AW processes, is to recognize these several degrees of freedom, to find stationary solutions, and to investigate their stability and transitions between them.

3.1 Basic stages of investigation

The choice of a particular method in attempting to find solutions depends on the properties of the system being investigated. Most nonequilibrium systems under consideration have homogeneous stationary states. The nonequilibrium properties and diffusion coupling between the points of the system can lead to the existence of several inhomogeneous stationary solutions, which are either structures that do not change with time or stationary waves that propagate at a constant speed. Some methods of searching for such stationary solutions will be discussed in Section 3.2. One of the main problems arising here is choosing the basic characteristics of these AW processes: front velocity, TP width, and others.

The next step is to determine the stability of the stationary solutions (Section 3.3) by the small perturbation method. For small nonlinearities (or quasi-harmonic systems), concepts of increments or decrements of harmonics, dispersion, etc. are used.

A consideration of solutions that can occur in strongly nonlinear systems needs the knowledge of their stability and inhomogeneous stationary states including stationary waves. In this case, the stability of each stationary solution is determined for the whole set of the perturbation modes. In such an analysis, data on the unstable stationary solutions are also important. The well-known examples of AWPs indicate that the unstable structures are typically boundaries between the regions attracting stable stationary structures.

In a local system of the second order, knowledge of the stationary states, the type of their stability (centre, saddle, node, focus and complex states) and the law that the phase trajectories do not intersect allow for a complete qualitative representation of the behaviour of the system. In a distributed system, the part of the stationary states is played by stationary waves or stationary structures. Their stability is determined in a much more complicated manner. The type of stability (focus, saddle, node, etc.) depends on the perturbation wavenumber. However, it is still more difficult to predict transitions from unstable stationary states for each perturbation type to other, more stable stationary states or to another attracting set. Typical solutions, for which such an analogy to qualitative methods for local systems is physically relevant, will be given in the subsequent chapters. Note also that most of the known methods for analyzing AWPs, although seemingly different, have certain features in common: A description of AWPs can, to a certain extent, be divided into the following stages:

a) distinguishing characteristic structures; typically, these are DSs or stationary waves (self-simulating solutions) but only simplified characteristics of these structures are determined in this step;
b) investigating structures for stability;
c) defining transitions from unstable to stable states and transitions from stable states to other states.

All of these stages are, in a sense, included in the axiomatic description in Section 3.5 and in the method of successive approximations (nearly the same as the method of iterations) in Section 3.7 and in the methoas of dividing space-time processes into fast and slow stages in Section 3.7. An important qualitative concept of solutions in distributed systems as well as of the numerical counting problem in Section 3.9 is furnished by the simplest discrete models in Section 3.6. In addition, permissible symmetries of AW solutions in space can be listed by means of a group-theoretical approach (Section 3.8).

3.2 A typical qualitative analysis of stationary solutions in the phase plane

One-dimensional stationary structures and waves are described by mathematical models in the form of sets of differential equations in ordinary derivatives. Such models are obtained from the input equations in partial derivatives by introducing a self-simulating variable. Typically, $\eta = r \pm vt$ is used as a self-simulating variable. It appears that the velocity v is the eigenvalue of the nonlinear boundary-value problem. Such solutions are generally referred to as self-simulating solutions of the second kind (Barenblatt, Zeldovich 1971). The ways of solving nonlinear equations in ordinary derivatives have now been developed well enough (see, e.g., Andronov, Vitt, Khaikin 1981). As a rule, a qualitative consideration of possible solutions is performed in the phase space. Many examples of such solutions will be listed below in Chapters 4, 5, 6 and 8. As an illustration of such a qualitative analysis, we will consider a simple AW system the mathematical model for which is defined as Let

$$\varepsilon \frac{\partial x}{\partial t} = \frac{\partial^2 x}{\partial r^2} + F(x, y) \tag{3.1 a}$$

$$\frac{\partial y}{\partial t} = \varphi(x, y), \tag{3.1 b}$$

the function $F(x, y) = -y + x - x^3$ and $\varphi(x, y)$ be a piecewise linear function The null isoclines of set (3.1) are shown in Fig. 3.1. Here, $\varphi(x, y) > 0$ on the right of the y-null isocline and $\varphi(x, y) < 0$ on the left. In a local system with such isoclines autooscillations occur.[1])

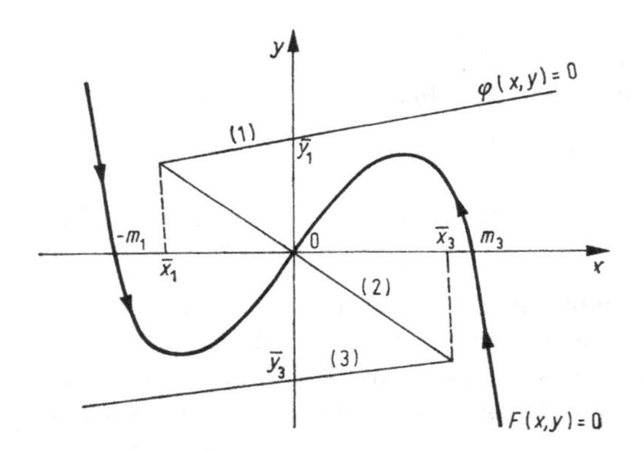

Fig. 3.1
The null isoclines of "local" equations corresponding to the sets (3.1) and (3.2)

Let us demonstrate that two types of fixed stationary solutions can exist in set (3.1). Assume that the solutions are time independent and consider the equations

$$\frac{d^2 x}{dr^2} - y + x - x^3 = 0, \quad \varphi(x, y) = 0 \tag{3.2}$$

for $dx/dr|_{r=\pm L} = 0$ (where $\eta = r$). The solutions of (3.2) will be analyzed in the phase plane x, dx/dr.

The first type of the stationary solutions occurs when the x and y variables change near the zero values ($\bar{x}_1 \leq x < \bar{x}_3$; $\bar{y}_3 < y < \bar{y}_1$ (see Fig. 3.1)). Thus, in trying out solutions, only the branches corresponding to the falling regions of the null isoclines $F(x, y) = 0$ and $\varphi(x, y) = 0$ are used. The stationary solutions correspond to the

[1]) It will be shown in Section 6.3 that in a distributed system, homogeneous autooscillations can lose their stability and inhomogeneous structures (DSs) can therefore arise.

limited trajectories inside a separatrix loop (Fig. 3.2). It is seen that a different number of solutions can exist depending on the length of the segment $2L$ being considered. As an example, three stationary solutions are shown by heavy lines in Fig. 3.2a. The basic distinctive feature of such solutions is a smooth variation of the x and y variables in space (Fig. 3.2b).

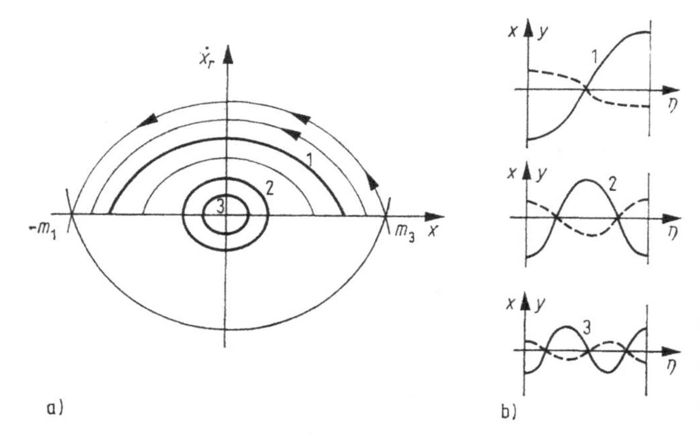

Fig. 3.2
Smooth stationary solutions of (3.2): a) phase trajectories of stationary solutions near the equilibrium point ($x = 0$, $y = 0$); b) spatial form of smooth stationary solutions: $x(r)$ is shown by a solid line, $y(r)$ by a dashed line ($\eta = r$)

Characteristic of the stationary solutions of the second kind is a jump discontinuity of the y variable in space. Such solutions are obtained by matching two solutions. One of them corresponds to branch (1) of the null isocline $\varphi(x, y) = 0$ and the other to branch (3). A transition from branch (1) to branch (3) and, therefore, the jump

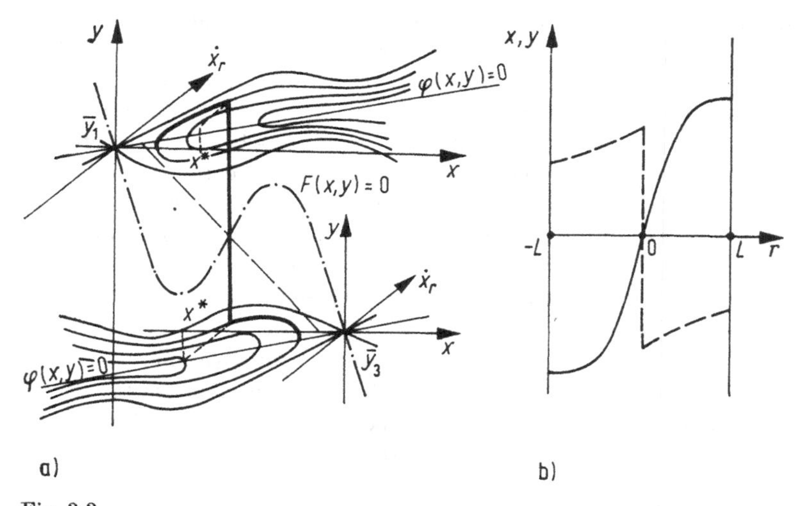

Fig. 3.3
Discontinuous stationary solutions of (3.2): a) the heavy line denotes a discontinuous (in y) stationary solution in the phase space (the dash-dotted line denotes the x-null isocline, and the dashed line, the y-null isocline); b) the change of x variables (solid line) and y variables (dashed line) in space for discontinuous stationary solutions

of the y variable, occur at a certain value x^*, which is chosen from the continuity conditions for x and its variable. Such a trajectory corresponding to a stationary solution with the jump of the y variable is shown in Fig. 3.3a in the phase space x, y, $\mathrm{d}x/\mathrm{d}r$. The solution itself in space is given in Fig. 3.3b. Thus, strongly nonlinear AW solutions are obtainable by purely qualitative methods, although the stability of these solutions is still to be investigated.

3.3 Study of the stability of stationary solutions

In this section, we will consider the following problems:

A. Study of stability "in the small" in the dispersion equation.

B. Stability of the homogeneous states of two-component systems.

C. Stability of three-component systems.

A. Study of stability "in the small" and of the stationary structures and states is the second necessary stage in the qualitative analysis of solutions that are possible in the AW system. Consider the definition of stability of the stationary states in more detail. Particular cases of stability of the inhomogeneous stationary structures will be investigated in the corresponding chapters.

The study of stability is based on Lyapunov's method (Demidovich 1967; Zubov 1961). An appropriate choice of Lyapunov's functions permits one, in principle, to investigate the stability with respect to arbitrary perturbations, not only small ones. However, regular means of constructing global Lyapunov's functions are still unknown and, therefore, skill is needed in such investigations. Specifically, characteristic features of particular systems and processes should be taken into account. As a rule, one has to confine oneself to qualitative results or, at most, to provisional estimates or, rather frequently, to a mere statement that hard transitions between the states of the system take place (see also stationary DSs, LCs and other regimes). In determining the evolution of arbitrary perturbations, a mathematically rigorous solution of the problem has not yet been obtained even as to the appearance of TFs and TPs, problems which are otherwise rather simple and have been investigated well enough. There are certain types of systems for which only perturbation thresholds have been found (see, e.g., the review by A. Scott 1975, Mornev 1981).

For small perturbations of the homogeneous states, the second Lyapunov method was developed by Prigogine et al. in their papers on the thermodynamics of irreversible processes (P. Glensdorf, I. Prigogine 1971; G. Nikolis, I. Prigogine 1977). Here, the entropy production is used as Lyapunov's functions. Study of the stability "in the small" is essentially the problem of definiteness of quadratic forms, which, in turn, is equivalent to the problem of the eigenvalues of the matrices obtained by the first Lyapunov method given below. We stress that both approaches lead to the same results. We prefer to apply the first method merely because of its simplicity and the absence of the stage of constructing Lyapunov's functions (functionals for inhomogeneous states), which is not always a trivial procedure.

Let the solution of set (1.3) be represented in the form

$$x_i(t, r) = \tilde{x}_{im}(t, r) + x'_{im}(t, r), \tag{3.3}$$

where $\tilde{x}_{im}(t, r)$ is the m-th stationary solution being investigated. Specifically, it can be a homogeneous solution $\overline{x}_{im}$. Substituting (3.3) into (1.3) we obtain the equations for x'_{im}, which, in terms of the theory of stability, should be called reduced equations, and represent them in the form

$$\frac{\partial x'_{im}}{\partial t} = M'_{ij} x'_{im} + F_i'\left(x'_{km}, x'_{qm}, x'_{km}, \frac{\partial x_{qm}}{\partial r}\right), \tag{3.4}$$

where M'_{ij} is the linear operator and F_i' does not contain terms linear with respect to x_{im}. The solutions of (3.4) should also satisfy the reduced boundary conditions. Study of the nontrivial solutions of the linearized set (3.4) determines the stability "in the small" of the solution $\tilde{x}_{im}$ for (1.3). Such an investigation is readily performed for homogeneous stationary states $\overline{x}_{im}$ when the matrix M_{ij} has constant coefficients. In this case, any solution can be represented as a superposition of waves of form $A_{imk} \exp\{p_{mk}t\}\, \Psi_{mk}(r)$, where $\Psi_{mk}(r)$ is the k-th eigenfunction of the operator M_{mk}; the spectrum is a discrete one because of the boundary conditions.

We insert the following series into the linearized reduced system

$$x'_{im} = \sum_{k=0}^{\infty} A_{imk} \exp\{p_{mk}t\} \exp\left\{j\, \frac{\pi k_m}{L}\, r\right\}, \tag{3.5}$$

where k_m is the wave number determining the wavelength $\lambda_{mk} = 2L/k_m$. Elsewhere below, where this does not lead to a misunderstanding, we will omit the index "m" corresponding to the number of the state being investigated. Multiplying the equations obtained here by the eigenfunctions $\exp\{j\pi kr/L\}$ and integrating from 0 to L we obtain conditions for the existence of nontrivial solutions ($A_{ik} \neq 0$):

$$\det\left\{M_{ij} - \frac{\pi^2 k^2}{L^2} D_{ij}(x) - p_k \delta_{ij}\right\} = p_k{}^n + q_{n-1}(k^2)\, p_k{}^{n-1} + \cdots + q_0(k_0{}^2) = 0 \tag{3.6}$$

where $\det\{\ldots\}$ is the matrix determinant, δ_{ij} is the Kronecker delta. By Eq. (3.6), the complex frequencies $p_k = \delta_k \pm j\omega_k$ are related to the wavelengths $\lambda_k = 2L/k$ and coefficients of set (1.3). This equation is generally referred to as a dispersion equation or a characteristic equation.

Note that the dispersion equation (3.6) does not change its structure in transition to the discrete analogues of the AW systems, which will be considered in Section 3.6. Taking into account the discreteness of the model leads to the following substitution in (3.6)

$$\hat{D}\left(\frac{\pi k}{L}\right)^2 \div \hat{D}_{\text{discr}}\, 2\left(1 - \cos\left(\frac{\pi k}{N}\right)\right), \qquad k = 0, 1, 2, \ldots, N - 1, \tag{3.7}$$

where $\hat{D}_{\text{discr}}$ is the diffusion matrix of a discrete system composed of diffusion-coupled elements. If the discrete model is taken as an approximation of a continuous model, then the diffusion matrix elements are related by an equation $\hat{D}_{discr} = DN^2/L^2$. It is easy to see that for any mode (its number k is fixed) (3.7) becomes an equality as $N \to \infty$. For small N the difference is great: 100% at $N = 3$, 10% at $N = 10$, and so on. This fact should be taken into account in numerical experiments when determining the positions of the bifurcation points. However, qualitatively, the same result is yielded by the dispersion equations of discrete and continous models.

Stability "in the small" of the trivial solution of reduced set (3.4) is adequately described by the root signs of the dispersion equation. According to the simplest classification of instabilities, two types can be distinguished:

1. oscillatory instabilities and
2. DS instability (also referred to as Turing instability).

Such a classification of instabilities corresponds to two types of AW solutions (e.g., standing waves) and spatially inhomogeneous stationary solutions (DSs) (see Chapter 7). The instability type is determined by the number of roots with positive real part in a manner similar to that for lumped systems. The number of such roots is even in the case of oscillatory instability for the k-th mode and odd for DS instability. Another indication of DS instability is a negative free term of the dispersion equation $\left(q_0(k^2) < 0\right)$. It is pertinent to note that different instability types can arise for different modes.

We now consider typical forms of dispersion dependence. Particular emphasis will be placed upon the relation between the form of this dependence and the number of components in the system, the singularity type corresponding to a lumped model and the presence of nondiagonal terms in the diffusion matrix.

B. The dispersion equation coefficients of a two-component system (1.3) ($n = 2$) are given by

$$\begin{aligned} q_1(k^2) &= -(M_{11} + M_{22}) - (D_{11} + D_{22})\,(\pi^2 k^2/L^2), \\ q_0(k^2) &= \det\{M_{ij}\} - [(D_{11}M_{22} + D_{22}M_{11} - D_{12}M_{21} - D_{21}M_{12}) \\ &\quad + \det\{D_{ij}\}\,(\pi^2 k^2/L^2)] \left(\frac{\pi k^2}{L^2}\right), \end{aligned} \tag{3.8}$$

where $\det\{M_{ij}\} = q_0(0) = M_{11}M_{22} - M_{12}M_{21}$; $\det\{D_{ij}\} > 0$ for systems with active transfer and is less than zero in any other cases.

The dependence $\mathrm{Re}\, p_k = \delta_k$ on wave numbers is given in Figs. 3.4 and 3.5, where I is the oscillatory instability region, II is the DS instability region and III is the stability region.

The oscillatory instability in two-component systems occurs only if the singularity is an unstable focus (Fig. 3.5e, f) or a node (Fig. 3.5a). Such an instability is caused by an autocatalysis, i.e., by a local feedback ($M_{11} > 0$ or $M_{22} > 0$). The self-excitation conditions can be satisfied for a finite number of waves with an upper limit $R < R_{cr} = (M_{11} + M_{22})\,L^2/\pi^2(D_{11} + D_{22})$. Here, the increment δ_k decreases monotonically with decreasing wavelength (Fig. 3.4). Assuming that δ_0 is a small parameter (quasi-harmonic regime) it is easy to show that only oscillations with the maximum wavelengths ($k = 0$ for the second boundary-value problem and $k = 1$

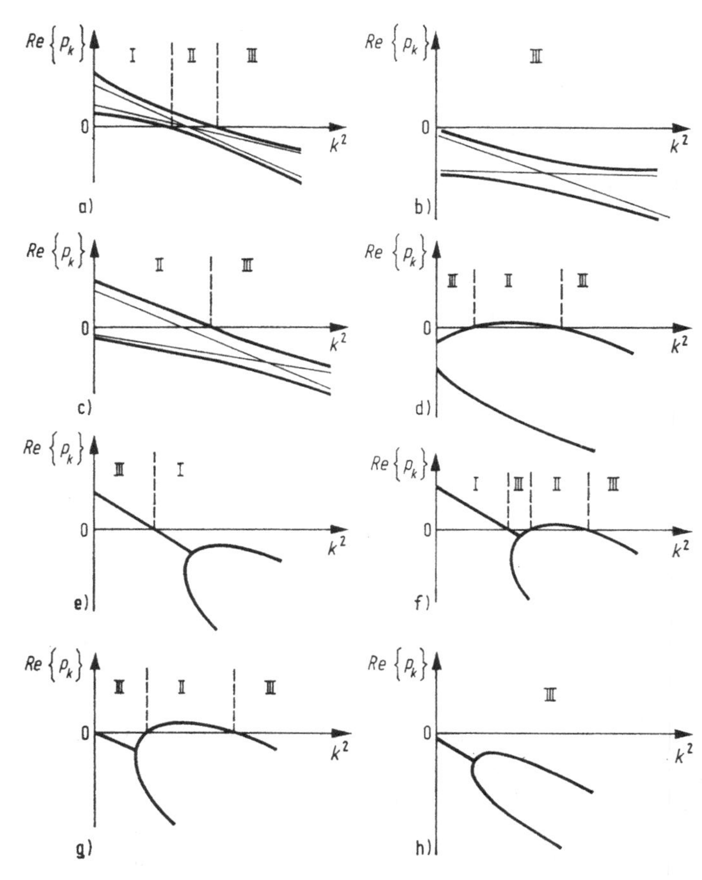

Fig. 3.4
The real parts of the roots of the characteristic equation plotted against the wavenumber for systems with $n = 2$, $D_{12} = D_{21} = 0$, $D_{11} \neq 0$, $D_{22} \neq 0$. I is the oscillatory instability region, II is the DS instability region, and III is the stability region.

for the first one) are stable. The situation is much more complicated in relaxation-type systems (Fig. 3.4a) (see Chapters 5 and 6).

In systems with the diagonal diffusion matrix, the DS instability (regions II in Figs. 3.4 and 3.5) occurs only in the presence of an autocatalysis ($M_{11} > 0$ and $M_{22} < 0$). From the viewpoint of the finite state, the dispersion dependence shown in Fig. 3.4d,e is interpreted most easily. If the dimensions of a system with physically relevant local kinetics are such that at least one mode corresponds to region II, then stationary DSs can occur. We recall that such DSs were first investigated in systems with this dispersion dependence (Turing 1952). The case when there are two types of instability (Fig. 3.5a, f) is not so definite: the DSs exist but their stability, i.e., occurrence, is still to be investigated (see Chapter 7).

Systems with a saddle-type singularity (trigger local kinetics: $q_0(0) < 0$, see Fig. 3.4c) always have a region II of DS instability. Central to these systems is an AWP by which the waves of a transition (TF) (see Chapter 4) propagate from one stable state to another, with dispersion dependence such as shown in Fig. 3.5b, h.[1])

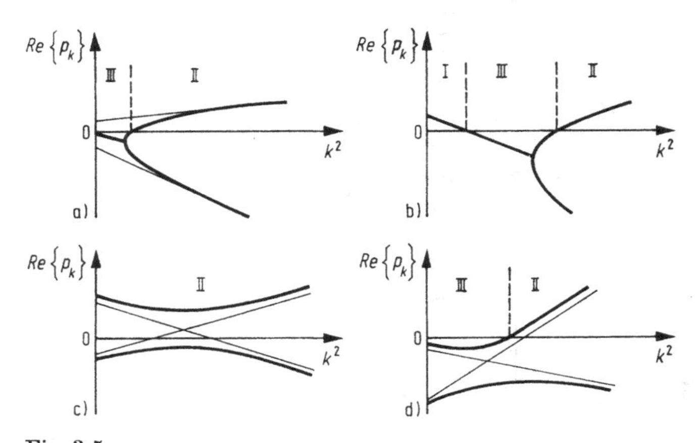

Fig. 3.5
The real parts of the roots of the characteristic equation plotted against the wave number for systems with $n = 2$, $D_{12} = D_{21} = 0$ and $D_{11} \neq 0$ (or $D_{22} \neq 0$). The designations I, II, and III are the same as those in Fig. 3.4.

Finally, the cases shown in Fig. 3.5b, h correspond to an instability with respect to small perturbations of a homogeneous stationary state. By these graphs, which are readily interpreted and need no additional comments, types of dependence $\delta_k = \text{Re}\, p_k(k^2)$ that occur in systems with $D_{11} \neq 0$ and $D_{22} \neq 0$ are exhausted. Note that eight types of the graphs (Fig. 3.4) can be obtained from two of them (e.g., "a", "e") by means of a translation. This corresponds to two types of local kinetics near a stationary state: aperiodic (Fig. 3.4a—e) and "quasi"-periodic (Fig. 3.4e—h).

It is important to note that in systems with $D_{12} = D_{21} = 0$, an instability can arise only in the presence of a local positive feedback (autocatalysis): $M_{11} > 0$ and/or $M_{22} > 0$. In such systems, the diffusion "reveals" a potential instability of the local kinetics in the system. In systems with the mutual diffusion of the components $D_{12} \neq 0$ and $D_{21} \neq 0$, det $\{M_{ij}\} < 0$, the DS instability is possible even if $M_{11} < 0$ and $M_{22} < 0$, i.e., if the singularity is stable at all parameter values. Here, the dispersion characteristic has a form analogous to that shown in Fig. 3.4d. We emphasize that the instability is possible even for such coefficients D_{12} and D_{21} which are small compared to D_{ii}.

Also, the dimensions of the instability region, or the number of self-excited modes, are important for AW systems, since these are the factors which often determine the

[1]) Stationary DSs in the two-component systems being discussed are unstable. However, stable DSs can occur in analogous systems with a large number of components, e.g., for $n = 4$ (Grigorov, Polyakova, Chernavskii, 1967; Romanovskii, Stepanova, Chernavskii, 1975, 1984).

properties of the solutions. Thus, Fig. 3.5 indicates that the DS instability region increases infinitely in systems with $|D_{11}M_{22}/D_{22}M_{11}| \to 0$. Characteristic of such systems (see Chapter 7) are solutions in the form of contrasting (discontinuous) DS. An infinitely increasing number of modes are excited simultaneously although not all of them occur in reality. Hence, the width of the DS instability region can be thought of as an equivalent to the relaxation degree in the theory of lumped systems.

The dispersion dependence in the region of the oscillatory instability of a homogeneous state (denoted by "1" in Fig. 3.4) indicates that $p_k > p_{k+1}$ for oscillatory modes. Therefore such an instability can occur only in systems with autooscillatory local kinetics. A linear analysis shows that homogeneous oscillations are always stable at $D_{11} \neq 0$ and $D_{22} \neq 0$. Nontrivial oscillatory regimes are asymptotically unstable on the segment L. However, this conclusion, which holds for two-component systems, can be inadequate for three-component systems.

C. We now discuss the specific features of the dispersion dependence at $n = 3$ (see Vasilev, Romanovskii 1975, 1976). We will not consider the cases in which the properties of these systems are analogous to those considered above ($n = 2$). First, we assume that $D_{ij} \to 0$ $(i \neq j)$ and then note the new properties arising in a system with the nondiagonal matrix D_{ij}. The necessary coefficients can be taken from the dispersion equation (3.6)

$$q_2(k^2) = \sum_{i=1}^{3}{}' \left(M_{ii} + \frac{\pi k^2}{L^2} D_{ii}\right), \tag{3.9a}$$

$$q_1(k^2) = \sum_{i=1}^{2} M_{ii} + \frac{\pi^2 k^2}{L^2} [D_{11}(M_{22} + M_{33}) + D_{22}(M_{11} + M_{33}) + D_{33}(M_{11} + M_{22})] + \frac{\pi^4 k^4}{L^4} [D_{11}D_{22} + D_{11}D_{23} + D_{22}D_{33}], \tag{3.9b}$$

$$q_0(k^2) = -\det\{M_{ij}\} + \frac{\pi^2 k^2}{L^2} \sum_{i=1}^{3} D_{ii}\overline{M}_{ii} - \frac{\pi^4 k^4}{L^4} (D_{11}D_{22}M_{33} + D_{11}D_{33}M_{22} + D_{11}D_{22}M_{33}) + \frac{\pi^6 k^6}{L^6} D_{11}D_{22}D_{33}, \tag{3.9c}$$

where M_{ii} and M_{ii} are the diagonal elements and minor subdeterminants of a local system. For systems with a sole singularity, the inequality $\det\{M_{ij}\} < 0$ is satisfied, since otherwise the nonlinear system has indefinitely increasing solutions. We then assume that $q_0(0) > 0$.

The DS instability condition $q_0(k^2) < 0$ (3.9b) can be met for appropriate D_{ii} if only one of the M_{ii} is positive or one of the M_{ii} is negative. In the first case, the positive feedback in the local kinetics is caused by an autocatalysis ($M_{ii} > 0$) and in the second case, by cross-catalytic circuits ($M_{ij}M_{ji} > 0$, $i \neq j$). In two-component systems, the second case is possible only if there is mutual diffusion between the components. This is one of the distinctions between systems with $n = 2$ and those with $n = 3$.

There is also another difference. If autooscillations are possible in a system with $n = 2$, then there are values of D_{ii} at which a DS instability occurs. In systems with

$n = 3$, cases for which all M_{ii} are negative and $\overline{M}_{ii}$ are positive but $[q_0(0) - q_2(0)\, q_1(0)] > 0$ can occur. There are no positive values of D_{ii} at which such systems have a DS instability region but autooscillations arise there by soft excitation (Fig. 3.6a). As an illustration of such a system the model in (3.34) can be used. Here, the mechanism of excitation of autooscillations is based on the delay phenomenon in the cross-catalytic circuit. There are no parameter values at which these systems reduce to those with $n = 2$ without losing their autooscillatory properties. This case exhausts the difference with respect to DS instability between systems with $n = 3$ and those with $n = 2$.

We now consider the oscillatory instability regions. The roots of the third-order algebraic equation are given by the Cardano formulas but these are not so convenient. We therefore give approximate expressions, which hold near the bifurcation points.

In the region of oscillatory instability, $q_0(k) > 0$ and, therefore, there is one real negative root at least. By virtue of the Vieta theorem this root is equal to

$$p_{0k} = -[q_0(k^2) + \delta_k]. \tag{3.10}$$

Assuming that the real parts of the two other roots are small ($\delta_k \ll 1$) we find, correct to terms of order δ_k^2, that

$$\delta_k \simeq [q_0(k^2) - q_2(k^2)\, q_1(k^2)]/[q_2(k^2) + q_1(k^2)], \tag{3.11}$$

$$\omega_k^2 = q_0/\lambda_{0k} + \delta_k^2 \simeq q_0(k^2)/[q_0(k^2) + \delta_k]. \tag{3.12}$$

In general, these expressions are also too cumbersome. We therefore explore the question of whether different types of the dispersion dependences can occur at typical relations between the D_{ii}.

Let $D_{11} = D_{22} = D_{33} = D/(\pi k^2/L^2)$. Substituting these values into (3.11) we find:

$$\delta_k = \frac{\delta_0 - 2[q_1(0) + q_2^2(0)]\, D - 4q_2(0)\, D^2 - 8D^3}{q_1(0) + q_2(0) + 8q_2(0)\, D + 12D^2}. \tag{3.13}$$

From this relation it follows that if the singularity is stable ($\delta_0 < 0$, $q_0(0) > 0$, $q_1(0) > 0$), then the homogeneous state of the distributed system is also stable. Differentiation of (3.13) with respect to $D \sim k^2$ yields:

$$\left.\frac{\mathrm{d}\delta_k}{\mathrm{d}k^2}\right|_{k^2 \to 0} = -\frac{2 + 8\delta_0 q_2(0)}{q_1(0) + q_2(0)}. \tag{3.14}$$

If $q_2(0)$, then the dispersion dependence (Fig. 3.6b) has the same form as that in two-component systems (Fig. 3.4e). The quantity $\mathrm{d}\omega^2/\mathrm{d}k^2$ can also be either positive or negative. This was to be expected, since in the case $q_2(0) > 0$, the autooscillation generation mechanism (the same as that for $n = 2$) is based on an autocatalysis. But if $q_2(0) < 0$ and $\delta_0 > 0$, then $\mathrm{d}\delta_k/\mathrm{d}k^2$ can be positive (Fig. 3.6a). In this case, $\mathrm{d}\omega_k^2/\mathrm{d}k^2 < 0$. This is a factor stabilizing homogeneous autooscillations.

For the theory of AWPs, systems in which one of the D_{ii} is substantially larger than the rest are of particular interest. Let $D_{11} = D_{22} = 0$, $D_{33} = D_3 \Big/ \left(\dfrac{\pi^2 k^2}{L^2}\right)$.

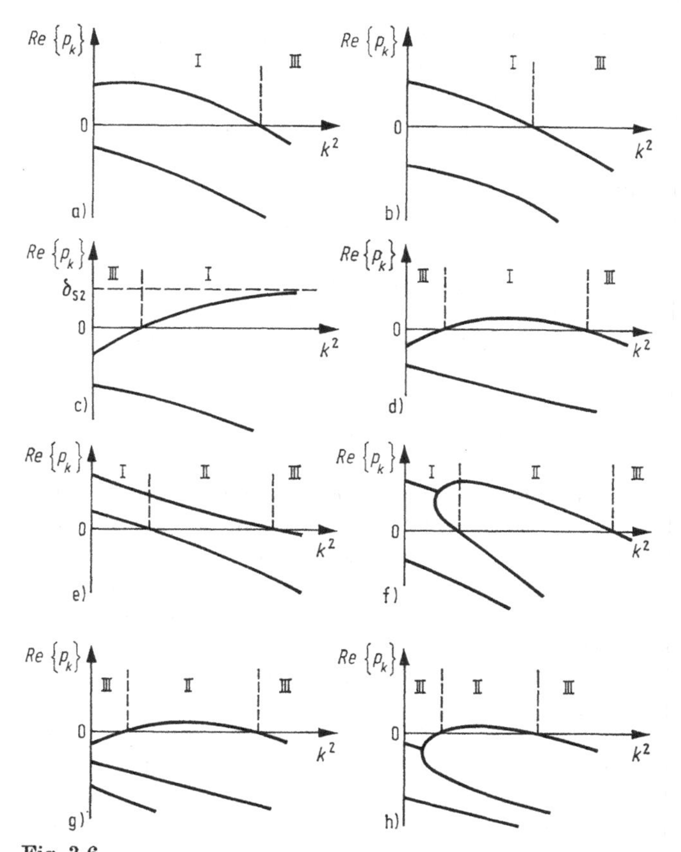

Fig. 3.6
The real parts of the roots of the characteristic equation plotted against the wave number for systems with $n = 3$ without mutual diffusion. The designations I, II, and III are the same as those in Fig. 3.4.

We denote $\delta_{s2} = M_{11} + M_{22}$. The quantity δ_{s2} has a simple meaning: This is a delay decrement of a local two-component subsystem, which occurs at a fixed third variable: $\bar{x}_3 = 0$.

For $D_3 \to \infty$ the roots of the dispersion equation are given by

$$p_{0k} \simeq -D_{33}; \quad p_{1,2;k} \simeq \frac{1}{2} \left(\delta_{s2} \pm \sqrt{\delta_{s2}^2 - 4\overline{M}_{33}}\right). \tag{3.15}$$

It is seen for that $D_{33} \to \infty$ and/or small wavelengths $[\pi^2 k^2/L \to \infty)$ two roots coincide with the characteristic numbers of a local two-component subsystem. Expression (3.15) shows the asymptotic behaviour of the dispersion dependence for $k^2 \to \infty$.

At $D_3 \to 0$, we find another asymptotic form

$$\delta_k = \frac{\delta_0 + \big(M_{33} - q_1(0) + \delta_{s2} q_2(0)\big) D_3 + \delta_{s2} D_3{}^2}{q_0{}^2(0) + q_1(0) + \big(2q_2(0) - \delta_{s2}\big) \delta_{s2} + \delta_{s2} D_3{}^2}. \tag{3.16}$$

It is easy to see that the oscillatory instability ($\delta_k > 0$) occurs even if the self-excitation conditions are not satisfied in the local system. The dispersion dependence in Fig. 3.6c is now fully conceivable. The upper limit on the spectrum of the self-excited modes vanishes if D_{11} and D_{22} are small but are not exactly equal to zero (Fig. 3.6d).

If there are autocatalytic circuits ($\overline{M}_{33} > 0$) in a local system ($\delta_0 < 0$) but $\delta_{s2} < 0$, then an upper limit is placed on the spectrum of the self-excited modes (Fig. 3.6a). We recall that the DS instability conditions in such systems $\left(q_0(k^2) = q_0(0) + \overline{M}_{33}D_3\right)$ are not at all fulfilled. For the oscillation frequencies of the modes we have: $\mathrm{d}\omega_k{}^2/\mathrm{d}k^2 > 0$ at small k^2. Thus, the cases considered here differ radically from those with $n = 2$. Typical of such systems are space–time oscillation regimes, specifically standing waves (see Section 5.5).

We propose the following interpretation of the results obtained here which seems to be illustrative enough. Assume that there is a local generator ($n = 2$, $\tilde{x}_1$ and $\tilde{x}_2$; $\tilde{x}_3 = 0$; δ_{s2}, $\overline{M}_{33}$). When included in a system with $n = 3$, such a generator will have an additional delay feedback through $\tilde{x}_3$. This feedback is able to suppress the autooscillations or, on the contrary, contribute to their excitation. In a distributed system, the diffusion $\tilde{x}_3$ for modes with $k^2 \neq 0$ smoothes out the feedback distribution and decreases the feedback oscillation amplitude, or, so to speak, reduces the action of the feedback. For large $D_3 = (\pi^2k^2/L^2)\, D_{33}$, the system is essentially a generator with δ_{s2} and $\overline{M}_{33}$; for small D_3, this is a local system with $n = 3$; and at intermediate values of D_3, the additional delay in feedback introduced by the diffusion is important. These mechanisms acting on the feedback contribute to the mode excitation. The oscillatory instability $\left(q_0(k^2) > 0,\ \delta_k > 0 \text{ but } \delta_0 < 0\right)$ is possible only for systems with the potential autooscillatory kinetics. This means that as the quantities M_{ij} vary in magnitude (but not in sign), the local system becomes an autooscillatory one. It appears that for the zero coefficients of mutual diffusion, the oscillations can be enhanced in a distributed system even if the autooscillatory local kinetics cannot occur. We emphasize that in this respect, systems with $n = 3$ behave as those with $n = 2$ with regard to DS instability. Such a mechanism of oscillation enhancement can be easily understood if we make a substitution $M_{ij} = M_{ij} - (\pi^2k^2/L^2)\, D_{ij}$. Then, M_{ij} can change sign and, therefore, the oscillations can be enhanced through cross-catalysis. This mechanism can be called "spatial cross-catalysis" (it is understood that the oscillations are driven by an external force: the smaller D_{ij} is, the more additional energy is needed).

Systems with $n = 3$ are, of course, more diverse in dispersion dependence as those considered above. Some of them are listed in Fig. 3.6e, g. It is seen that these forms of dependence are qualitatively the same as those for systems with $n = 2$. Note also that the graphs of Fig. 3.4 are suited for systems with $n = 3$ but a monotonically decreasing curve of the third root is added in this case. This last note having been taken into account, the analysis of the dispersion dependences of systems with $n = 3$ seems to be exhausted.[1])

---

[1]) Note that new types of the graphs are not needed for "simple" systems with $n > 3$. We mean those systems whose local kinetics has only one autooscillatory subsystem, a simple, rough singularity, etc.

3.4 The small-parameter method

The construction and analytical study of AW solutions can be much simplified if certain parameters of the basic models are treated as small parameters. Such parameters are not necessarily present in the explicit form of the basic model. For instance, the distance defined in the parameter space with respect to the bifurcation point (point of solution branching), measure of proximity of the desired solution to the known one, etc. can serve as small parameters.

We can distinguish two types of systems containing small parameters. First, the search for solutions is easier in systems with small parameters in the derivatives. Their AW solutions have discontinuities (step fronts), which are manifestation of the nonlinear properties. Second, there are weakly nonlinear systems close to the bifurcation points, those with small parameters in the nonlinear functions. Essentially, an expansion of solutions in a small parameter is performed in both cases but the methods applied are slightly different.

If the small parameter is before a derivative, it means that there are several scales for the simulated processes in the system. Thus, the terms having derivatives of the form $\partial/\partial t$ and $\varepsilon\, \partial/\partial t$ yield two time scales t and εt, and the terms $\partial^2/\partial r^2$ and $\varepsilon\, \partial^2/\partial r^2$ yield two space scales, r and $\sqrt{\varepsilon}\, r$. Therefore the solutions have regions of fast and slow variations. The equations of motion in these regions account only for the changes of those variables which have the corresponding scales. They are simple enough and, therefore, it is possible to construct families of solutions for each region. The resulting solutions are matched by the continuity condition. Such a solution can be treated as a zero approximation. The patterns by which the subsequent approximations are constructed are rather simple but cause technical difficulties.

Essentially, such patterns follow from the theory of relaxation-type autooscillations (Andronov, Vitt, Khaikin 1981). A rigorous substantiation and conditions for applicability of such patterns for ordinary differential equations were obtained by A. N. Tikhonov (1952). These conditions require in addition to a small ε, that the Jacobian eigenvalues corresponding to the reduced variables are negative. For distributed systems, such general theorems are still unavailable. However, it is evident that for AWPs, the basic model can be reduced to a truncated system, since the fast variables reach their stationary distributions. Tikhonov's theorem applies, in a sense, to the case of distributed systems. First, a consideration of stationary solutions (which is one of the basic stages in the analysis of distributed systems) is associated with the solution of ordinary differential equations for which this theorem is valid. Second, it is possible to consider a distributed system in the form of its discrete analogue composed of N elements. It is natural to assume that for large N, the solutions of these elements correspond, in the limit, to the solution of a distributed system. Of course, such arguments are not rigorous but there is an indication that the reduction conditions should not be weaker than those in Tikhonov's theorem. Note that here, the necessary condition is not only the presence of variables of different scales but also the existence of attracting sets of reduced variables in certain regions.

We now consider ways of obtaining "weakly nonlinear" periodic solutions. The basic concepts of such methods were also first developed in the oscillation theory for local systems (Bogolyubov, Mitropolskii 1974). Then methods of deriving truncated equations for distributed systems and the theory of solution branching (bifurcation theory) were developed. The first step is, as a rule, a transition to self-simulating variables determining the type of a desired AW solution. Thus, a substitution $\eta = r - vt$ is used for the investigation of plane waves, a substitution $\eta = t - f(r, \varepsilon)$ for the study of sources of TPs (LCs), $x_i = A_{ik}(t) \cos(\pi kr/L)$ for standing waves, $x_i = A_{ik} \cos(\pi kr/L)$ for dissipative structures, etc. After the variables are substituted, the perturbation theory procedure is applied. The solutions are presented in the form of a series expansion

$$x_i = \sum_{j=0}^{\infty} x_{ij}\varepsilon^j, \qquad B = B_0 + \sum_{j=1}^{\infty} B_j\varepsilon^j, \tag{3.17}$$

where B_0 is the value of the parameter B at the point of branching of a desired solution, for example of the autooscillation frequency in a local system or of a reaction rate at which a DS instability arises, etc.

Expressions (3.17) are substituted into the equations containing terms with ε. Terms with equal powers of ε (ε for $\tilde{x}_{i1}$, ε^2 for $\tilde{x}_{i2}$, etc.) are retained. The second expansion in (3.17) is needed to exclude secular terms from $\tilde{x}_{ij}$ by applying the conditions of orthogonality for the "remainings" $\{\Phi(\tilde{x}_{ij}, \ldots)\}$ of the preceding approximation. If the system has more than two characteristic scales, then further a simplification is allowed, i.e., the stepwise integration method proposed by R. V. Khokhlov (Ilinova, Khokhlov 1963) applies.

At present, there are many patterns by which the truncated equations are constructed. Basically, these patterns differ in the particular means of excluding secular terms [e.g., the Floke theory (Ortoleva, Ross 1974) or the averaging method are also useful]. It seems that the choice of a pattern is purely intuitive rather than depending on the problem to be solved. All these patterns yield similar results and have the same regions of applicability. Their basic advantage is simplicity. The disadvantage is that technical difficulties arise when complicated solutions are investigated. Hence it is no easy matter to construct a set of truncated equations describing the interaction between solutions of different types. The stability of solutions suggested by the truncated equations is a necessary but an insufficient condition for these solutions to be stable in a complete set. Some patterns fail to give even such poor information on the stability of approximate solutions.

The analysis of solutions for stability in a complete set involves a variety of complicated problems many of which are not yet resolved for the AWPs discussed below. Sometimes simple qualitative considerations of the form of "dangerous" perturbations or "dangerous" mode interaction make it possible to estimate the stability and to conjecture which path the evolution of the transient processes will take. However, at present, only numerical experiments can give an answer to the question of what form this or that stable stationary AW solution should evolves into.

We stress once again that the efficiency of the small parameter method depends, in the final analysis, on the rate of convergence of series like, those in (3.17).

3.5 The axiomatic approach

Historically, along with nonlinear wave theory, which is described by models like those in (1.1), an "axiomatic" theory of excitable media has been developed. A study of these problems was pioneered by N. Wiener and A. Rosenbluth as early as in 1946. I. M. Gelfand and M. L. Tsetlin (1960) were the first Soviet researchers in this field. Since then, the axiomatic theory has been developed successfully in the Soviet Union (Krinskii 1968; Fomin, Berkinblit 1973; Bogach, Reshodko 1979).

There are different levels of describing an object by the axiomatic theory method. In the simplest case, it is assumed that the active medium is composed of discrete elements, or finite automata, which can be only in two states: excited and refractory. The axiomatic theory can be applied without detailed knowledge of the kinetics of real objects. Another important advantage of such an approach is the possibility of considering a broad spectrum of general problems and the simplicity of computer experiments.

Essentially, the axiomatic approach embodies the strongest idealization of the methods applied in nonlinear wave analysis. It has to be admitted that in such a system, the element driven by an external force can generate a standard pulse of duration τ, with the fronts of zero duration, which returns to its equilibrium state over a time $R - \tau$, where R is the refractory time. The rate v of transfer of excitation is treated as being given. The relationship between the axiomatic approach and the characteristics of the nonlinear wave theory is shown in Fig. 3.7. Typical of the axiomatic approach as a whole are the following features:

a) Only one excitation pulse is chosen as a characteristic stage. As was mentioned above, axiomatic quantities are introduced which serve as the characteristics of such an elementary process. When taking an axiomatic approach, it is enough to distinguish regions in which the wave is in the excited state (shown black in Fig. 3.7 c, d) and in the refractory state (shown dashed);
b) The pulse is assumed to be stable;
c) The transient processes by which the pulse is formed are not considered although their existence is implied when dealing with the excitation threshold.

The initial values of the variables, which appear to be less than a threshold value, convert into the equilibrium state. Once the threshold is exceeded, a stable stationary pulse is formed.

Also, the axiomatic approach was used in studying nerve tissue (Fomin, Berkinblit 1973), cardiac muscle (Krinskii 1968; Ivanitskii, Krinskii, Selkov 1978), a chemical system (Zhabotinskii 1974) and genetic networks. New autowave structures were predicted and qualitatively described: There are sources of pulses of the "echo" type (mutual retriggering of the neighbouring elements of the medium) (Krinskii 1968; for details see Section 5.1) and spiral waves, or reverberators (Selfridge 1948; Balakhovskii 1965; Krinskii 1968).

For the sake of simplicity, we mention only the simplest system of postulates for active media. Sometimes such a system is insufficient for a reasonable description of the phenomena and, therefore, one has to introduce, e.g., a variety of stable states of the elements and of the matrices of transitions between them, randomness in the transfer of interaction between elements, etc. multi-purpose applied program sets have been elaborated for the investigation of such models (see, e.g., Bogach, Reshedko 1979), which allow the complex networks of finite automata to be imitated.

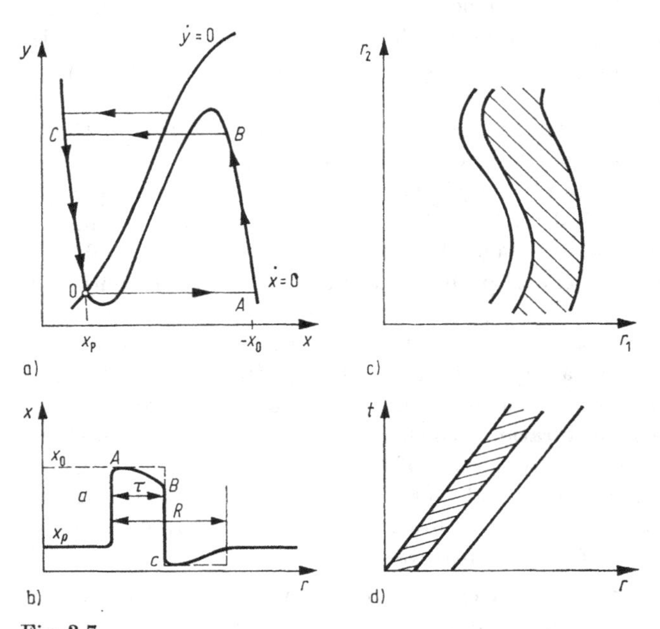

Fig. 3.7
An axiomatic description of a TP: a) Segments of trajectories in the phase plane are shown, which correspond to an axiomatic description of the excited state (*AB*), refractory state (*ABCO*), state of rest (small vicinity of the equilibrium state 0); b) the form of the TP; the dashed line denotes a TP described axiomatically; c) schematics of a TP in the plane r_1, r_2; the excited regions are shown white and the refractory regions are shown crosshatched; d) schematics of a TP in the plane r, t

Undoubtedly, the axiomatic approach is useful in AW theory and its further development is necessary. However, there is the problem of adequacy in describing phenomena in terms of axiomatic and dynamic models. Mathematically, this problem of adequacy is solved by a number of theorems, which, in particular, assert that any discrete automaton can be represented in its dynamic form; the inverse of this assertion is, generally, incorrect. In practice, it is necessary to postulate the properties of discrete automata from the dynamic properties of the local system and knowledge of the matrix of the diffusion coefficients D_{ij} (Lindenmayer, 1971). The Bulgarian researchers Sendov and Tsanev proposed models which combine discrete and kinetic methods for describing associations of automata (Tsanev, Sendov 1966; Sendov 1976).

3.6 Discrete models

Certain specific idealizations are adopted in each research method. Thus, in the axiomatic approach, the simplification is the assumption that the variables of the system change in space and time in a discrete manner. The next step in describing the input systems in more detail is to take into account the continuous change of the variables with time, the change in the space variable being thought of as being discrete. When taking such an approach, the AW system is represented as a sequence of a finite number of interacting elements. In the simplest cases, such discrete models have the form

$$\frac{d}{dt}\,{}^jx_i = {}^jF_i({}^jx_1, {}^jx_2, \ldots, {}^jx_n) + ({}^jI_i - {}^{j+1}I_i)$$

$$= {}^jF_i({}^jx_1, \ldots, {}^jx_n) + d_i({}^{j-1}x_i - 2\,{}^jx_i + {}^{j+1}x_i); \tag{3.18}$$

where $i = 1, 2, \ldots, n$ is the number of the variable, $j = 1, 2, \ldots, N$ is the number of the element in the chain and F_i is the local kinetic function. The right-hand side of Eq. (3.31) is written for a discrete representation of the flow in accordance with the Fick procedure: ${}^jI_i = d_i({}^{j-1}x_i - {}^jx_i)$. There is a relationship between the coupling coefficient d_i of the elements and the diffusion D_{ii} in a distributed model, which reads $D_{ii}N^2/L^2 = d_i$. For a discrete description the boundary conditions are given by the corresponding values of the flows and of the quantities themselves at the boundaries. For example, the impermeability of the chain boundaries can be expressed in the form ${}^0x_i = {}^1x_i$ and ${}^{N+1}x_i = {}^Nx_i$) (this corresponds to the conditions of the second kind (${}^1I_i = {}^{N+1}I_i = 0$).

The right-hand sides of Eqs. (3.31) are one of the possible finite-difference approximations which are used for diffusion systems and yield an accuracy of order $O(h^2)$, where $h = L/N$ (Samarsky 1977). The boundary conditions of form ${}^0x_i = {}^1x_i$ restrict the accuracy to $O(h)$. The accuracy of order $O(h^2)$ is ensured by the condition ${}^{-1}x_i = {}^1x_i$. The sum of all the right-hand sides of Eq. (3.18) is no longer exactly equal to zero. Such a situation is not desirable for AW systems; thus, the conditions ${}^0x_i = {}^1x_i$ are recommended: the use of three (or more) local patterns for the representation of the flows leads to increased complexity of the models and makes them less illustrative.

If the number N of elements is large, then a general investigation of a discrete model is not simpler than the analysis of a continuous model. The minimum number of elements for AW systems depends on the characteristic scales of the AWP.

The choice of model type is determined by the particular problems to be solved and the processes investigated. Typically, in order to gain a deeper insight into the nature of motions that occur in a distributed object, a model with too large an N is chosen, so that one has to use a computer. However, the simplest discrete models ($N = 2, 3 \ldots$) are useful when one needs to predict whether this or that AWP will occur under specified conditions. Such models are rather easy to investigate and help understanding a variety of important problems as well as gaining a better intuition with respect to the object under study.

When using these simplest discrete systems, the following assertion is valid: "If a continuous model describes a certain AWP, then at appropriate values of d_i, the simplest discrete model will have a solution inhomogeneous in space with a corresponding time dependence." The alternative assertion is, generally, incorrect, i.e. not all the types of solutions of the discrete models remain stable in continuous models. Such a statement can be illustrated by the fact that there is a correlation between the solutions of the simplest discrete ($N = 2$) and continuous models:

1. Stationary inhomogeneous states ${}^1x_i \neq {}^2x_i$ ($\mathrm{d}^jx_i/\mathrm{d}t = 0$) correspond to DSs;
2. Antiphase oscillations ${}^1x_i$ and ${}^3x_i$ correspond to standing waves;
3. Permanent advance in phase of one of the elements corresponds to LCs;
4. Retriggering of the elements corresponds to "echo", etc.

Thus, all the basic types of AWPs have images even in the simplest discrete models.

We now consider some characteristic features of describing the simplest discrete model ($N = 2$):

$$\frac{\mathrm{d}^1x_i}{\mathrm{d}t} = F({}^1x_1, \ldots, {}^1x_n) + d_i({}^2x_i - {}^1x_i),$$
$$\frac{\mathrm{d}^2x_i}{\mathrm{d}t} = F({}^2x_i, \ldots, {}^2x_n) + d_i({}^1x_i - {}^2x_i), \tag{3.19}$$

where $i = 1, \ldots, n$. Specifically, such systems are useful when simulating living tissues.

Let us introduce, as is usually done in the oscillation theory (Strelkov 1964), normal coordinates

$$\Delta_i = ({}^1x_i - {}^2x_i)/2, \quad S_i = ({}^1x_i + {}^2x_i)/2. \tag{3.20}$$

Here, the linear part of the model in (3.18) will have a cellular diagonal matrix. The coupling between in-phase (S_i) and anti-phase (Δ_i) modes occurs through linear interactions between these modes. The solutions of a continuous model can be represented in the form of a series expansion in harmonic modes: $x_i = \sum_{i=1}^{\infty} A_{ik}(t) \cos(\pi kr/L)$. A direct substitution shows that the equations for A_{i0} and S_i, as well as for A_{ik} and Δ_i, have the same structure (see Section 5.5). The parameter dependence of the solutions and the character of their stability in both "two-mode" approximations are similar, so that the qualitative results should coincide. For a "two-box" model, however, many results can be obtained and interpreted in a rather illustrative manner.

Consider the use of a phase plane in the analysis of AWPs. For definiteness, we will study the stationary states and assume $F_1 = x_2$ and $F_2 = -x_1 + \delta_1 x_2 - \delta_3 x_2^3$ ($\delta_1 > 0$, $\delta_3 > 0$). The null isoclines (lines $F_1 = 0$ and $F_2 = 0$) for such kinetics and the direction of the integral curves are shown in Fig. 3.8. There are regions in this field of integral curves in which they are oppositely directed, i.e., the representative points (${}^1x_1$, ${}^1x_2$) and (${}^2x_1$, ${}^2x_2$) subject to "local forces" ${}^1F_{1,2}$ and ${}^2F_{1,2}$ should diverge both along the O^1x- and O^2x-axes. But the "diffusion" acts as the distance between the representative points decreases. The fact that the "local" and "diffusion" forces are equal leads to the existence of stationary inhomogeneous solutions. We will not consider the images of other AWPs but note only that the simplicity of our reasoning in terms of a phase plane follows from the basic property of diffusion coupling, by virtue of which the representative points converge along each direction in the phase space. Specifically, this implies that there are no stationary noncoincident positions for two representative points in the phase space. Then, of course, there should be no such positions for three, four or more points, or for a continuous line, which is the image (projection) of the stationary solution of a continuous model.

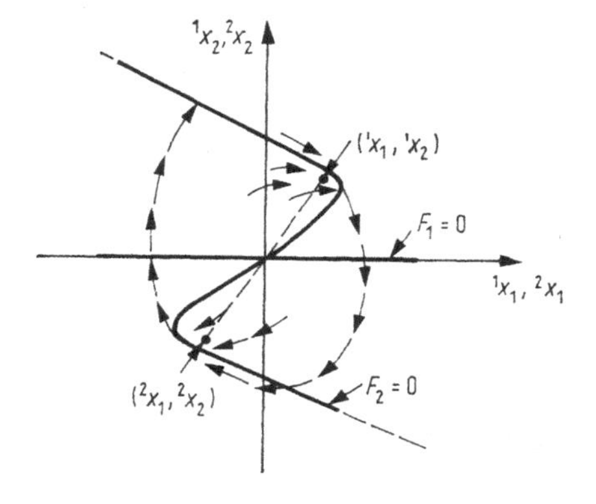

Fig. 3.8
The stationary states of a discrete model ($N = 2$) in the phase plane x_1, x_2. The arrows show the directions of motion along the phase trajectories in a local system.

As an example, we mention one of the models for oscillatory chemical reactions (Zhabotinskii, 1974):

$$F_1 = k_2x_1 - k_3x_1x_2; \quad F_2 = k_1x_1 - k_2x_2; \quad F_3 = k_2x_2 - k_3x_3, \tag{3.21}$$

where k_1, k_2, and k_3 are positive. Nontrivial AW regimes can occur in a system with such F_i. However, a rigorous analysis of this system is rather difficult. For a "two-box" representation, many positions are obtained most easily, since there are stable out-of-phase oscillation regimes but no stationary homogeneous solutions ($\Delta_1(\Delta_1^2 + a^2) = 0$ (3.20)). Thus, the general property of the model is obtained in such an involved situation.

As a rule, the conclusion that some AW phenomenon is unrealistic can be easily generalized to a continuous model but a generalization of the conclusion on the regime stability encounters essential difficulties. The point is that the discrete models possess additional threshold properties. Thus, it is well known (Chaffee 1974) that the stationary inhomogeneous solutions are unstable in a continuous one-component model ($n = 1$). Figure 3.9 shows the phase plane of such a model ($F_1 = x_1 - x_1^3$).

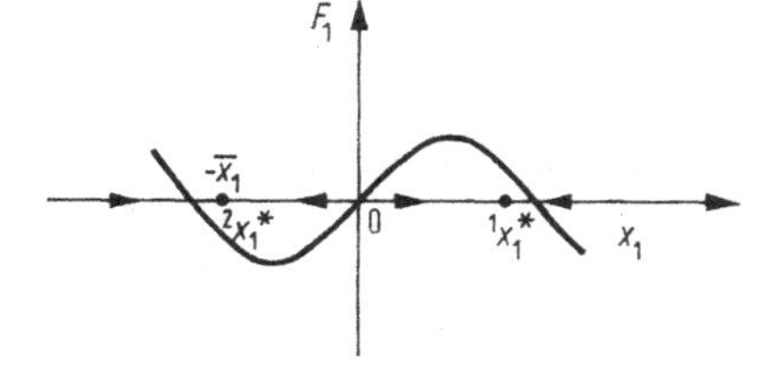

Fig. 3.9
The phase plane of a first-order model ($n = 1$); $^1x_1^*$ and $^2x_1^*$ are the stable stationary states in a discrete system ($N = 2$). The arrows show the directions of motion in a local system.

It is seen that even the simplest discrete model ($N = 2$) has a stable homogeneous solution. Moreover, such solutions are allowed until N is increased up to a certain critical value. At large N, the threshold disappears and all the points revert to one of the equilibrium states ($\bar{x}$ or $-\bar{x}$).

Similar threshold properties are inherent in systems with trigger local kinetics for $N = 2$, as well as in systems where one of the D_{ii} vanishes. The threshold properties are observed not only for stationary solutions but also for oscillatory ones. We emphasize that in a numerical experiment, one often has to deal with discrete models (although at large N) and, therefore, additional threshold properties can manifest themselves as solutions, which can occur in a numerical experiment but are unstable in continuous systems. Even a tentative analysis of the simplest discrete models permits one to predict whether or not such thresholds are possible in particular systems.

Thus, the investigation of the simplest discrete models is rather easy. Also, this is an effective mean of studying AW systems. In many cases such a study helps understanding the behaviour of complex systems.

3.7 Fast and slow phases of space–time processes

Present day methods for a qualitative study of AW systems in which the motion phases with radically different space–time parameters can be distinguished are well developed. Each of these phases is described by the simplified equations. Moreover, the solutions of these equations are asymptotically stable and, therefore, in the transition from one phase to another, the separate solutions can be combined into one general solution for the whole AWP in a most natural way. We note two versions of such an approach using the following "basic" model for an AW system:

$$\tau_x \frac{\partial x}{\partial t} = D_x \frac{\partial^2 x}{\partial r^2} + P(x, y), \tag{3.22a}$$

$$\tau_y \frac{\partial y}{\partial t} = D_y \frac{\partial^2 y}{\partial r^2} + Q(x, y). \tag{3.22b}$$

There are two parameters in these equations: $\varepsilon = \tau_x/\tau_y$ is the parameter which determines the relationship between characteristic time intervals, and $\alpha = D_x/D_y$ is the parameter which determines the relationship between space intervals.

The first approach consists in assigning the form of the nonlinear function $P(x, y)$ directly in terms of the space and time variables, i.e. $P = P(r - vt)$ from Eq. (3.22a). In this case, there are two stationary TPs which differ in form. The TP with the larger amplitude moves at a higher velocity and is stable with respect to small perturbations. The TP having the smaller amplitude is unstable. The form and velocity of TPs were first calculated in such a manner for a model of nerve tissue (where the function $P(x, y)$ corresponds to the current that flows through the membrane). That is why such an approach was called "investigation of a model with the assigned current generator". Note that all three stages of searching for AW solutions mentioned in Section 3.1 are present in this approach. Indeed, the solutions in the form of pulses correspond to the stage at which stationary solutions are sought; two types of the pulses are differentiated by their stability to small perturbations in the second stage and, finally, the unstable pulse is thought of as a boundary which separates perturbations leading to the formation of a stable TP in the third stage. The calculation procedure for a model with the assigned current generator was reported in ample detail by Markin, Pastushenko, Chizmadzhev 1977, 1981.

The second approach, which is associated with the investigation of Eqs. (3.22), will be demonstrated for the case of a strong-relaxation system ($\varepsilon \ll 1$) without taking into account slow variable diffusion ($1/\alpha \to 0$). Here, instead of Eqs. (3.22), we introduce approximate characteristics, which enable a qualitative description of actual nonlinear AW structures to be made (see Chapters 4 and 5). Such characteristics are obtained from very simple considerations. The available solutions of the set (3.22) (Zhabotinskii 1974; Ivanitskii, Krinskii, Selkov 1978; Vasilev, Romanovskii, Yakhno 1979) indicate that only three phases of motions can exist: a) a phase of fast motions in time and abrupt changes in space (TF propagating at a velocity proportional to $1/\varepsilon$); b) a phase of slow motions in time and sharp changes in space (immobile front); c) a phase of slow motions in time and smooth changes in space (pulse plateau).

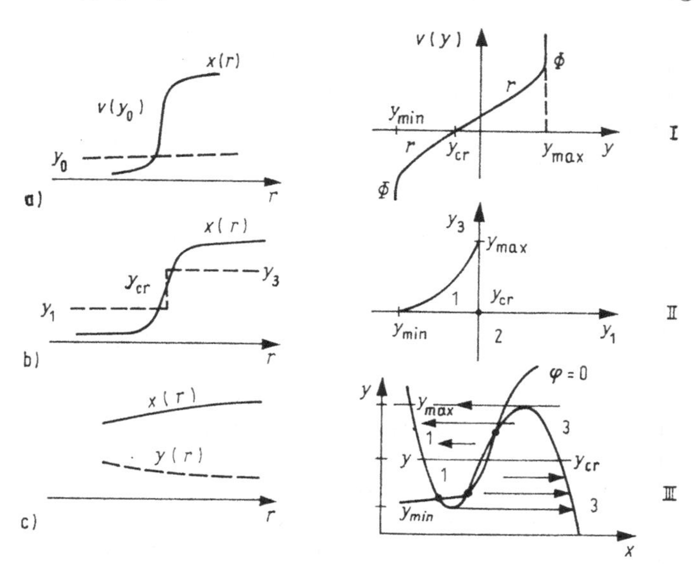

Fig. 3.10
The characteristic planes of space-time motions in a relaxation system at $\varepsilon \ll 1$ and the characteristics describing the dynamics of these motions: a) The TF is described by characteristic I (dependence of the front velocity on the slow variable): I is the region corresponding to "trigger" waves, Φ is the region corresponding to "phase" waves; b) a stopped TF in the discontinuity (from y_1 to y_3) of the slow variable. Characteristic II determines the region of existence of an immobile front in the variables y_1 and y_3; c) The slow motions of the smooth regions are qualitatively described by characteristic III (null isocline of set (3.1) at $\partial x^2/\partial r^2 = 0$; y_{cr} corresponds to the zero velocity of the TF. (Yakhno 1981)

These phases are shown on the left in Fig. 3.10. A particular approximate characteristic is used for each phase of motions (Fig. 3.10a, b, c). We introduce the following designation for these characteristics: I indicates the dependence of the front velocity on the slow variable value; II is the parameter region of a stepwise change of the slow variable y_1, y_3, which corresponds to an immobile front, and III the null isoclines of set (3.22). A description of the solutions by means of the approximate characteristics will be given in Chapters 4 and 5. We now briefly define the characteristics for each phase.

a) *The dependence of the front velocity on the slow variable value*

Characteristic I is obtained if the "fast" motion occurs in the region of a smooth change of the slow variable (the characteristic size of the inhomogeneity is much larger than the diffusion length $l_{\text{front}} \simeq \sqrt{D_x \tau_x}$, i.e. it is used in the absence of sharp changes of the y variable in space. In this case, the fast motion equation (3.35a), which describes the motion of the excitation front, can be substituted by the algebraic dependence of the front velocity on the rate of change of the slow variable in place of the front $\big(v = v(y)\big)$ (Ostrovskii, Yakhno 1975). The fact that such a substitution is possible follows from the specific features of the solutions of Eq. (3.22a). The arbitrary initial solution x (at $y = \text{const}$) in the form of a slope tends asymptotically (characteristic time $\tau \simeq \varepsilon$) to a stationary wave that moves at a certain velocity (Kanel 1962, 1964). The method for determining the velocity of such a stationary wave consists in solving the eigenvalue problem. This method will be discussed in Section 4.1. Note only that at $y_{\min} < y < y_{\max}$, if there are three roots in the equation $P(m_i)|_{y=\text{const}} = 0$, the velocity has a unique value (sometimes such a wave is referred to as a "trigger" wave). At $y = y_{\max}$ or $y = y_{\min}$ the front velocity has a continuum of values exceeding a certain value (see Section 4.1). These are so-called "phase" waves. A possible dependence $v = v(y)$ is listed in Fig. 3.10a. For some functions $P(x, y)$ the "trigger" wave velocity can be calculated analytically (see Table 3.1). The table indicates that for the chosen nonlinear functions, the front velocity is governed by the difference between the threshold values $(m_3 - m_1)$ and $(m_2 - m_1)$, which correspond to the stable states of a local system. It appears that the wave moves in this direction so as to cause the system to be converted from the state with the lower threshold to the state with the higher threshold. However, generally, the velocity direction depends on the relation between the positive and negative values of the function $P(x, y = \text{const})$. In an inhomogeneous, smoothly varying medium, the dependence $v = v(y)$ is also useful. Note that this characteristic is applicable at any value of the parameter $\alpha = D_x/D_y$; it is only necessary that the characteristic size of the space inhomogeneity of the slow variable is much larger than $l_{\text{front}} \simeq \sqrt{D_x \tau_x}$.

b) *The parameter region corresponding to an immobile front*

The excitation front can be immobile at a given space step of the slow variable $y(r)$, if the values of y are larger than y_{cr} on one side of the step and are less than y_{cr} on the other. For definiteness, we designate the slow variable values in the nonexcited region by y_1 and that in the excited region, by y_3 (see Fig. 3.10b). Characteristic II

Table 3.1

Analytical equations for the front velocity

form of a nonlinear function	equation for the front velocity
1) 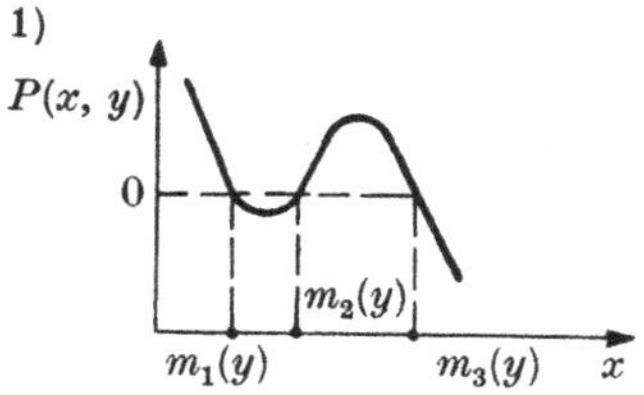 $P(x, y) = -\gamma(y)\,[x - m_1(y)]\,[x - m_2(y)] \times [x - m_3(y)]$	$v(y) = \pm\sqrt{\gamma(y)/2}\,[m_1(y) + m_3(y) - 2m_2(y)]$
2) 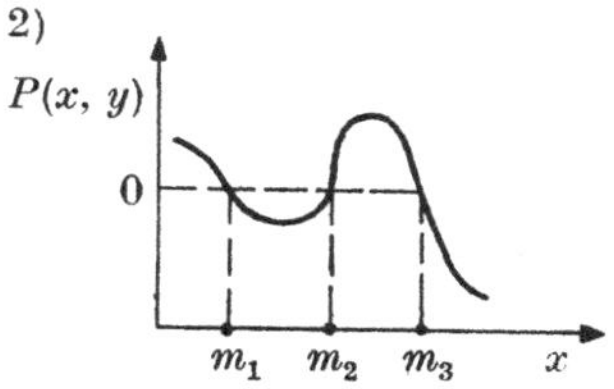 $P(x, y) = -\gamma(x - m_1)\,(x - m_2)\,(x - m_3) \times (1 + \beta x + \alpha x^2)$	$v(y) = \sigma\left[m_1(y) + m_3(y) + g m_1 m_3 - \frac{\gamma(y)}{\sigma^2}\, m_2(y)\right]$, where $\sigma = \pm\sqrt{\frac{\gamma(y)}{2 + g(2m_1 + 2m_3 - 3m_2)}}$
3) 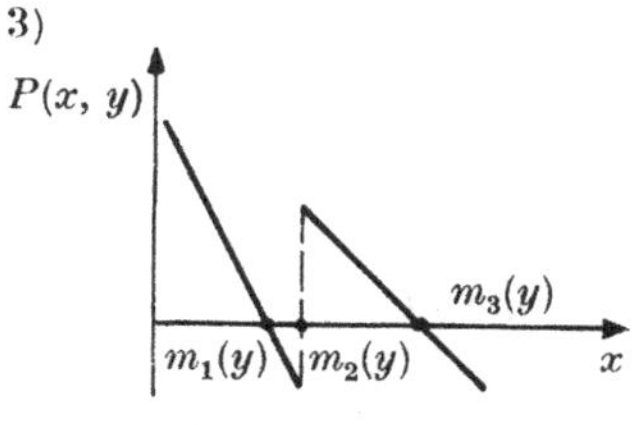 $P(x, y) = -\lambda_i(x - m_i(y)]$, $i = 1$, if $x < m_2(y)$, $i = 3$, if $x > m_2(y)$	$v(y) = \pm\frac{\lambda_1(m_1 - m_2)^2 - \lambda_3(m_3 - m_2)^2}{k}$, where $K = \{(m_3 - m_2)\,(m_2 - m_1)\,(m_3 - m_1) \times [\lambda_3\,(m_3 - m_2) + \lambda_1(m_2 - m_1)]\}^{1/2}$
4) 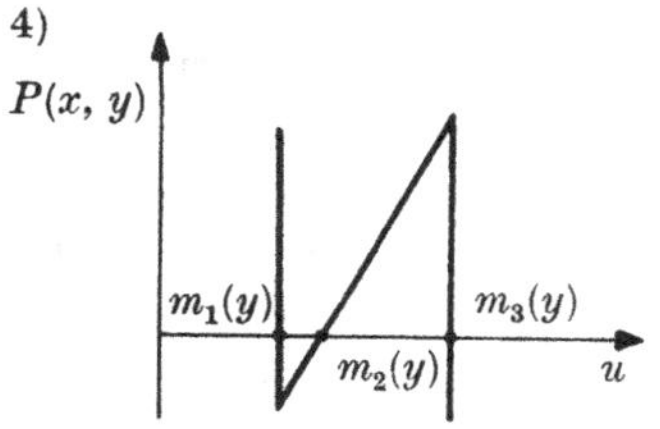 $P(x, y) \doteq \gamma(y)\,[x - m_2(y)]$, $m_1(y) \leqq x \leqq m_3(y)$	$v(y) = 2\left\{\gamma(y)\ln^2\frac{m_2 - m_1}{m_3 - m_2}\Big/\left[\pi^2 + \ln^2\frac{m_2 - m_1}{m_3 - m_2}\right]\right\}^{1/2}$

is a region in the plane y_1, y_3 which corresponds to those parameter values y_1 and y_3 of the step for which an immobile excitation front can exist. In order to define the boundaries of such a region, we will consider the stationary solutions corresponding to an immobile front. Fast motions are absent and, therefore, the time derivative in (3.35) can be neglected.

We will perform an approximate consideration of the solutions assuming that the stepwise distribution of the slow variable $y(r)$ for which $y(r) = y_3$ in the excited region and $y(r) = y_1$ in the nonexcited region is given. This is allowed only in systems with $\alpha \gg 1$ ($D_x \gg D_y$) (only in this case the y variable can change faster in space than the x variable).

The solutions for the front can then be determined from the equation

$$D_x \frac{\mathrm{d}^2x}{\mathrm{d}r^2} + P\big(x_0(r), y(r)\big) = 0 \tag{3.23}$$

with the boundary conditions

$$\left.\frac{\mathrm{d}x_0}{\mathrm{d}r}\right|_{r\to\pm\infty} = 0; \quad x_0|_{r\to\infty} = m_3; \quad x_0|_{r\to-\infty} = m_1\,. \tag{3.24}$$

The conditions for the existence of stationary solutions can be most clearly demonstrated in the phase space y, x, $\partial x/\partial r$. Figure 3.11 shows how the desired solution of Eq. (3.23) is constructed from two solutions of this equation at the constant values of the slow variable ($y = y_1$ and $y = y_3$). Since our interest is in the wave of a transition between the states m_1 and m_3, the desired solution corresponds to the separatrix trajectories. Figure 3.11 shows how such separatrix trajectories are matched. We begin by moving from the point m_1 along the separatrix curve corresponding to the solution at $y = y_1$. We draw a vertical line from the representative point on the separatrix. We move the point O_1 until the vertical line crosses the separatrix trajectory corresponding to the solution of Eq. (3.23) for $y = y_3$ at the point O_1'. The values x_0' and $\mathrm{d}x_0'/\mathrm{d}r$, which correspond to the place where the separatrices are connected through the vertical line O_1O_1', determine the conditions for matching the solutions at the point of a stepwise change in the slow variable from y_1 to y_3. We obtain a solution corresponding to the phase trajectory $m_3(y_1)\,O_1O_1'm_3(y_3)$. The spatial form of the first solution $x_0(\eta)$ is given in Fig. 3.11b. If we continue to move the point O_1 along the separatrix in the plane $y = y_1$ (phase trajectory $m_1(y_1)\,O_1O^*O_2O_2'O_1'm_3(y_3)$), then we obtain the second solution shown in Fig. 3.11b. The values of the variables at the place of matching $x_{0\mathrm{left}} = x_{0\mathrm{right}}$ and $(\mathrm{d}x/\mathrm{d}r)_{\mathrm{left}} = (\mathrm{d}x/\mathrm{d}r)_{\mathrm{right}}$ correspond to the position of the line O_2O_2' shown in Fig. 3.11a. Thus, if the values y_1 and y_3 are assigned, then there are only two solutions of Eq. (3.23), which correspond to a transition from m_1 to m_3. These solutions are obtained for the case when the value $y(r)$ is larger than $y_{\mathrm{cr}}(y_3 > y_{\mathrm{cr}})$ in the excited region and less than y_{cr} ($y_{\mathrm{cr}} > y_1$) in the nonexcited region. But if $y_3 < y_{\mathrm{cr}}$ (we denote this solution by $\bar{y}_3$) and $\bar{y}_1 > y_{\mathrm{cr}}$, then there is a unique solution $x_0(r)$, which relates $m_1(\bar{y}_1)$ to $m_3(\bar{y}_3)$ (the trajectory $m_1(\bar{y})\,OO'm_3(\bar{y}_3)$) in Fig. 3.11a. The qualitative form of such a solution is shown in Fig. 3.11c. The change of the values y_1 and y_3 with the stepwise change of the slow

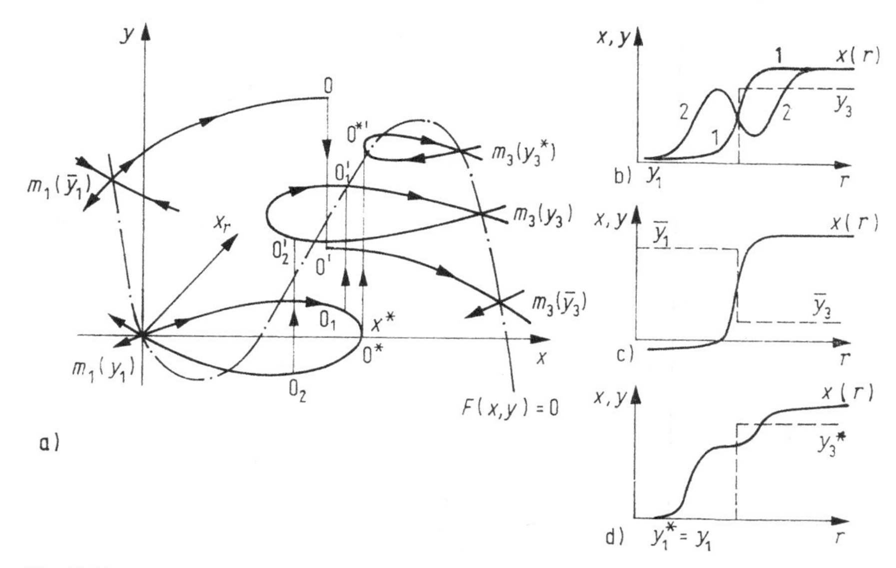

Fig. 3.11
The solutions of Eq. (3.23) for a jump-like distribution $y(r)$: a) matching of solutions in the phase space y, x, $\partial x/\partial r$; b) the form of solutions corresponding to the trajectories $m_1(y_1)\ O_1O_1{}'m_3(y_3)$ (curve 1) and $m_1(y_1)\ O_1O_2O_2{}'O_1{}'m_3(y_3)$ (curve 2); c) and d) are the form of solutions for the trajectories $m_1(y_1)\ O_1O_1{}'m_3(y_3)$ and $m_2(y_1)\ O_1O^*O^{*\prime}_{m_3(y_3^*)}$, respectively (Rozenblum, Starobinets, Yakhno, 1981).

variable leads to the change of the position of the equilibrium points in the solution $x_0(r)$ and of the values of x_0 at the point of matching. It is easy to see that there is a range of values y_1 and y_3 for which the matching point does not disappear and the solutions exist although a certain shift of the solution $x_0(r)$ with respect to the step $y(r)$ can occur.

Thus, there are ranges of values which correspond to the existence of solutions of Eq. (3.23). These ranges represent the desired characteristic II. Their own regions of existence of stationary solutions are obtained for each nonlinear function in (3.36). Examples of such regions are given in Figs. 3.10b and 3.12a. We note the peculiarities of the solutions in the different regions of characteristic II (see Fig. 3.12a). There are two such regions. The first one, denoted by 2, corresponds to a stable, and region 1 to an unstable immobile transition wave. The form of the solutions in region 2 is shown in Figs. 3.11c and 3.12a. A dissipative structure can be constructed from such stable fronts (see Section 7.4). The solutions corresponding to region 1 are shown in Figs. 3.11b and 3.12b. We stress that the points of the boundary $r \div r'$ of region 1 (Fig. 3.12a) correspond to a critical solution having a plateau with zero space derivative at the matching point (Fig. 3.11d) ($x_0 = x^*$, $\mathrm{d}x_0/\mathrm{d}r = 0$). When such a solution is formed, two other solutions (types 1 and 2 in Fig. 3.11b) merge into one solution. The critical solution in the phase plane corresponds to the trajectory $m_1(y)\ O_1O^*O^{*\prime}m_3(y_3)$, which is produced by the tangency of the separatrix loops.

We can judge the stability of the fronts in regions 1 and 2 (Fig. 3.12a) by the direction of those motions which arise in the initial deflection of the front from the stopping point to a distance exceeding its length. For the solutions corresponding to region 2, the front velocity is directed towards the stopping point and therefore the front returns to its initial position. For the solutions in region 1, the perturbation responsible for the shift of the solution (see $g_1(r)$ in Fig. 3.13) will increase (the slow variable values are such that the front velocity is directed away from the stopping

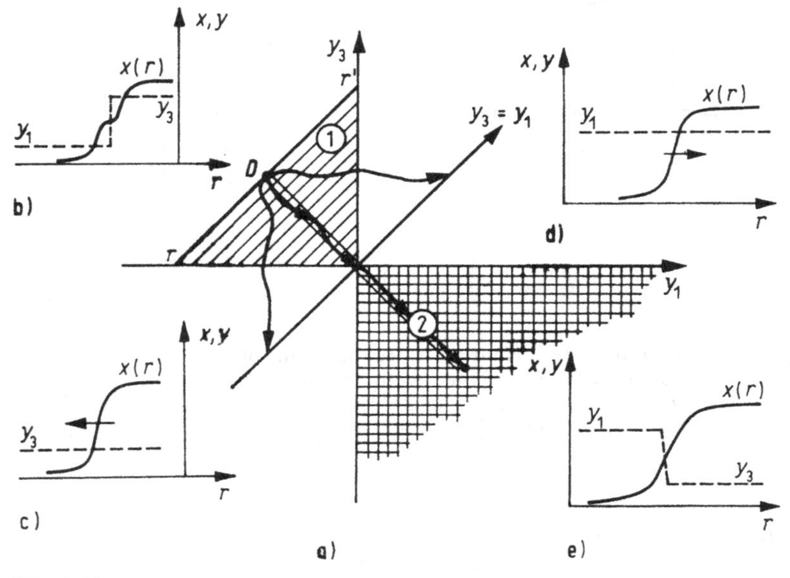

Fig. 3.12
a) The region of existence of a front with the zero velocity in the plane y_1, y_3. The front is unstable in region 1 and is stable for spatial shear perturbations in region 2. The front is unstable for the first "bending" perturbation mode, which causes the front to fission, at point D;
b) The spatial distributions of the x and y variables before the fission;
c), d) and e) are the distributions of the x and y variables after the fission of the front which differ in velocity.

point). The investigation of the stability with respect to small perturbations and numerical calculations (Rozenblum, Starobinets, Yakhno 1981) have confirmed these simple considerations. The increasing shear perturbation mode $g_1(r)$ forms a stable perturbation front that moves at a velocity $v(y)$. However, the different stationary solutions (denoted by 1 and 2 in Figs. 3.11b and 3.14) in region 1 differ in the character of stability with respect to the different perturbation modes. The first stationary solution in Figs. 3.11b and 3.14 is unstable only for shear perturbations, while the second stationary solution is unstable for two perturbation modes: a shear perturbation, $g_1(r)$, and the first bending perturbation mode, $g_2(r)$ (see Fig. 3.13). Both of these stationary solutions are stable for all other bending perturbation modes.

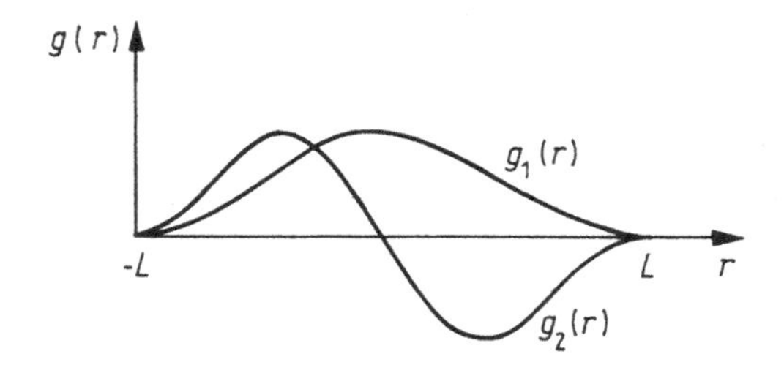

Fig. 3.13
The qualitative form of two first spatial perturbation modes: $g_1(r)$ is the "shear" perturbation and $g_2(r)$ is the first "bending" mode.

Conceivably, the second stationary solution is a sort of boundary that separates a bending perturbation superimposed on the first stationary solution. If the perturbation amplitude is small (shown by a dashed line between solutions 1 and 2 in Fig. 3.13), then the perturbation will disappear and the first stationary solution will not be violated by the bending mode. If the perturbation amplitude exceeds the difference between the second and the first stationary solutions (see Fig. 3.14), then the increasing bending perturbation mode will cause a break-up of the stationary solution. Basically, the processes by which the stationary front deteriorates were investigated by numerical calculation (Yakhno 1975; Zhislin, Yakhno, Goltsova 1976). It was shown

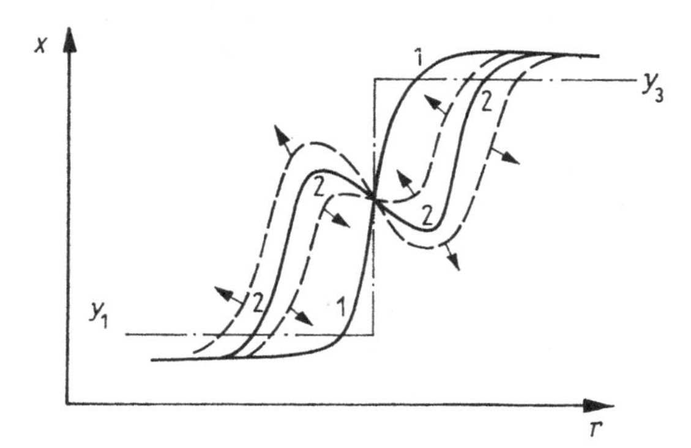

Fig. 3.14
Unstable stationary solutions (solid lines) in the discontinuity of the slow variable y (dash–dotted line). Solution 1 is unstable only for the shear perturbation mode. Solution 2 is unstable for the shear perturbation mode and the first bending perturbation mode (dashed lines).

that in this case fission of the front occurs.[1]) As the perturbations associated with the fast motions increase, the immobile front splits up into three new fronts (see Fig. 3.15). One of these new fronts remains immobile, while the other two fronts diverge at a velocity $v(y_3)$ in the region where $y = y_3$ and at a velocity $v(y_1)$ in the region with $y = y_1$. The formation of a pulse source in such a process will be considered in Chapter 5. The pattern by which the front fissions on characteristic II is represented in the form of three wavy arrows (see Fig. 3.12a). The lateral arrows that show the formation of two divergent fronts extend to the line $y = y_1$. The front velocities along this line are determined by means of characteristic I. The central arrow indicates that the new immobile front arising here enters region 2, which corresponds to a stable front. Note that the most favourable conditions for front fission are at

[1]) First, the front fission was obtained in the numerical calculations of Zaikin and Zhabotinsky, 1971.

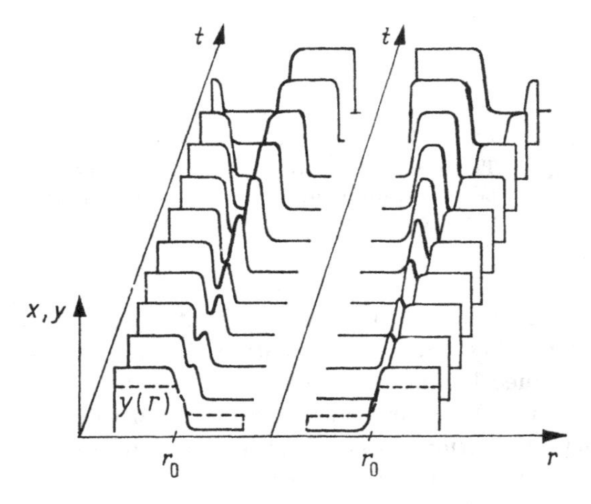

Fig. 3.15
The fission of a stopped front (x is shown by a solid line, y is shown by a dashed line)

the boundary of region 1, which is denoted by $r \div r'$ in Fig. 3.12a. At this boundary both of these stationary solutions (see solutions 1 and 2 in Fig. 3.11b) merge into one stationary solution having a plateau, which is unstable with respect to the infinitesimal bending perturbation mode. However, the question arises of whether this system can reach the boundary $r \div r'$ by slow motions. The point is that the increasing shear perturbations can move the excitation front away from the slow variable step y_1, y_3 and not allow this step to reach the boundary $r \div r'$. The numerical calculations show that such a process is realistic, i.e., such slow variations in y_1 and y_3 (and the trajectory in region 1, which corresponds to these variations) can occur, and do not cause any shear perturbations at the excitation front. It is natural to call such trajectories of slow motions in region 1 "separating" trajectories. A calculation of these trajectories has not yet been performed. However, it is seen from (3.23) that their form is determined by the nonlinear function $P(x, y)$, while the real slow motions are also governed by the function $Q(x, y)$. Therefore a qualitative analysis is possible only by a comparison of the slow motions of the representative point with the separating trajectories on characteristic II (see Fig. 3.12).

The description of characteristic II which we propose, is based on the calculations for a system with $\alpha = D_x/D_y \gg 1$. However, such a system seems to be useful in determining the characteristic features of the solutions in systems with $\alpha \gtrsim 1$.

c) *The null isoclines for a distributed system*

Characteristic III indicates the well-known null isoclines for a local system ($D_x = 0$, $D_y = 0$ in Eqs. (3.22)). However, there are some other distinctive features. In a local system, the transitions from one stable branch to another occur only at the values y_{max} and y_{min}, while in a distributed system, such transitions are caused by the excitation front. Therefore they can also take place for other values of the slow variable. Thus, for $y_{min} < y < y_{cr}$ a transition from branch 1 to branch 3 is possible and for $y_{cr} < y < y_{max}$ a transition from branch 3 to branch 1 is possible (see

Fig. 3.10c). These transitions are shown by arrows. Besides, for distributed systems with $\alpha = D_x/D_y \gtrsim 1$, one can judge the direction of the slow motions that occur in the system by the form of the null isoclines. But if $\alpha = D_x/D_y \ll 1$, then such an assessment will be incorrect, since even smooth inhomogeneities of the slow variable in space will strongly influence its variations in time. This can be illustrated by the mere existence of contrasting dissipative structures (see Section 7.4). The projections of the smooth regions of such structures are situated on the stable branches of the N-shaped null isocline far away from the equilibrium points of a homogeneous system but their positions do not vary at all with time (see also Sections 3.2 and 6.3).

In this approach, the considerations regarding both the fast and slow variable stages are, essentially, the same as in other methods. Common steps in the qualitative description are apparent. Stationary structures serve as a basis for AW solutions. Knowledge of such waves and structures permits one to predict the evolution of the transient processes in AW systems.

3.8 The group-theoretical approach

The diffusion-type equations without sources have a wide variety of space-time symmetries. The question arises which types of symmetries occur in the different nonlinear sources. The answer to this question can be given by group-theoretical investigations. A number of general results have already been obtained (Danilov 1980; Berman, Danilov 1981). We will not consider the methodological aspects of this problem but will only report some results following the papers cited above, since thus far the group-theoretical approach finds restricted applications in AW theory.

The set of transformations leaving the system invariant is called a group admitted by this system. The Lie theory enables one significantly to simplify the study of the groups of continuous transformations and to reduce the problem to the investigation of the so-called Lie algebra. Thus, for a basic model with $n = 2$ and $F_1 = F_2$, $D_{11} \neq D_{22}$, it is proved that the corresponding Lie algebra has a basis consisting of the following operators:

$$\xi_1 = \frac{\partial}{\partial t}; \quad \xi_2 = \frac{\partial}{\partial r}; \quad \xi_3 = r\frac{\partial}{\partial r} + 2t\frac{\partial}{\partial t};$$

$$\xi_4 = x_1\frac{\partial}{\partial x_1}; \quad \xi_5 = x_2\frac{\partial}{\partial x_2}; \quad \xi_6 = 2t\frac{\partial}{\partial r} - \frac{1}{D_{11}}rx_1\frac{\partial}{\partial x_1} - \frac{1}{D_{22}}rx_2\frac{\partial}{\partial x_2};$$

$$\xi_7 = tr\frac{\partial}{\partial r} + t^2\frac{\partial}{\partial t} - \frac{1}{4D_{11}}(r^2 + 2D_{11}t)\,x_1\frac{\partial}{\partial x_1} - \frac{1}{4D_{22}}(r^2 + 2D_{22}t)\,x_2\frac{\partial}{\partial x_2}. \tag{3.25}$$

For $D_{11} = D_{22}$ the Lie algebra basis increases

$$\xi_8 = x_2\frac{\partial}{\partial x_1}; \quad \xi_9 = x_1\frac{\partial}{\partial x_2}. \tag{3.26}$$

We emphasize that the whole set of symmetries of two equations with sources is exhausted by these transformations. Also, it was proved that the Lie algebra with a basis (3.24; 3.25) forms the widest group admitted by the basic model in a one-dimensional space for any choice of the sources, i.e., for AW systems, the number of symmetries is always smaller.

For an arbitrary choice of the sources, the dimension of the basis is equal to 2, i.e., there are only transformations of shear in time ξ_1 and shear in space ξ_2. The Lie algebra of dimension 3 (Danilov 1980) is used for

a) pure powers ($F_1 = x_1^s$; $F_2 = x_2^p$),

b) cross powers ($F_1 = x_2^p$, $F_2 = x_1^s$),

c) mixed powers ($F_1 = x_1^q x_2^p$; $F_2 = x_1^s x_2^p$),

d) exponential sources and

e) exponential power-law sources.

The similarity transformations produce self-simulating solutions dependent on $r/\sqrt{t}$. Transformations of the rotations are added when dealing with plane problems. Combinations of the transformations of rotations and shears yield solutions in the form of spiral waves, or reverberators.

The group-theoretical methods also permit one to obtain interesting results in other fields of AW theory. For example, theorems on the number and position of the homogeneous limit cycles of the basic model for $n = 2$ and quadratic nonlinearity have been proved (Danilov 1982).

Another advantage of these methods is the mathematical rigour and generality of the results obtained. The group-theoretical approach does not require detailed knowledge of the structure of the systems. Besides, knowledge of only one solution is sufficient to construct a family of solutions. It is no easy matter to do this by other methods. Seemingly, the range of application of these methods in AW theory will be increasingly wide.

3.9 The numerical experiment

Numerical methods hold a prominent place in studies of AWPs. The point is that for strongly nonlinear and nonconservative systems, exact solutions can be found only in extremely simple models. Only some typical calculations will be listed in Chapters 4 and 7. Thus, one has to restrict oneself to using approximate, qualitative methods for studying the models.

Usually, the qualitative results need to be checked for correctness on a computer.

Numerical methods today constitute a well-developed branch of applied mathemathics which for space reasons we cannot survey here. These problems are widely discussed in the literature at different levels of rigour and accessibility. We will only mention some specific features which are important in conducting a numerical experiment for AW systems. Typically, a "numerical experiment", is not just any use of a computer but only calculation of the basic models, which is aimed at substantiating the existence of the processes under study or their peculiarities. Of course, numerical experiments are widely used to solve other problems, e.g., to construct homoclinic trajectories, parameter dependence on model coefficients, etc. However, these are not experiments: such applications are only a continuation of the qualitative analysis of a dynamical system and can be treated as a subsidiary procedure available to a researcher.

A numerical experiment is conducted using models which are discrete in space. As was noted in Section 3.6, characteristic of such models are threshold effects for N (number of elements). Basically, it is recommended to increase N and compare the form of the resulting two solutions in a certain functional space. Such an approach can be called "quasi-linear", since it does not account for the threshold phenomena. These phenomena can be illustrated as follows.

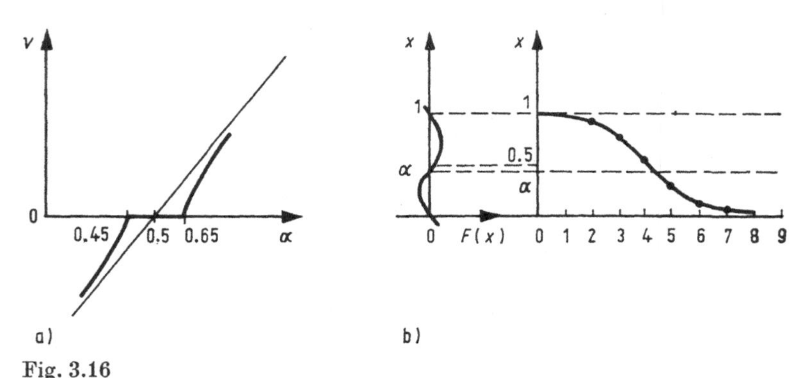

Fig. 3.16
The TF velocity plotted against the "threshold" for α:
a) A continuous model is shown by a thin line, a discrete model ($N_F = 5$) is shown by a thick line;
b) the graph of the function $F(x)$ and the profile of the stationary discontinuity

Figure 3.16 shows the dependence of the wave velocity in a one-component model with $F(x) = \varkappa(1 - x)(x - \alpha)x$. It is well known (Section 3.7) that the exact value of the TF velocity for $t \to \infty$ can be determined from the expression $v = (1 - 2\alpha)\sqrt{\varkappa D/2}$ ($v(\alpha)$ is shown by a thin line in Fig. 3.16a). The numerical experiment yields an alternative form of dependence: $v = 0$ for the range of values of α (thick line), i.e., there is a certain analogue of the "dry friction" region. A conventional three-point pattern was used for the approximation of the second derivative in our calculations: $\partial^2x/\partial r^2 \to {}^{j-1}I_i{}^{-1}I_i = ({}^{j-1}x_i - 2{}^jx_i + {}^{j+1}x_i)$. The number of elements was chosen such that the steep region of the front occupied not less than $N_F = 5$ cells. No doubt, the region analogous to "dry friction" decreases with increasing N but does not degenerate to a point in discrete models. For $N \gg N_F$, a transition from the immobile slope to the travelling wave regime occurs in a jumplike way, i.e., if $N' = 2N$, then the solution is nearly the same as that for N, while for $N'' = 2N + 1$, the immobile solution in the form of a front loses stability and a travelling wave arises.

It is well known that inhomogeneous stationary solutions are unstable in one-component systems. In a numerical experiment, however, there is a wide range of initial conditions which can lead to stable inhomogeneous stationary solutions. They are stable both as to a numerical count and in the parametric sense, i.e., minor variations of the model parameters yield minor variations of the solution. However, such solutions do not correspond to the initial distributed system and lose stability as N reaches a threshold value.

Similar effects are observed in multicomponent systems with $D_{xx} \sim \varepsilon > 0$ (x is the autocatalytic component). Thus, localized structures with unbounded space region defined by the initial conditions can occur in a discrete model for N less than a critical value (see Chapter 7 and Vasilev, Romanovskii, Yakhno 1979). The localization breaks up for large N. Note that in a distributed system, such solutions are possible only if D_{xx} is exactly equal to zero. Threshold effects are also observed for the nonperiodic stationary solutions of homogeneous models in numerical experiments (see Section 6.1).

Threshold effects arising in a quantized space occur also in other types of AWP but sometimes it is difficult to give account for such effects. Primary attention should be paid to those systems in which one of the D_{ii} is much smaller than the rest. These systems feature several space scales. In a numerical experiment, the length of the smallest scale should correspond to a number of elements such that their representative points overlapped "all the characteristic" phase space regions at every instant. In other words, it is necessary to combine a numerical experiment with a qualitative analysis of the models. If it can be shown that the threshold for N is exceeded, then traditional "quasi-linear" patterns of experiments can serve to check the solutions for accuracy.

Also, a computer experiment runs into some difficulties arising from the introduction of discrete time. Numerical patterns can be divided into two groups: explicit and implicit. Explicit numerical patterns have rather simple algorithms; thus,

$$^jx_i(t + \tau) - {}^jx_i(t) = \tau\left({}^jF_i(t) + D^j_{ii}\Delta_i(t)\right). \tag{3.27}$$

However, the need for count stability places an upper limit on the step in time

$$\tau < L^2/\left(2 \max_i \{D_{ii}\} N^2\right). \tag{3.28}$$

Such limitations are not necessary for implicit patterns but these are much more complicated:

$$^jx_i(t + \tau) - {}^jx_i(t) = \tau\left({}^jF_i(t + \tau) + D^j_{ii}\Delta_i(t + \tau)\right). \tag{3.29}$$

In this case, one has to solve complex sets of iteration equations in order to determine $^jx_i(t + \tau)$. Nevertheless, typically, implicit patterns are recommended for quasi-linear and nonlinear parabolic equations (Samarsky 1977) because a smaller number of operations are needed to achieve the required accuracy. However, when dealing with AWPs, additional restrictions should be placed on τ, which are mostly still more rigorous than those in (3.28). These restrictions depend on the form of the nonlinearity and are necessary both for an explicit and an implicit pattern. Thus, here, explicit patterns are believed to be more profitable.

The meaning of the additional restrictions on τ can be understood from the calculations of a local system. Indeed, for $N = 1$, the explicit pattern (i.e., the Euler broken line method) yields a limiting cycle larger in size than the real one, while the implicit pattern yields a smaller limiting cycle. The relative errors are approximately the same. The admissible accuracy can be achieved only for small values of τ. For example, for the Van-der-Pol equations, which were studied in connection with synchronization problems (see Section 6.2), it appeared that when the error was within 10%, a value of τ 20—40 times less than that implied by (3.28) for $N = 30$ is needed. An explicit pattern has therefore been chosen. Note that for relaxation-type systems, much more rigorous restrictions can be imposed on τ. If this is the case, then it can be more profitable to integrate by explicit patterns using methods for increasing accuracy as applied to ordinary differential equations (see, e.g., Runge-Kutta, Adams or Heming methods).

Discrete representations of a distributed system can be not only in the form of space partition by a grid but also in the form of expansions of solutions in other basis sets of functions, e.g. harmonic functions. If the number of terms in the series is restricted, then it is possible to obtain a set of ordinary equations the numerical integration of which presents no difficulties. However, such an approach is reasonable only if a much smaller number of terms in the series than N is admissible. Otherwise, the machine time exceeds, as a rule, the time required for grid discrete models.

Along with a direct discretization of models, other methods of numerical experiment are used, for example a reduction of the basic model to integral or integral-differential equations for which the problems of the stability of numerical patterns or their matching are simpler. However, when using such methods, one has to do much preparatory work before performing a computer experiment and take into account the specific features of each particular model.

When studying AWPs in large-dimension systems, many difficulties can be overcome by means of hybrid, digit–analog computers. For example, the circulation of an excitation in a plane was investigated (Zykov 1984). Seemingly, hybrid computers will not find wide application in AWP studies.

Multiprocess digital complexes for highdimensional problems seem to be more promising.

In conclusion, we emphasize that the tendency to take computer results on face value is clearly not justified. So far, conducting a numerical experiment belongs to a certain extent, to the realm of art.

Chapter 4
Fronts and pulses: Elementary autowave structures

We will consider models which are most frequently encountered in experiments with autowaves in the form of fronts or excitation pulses (travelling pulses (TP) and travelling fronts (TF)).

4.1 A stationary excitation front

A simple description of a solitary excitation TF enables one to use one equation (see Section 2.6)

$$\frac{\partial x}{\partial t} = \frac{\partial^2 x}{\partial r^2} + F(x), \tag{4.1}$$

where $F(x)$ describes the rate of change of x in a point system, and is usually represented by one of two curves shown in Fig. 4.1. Specifically, such a dependence

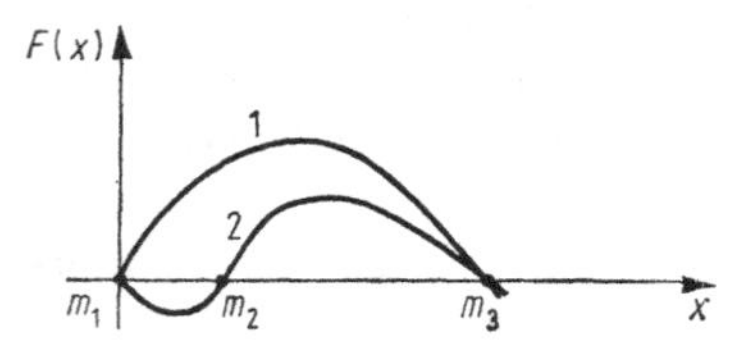

Fig. 4.1
Dependence of the nonlinear function in Eq. (4.1) for nonthreshold (1) and threshold (2, 3) cases of front propagation

follows from a nonlinear function $F(x, y)$ in (2.53) at $y = \text{const}$. Physical boundary conditions on an infinite segment correspond to a vanishing $\partial x/\partial r$ at those values of $x_i = m_i$ for which $F(x_i) = 0$. First we will consider stationary solutions (4.1). Introducing a self-similarity variable

$$\eta = r - vt \tag{4.2}$$

the front propagation velocity problem in (4.1) reduces to

$$W \frac{dW}{d\eta} + vW + F(x) = 0, \tag{4.3}$$

where $W = dx/d\eta$ and the velocity v are found from the solution of this nonlinear equation with the boundary conditions $W(m_1) = W(m_3) = 0$.

The existence of desired solutions can be clearly demonstrated in the W, x phase plane. Let us consider the case $F = F_1$ (the nonlinear function has only two zeroes (curve 1 in Fig. 4.1). Bounded solutions corresponding to physically relevant cases exist over the entire range $0 < v < \infty$.

For $0 \leqq v \leqq v_{min}$ only those solutions are bounded which oscillate about $x = m_1$. These solutions are unstable with respect to perturbations in the framework of Eq. (4.1) (see Section 4.2) and, moreover, do not satisfy the boundary conditions assigned here. But if the focus (point $x = m_1$) is converted into a node, then solutions in the form of a front without oscillations, those which satisfy the chosen boundary conditions, are possible. As is seen from Fig. 4.2, a saddle-to-node separatrix solution exists

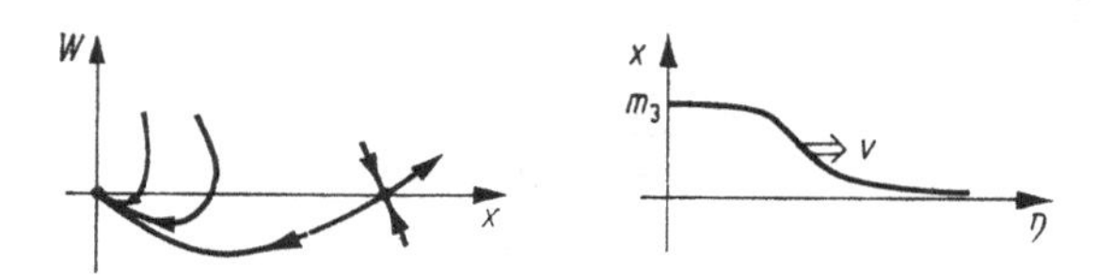

Fig. 4.2
Solutions of Eqs. (4.1) and (4.3) for the case where the nonlinear function has only two zeroes (denoted by "1" in Fig. 4.1) and $v > v_{min}$

over the entire range $v_{min} \leqq v < \infty$. Such a problem was first solved by A. N. Kolmogorov, I. G. Petrovskii and N. S. Piskunov back in 1937.[1]) For the nonlinear function with a monotonically varying first derivative $F_x'(x)$ it is easy to find $v_{min} = v_0 = 2\sqrt{F_x'(m_1)}$. This value corresponds to the transformation of a focus-type singularity to a node type. If the variation is no longer monotonic, then the first saddle-to-node solution is possible only for $v_{min} > v_0$. The presence of a range of velocities is caused by the instability of a homogeneous state $x = m_1$. In the framework of Eq. (4.1), stationary saddle-to-node solutions possess the highest stability. They are unstable only with respect to those perturbations which cause them to turn from a front with one velocity to a front with another velocity within the same range. These solutions are stable with respect to any other perturbations.

In the case $F = F_{II}$ (the nonlinear function has three zeroes) possible stationary solutions (4.1) are shown in Fig. 4.3. The only stable solution is a nonoscillating front (Fig. 4.3c), which represents a saddle-to-saddle separatrix (Kanel 1962, 1964). All other stationary solutions are unstable. Cases of unstable solutions that satisfy the boundary conditions are shown in Fig. 4.3a, b, c. Typical nonstationary processes associated with these solutions are given in Section 4.2.

Generally, the eigenvalue problem (4.3) is solved by numerical methods. However, sometimes an analytical solution is obtained (see Table 3.1 in Section 3.7). For this,

---

[1]) Nearly at the same time, the travelling front problems in the burning case were solved by Zeldovich and Frank-Kamenetskii (1938). Those investigations were generalized in monographs by Frank-Kamenetskii 1967 and Zeldovich et al. 1980.

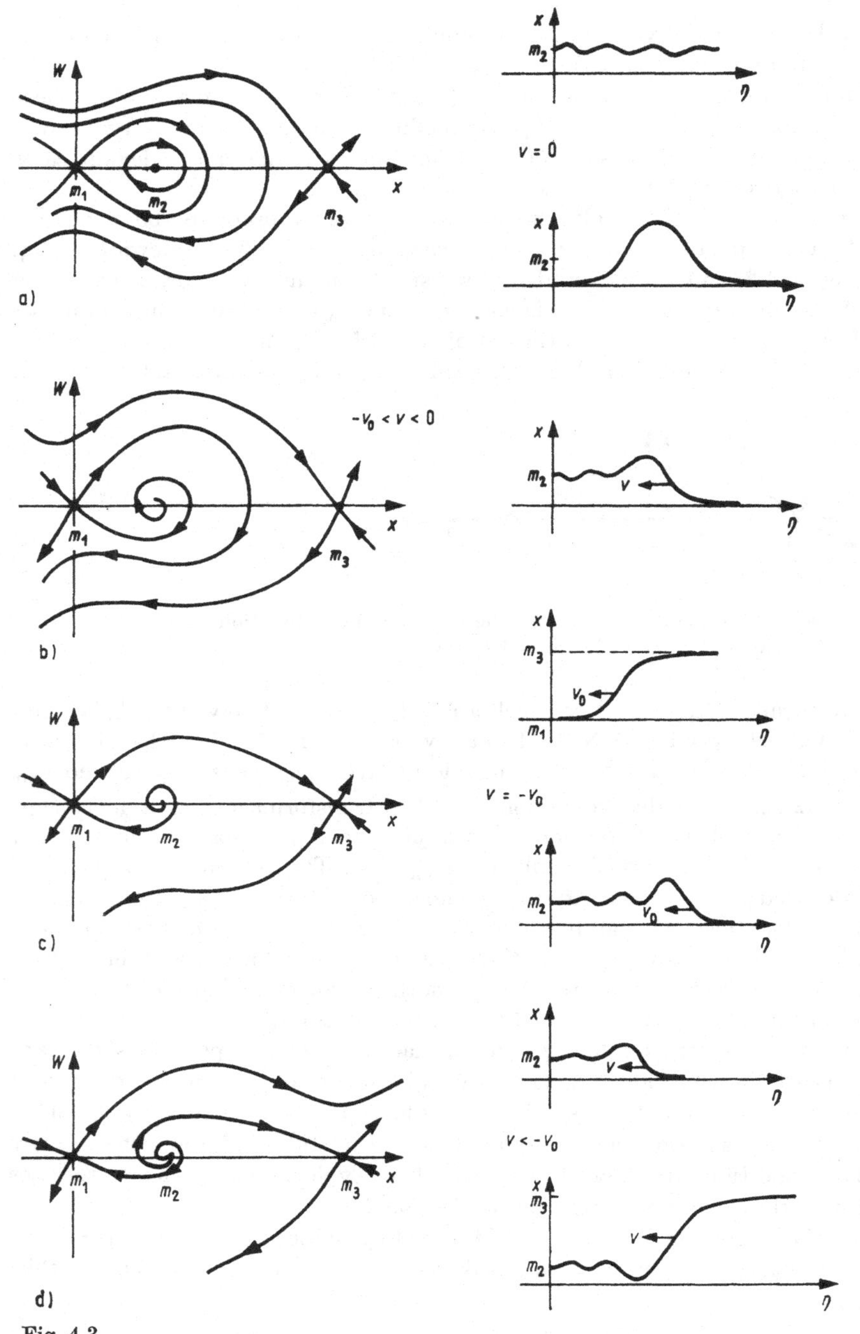

Fig. 4.3
Solutions of Eqs. (4.1) and (4.3) for the case where the nonlinear function has three zeroes (denoted by "2" in Fig. 4.1): a) $v = 0$; b) $-v_0 < v < 0$; c) $v = -v_0$; d) $v < -v_0$

for example, a piece-linear approximation for $F_{II}(x)$ is useful. But if $F_{II} = -\gamma(x - m_1) \times (x - m_2)(x - m_3)$, then substituting $W = \sigma(x - m_1)(x - m_3)$ into (4.3) in a manner similar to that in (2.11) (Section 2.1) and equating coefficients of x with equal powers we have:

$$v_0 = \sqrt{\gamma/2}\,(m_1 + m_3 - 2m_2), \quad \sigma = \sqrt{\gamma/2}. \tag{4.4}$$

Such a representation is very often used in theoretical studies (Frank-Kamenetskii 1967, Hunter, Hauchton, Noble 1965, Nagumo, Joshizawa, Arimoto 1975, Yakhno 1976, Ebeling and Czajkowski 1976; Ebeling, Schimansky-Geier 1979, 1980; Schimansky-Geier, Ebeling 1983). Also, there is a case where the function $F(x)$ has five or more zeroes. Then several stable TFs differing in amplitude can be excited in the system (4.1).

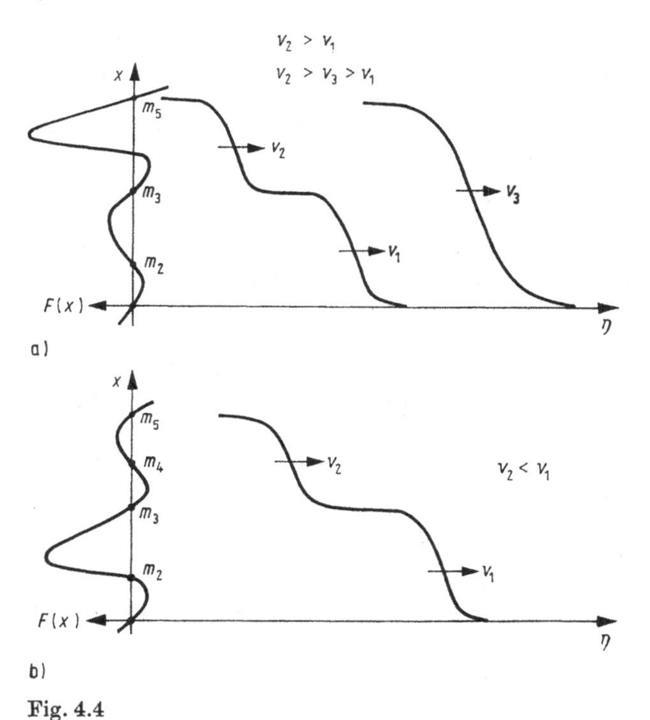

Fig. 4.4
Stable stationary solutions of Eq. (4.1) for the case where the nonlinear function has five zeroes (the graphs $F = F(x)$ are shown on the left): a) three types of stable TFs at $v_2 > v_1$; b) two types of stable TFs at $v_2 < v_1$.

Figure 4.4 shows a qualitative picture of possible stable stationary TFs. Two variants are possible:

a) TFs with small amplitudes propagate and converge eventually ($v_1 < v_2$ in Fig. 4.4); in this case, a front with a large amplitude and a velocity between v_1 and v_2 is possible;

b) TFs with small amplitudes diverge ($v_1 > v_2$ in Fig. 4.4); in this case, a large-amplitude TF cannot exist.

Using these data as a basis, it is easy to analyze the motions of a TF in the case where the function $F_{III}(x)$ has seven or more zeroes. Mathematically, the problems of the existence of solitary fronts are studied in ample detail by Volpert (1983).

4.2 A typical transient process

It is well known that in a local system of the first or second order, the analysis of the position of equilibrium points and of the behaviour of small perturbations near these points allows a qualitative description of all possible motions. Of course, we are tempted to construct a similar procedure for a qualitative description of the solutions in distributed systems. Let us consider the relevance of such an approach using Eq. (4.1), where $F(x) = -\gamma x(x - m_1)(x - m_3)$ (Starobinets, Yakhno 1983). The solutions will be considered on a segment $-L \leqq r \leqq L$, at the boundaries of which the conditions $\partial x/\partial r|_{r=\pm L} = 0$ are met.

It is natural to suppose that in a distributed system, stationary solutions are obtained from (4.1) (provided $\partial x/\partial t = 0$), and are analogous to the equilibrium points in a local system. The important part played by stationary solutions was repeatedly noted when nonlinear wave processes were investigated (Gaponov, Ostrovskii, Rabinovich 1970; Rabinovich 1974).

The next step is to determine the stability of stationary solutions. In a distributed system, small perturbations near the stationary solutions are presented, in contrast to a local system, in the form of a set of space modes. Therefore when analyzing the instability it is necessary to trace the "fate" of each space mode. Let us demonstrate the existence of a certain relation between the number of possible stationary solutions and the character of their stability with respect to space perturbation modes.

In the W, x phase plane, limited stationary solutions of Eq. (4.1) are shown by closed phase trajectories included inside the separatrix (see Fig. 4.3a). It is seen that the boundary conditions $\partial x/\partial r|_{r=\pm L} = 0$ are satisfied for not only solutions limited by the separatrix but also for the equilibrium points $x = m_1$ and $x = m_3$. As will be shown below, the number of solutions depends on the length of the segment $2L$. Let us introduce some designations: $\varkappa_i = \pi i/2L$ is the wave number of the perturbation mode (the index i denotes the number of zeroes for the given mode on the $2L$ segment), γ_{ij} is the instability increment for a perturbation mode having i zeroes on the segment and superimposed on a stationary wave having j zeroes.

Homogeneous stationary solutions $\bar{x} = m_1$ and $\bar{x} = m_3$ are always stable with respect to arbitrary infinitesimal perturbations. The stability of the central homogeneous solution $\bar{x} = m_2 = 0$ depends on the increment $\gamma_i = -\varkappa_i^2 + F_x'(0)$. Let us consider segments of different length:

a) $0 \leqq 2L < \pi/F_x'(0)$. In this case, only homogeneous stationary solutions occur in the system. The solution $\bar{x} = m_2 = 0$ is unstable only with respect to shear perturbations $\left(\gamma_{0\infty} = F'(0) > 0\right)$. All perturbations with the wave number

$\varkappa_i$ ($i > 0$) will be damped. The increments for the perturbation modes of a homogeneous stationary solution $\bar{x} - m_2 = 0$ are shown schematically in Fig. 4.5.

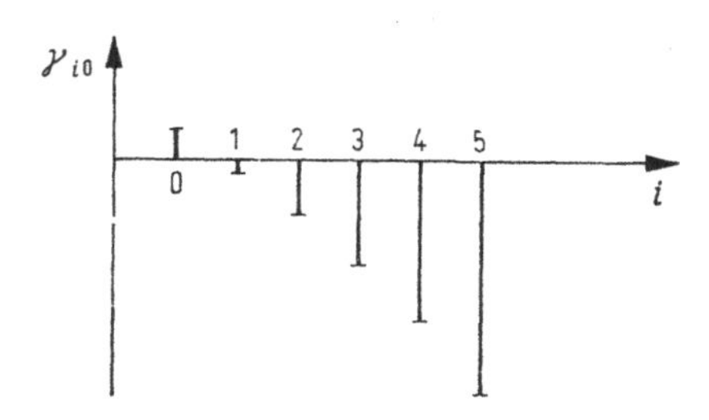

Fig. 4.5
Dependence of the instability increment on the perturbation mode number for the solution $x = m_2$ (the case $0 \leqq 2L \leqq \frac{\pi}{F_x'(m_2)}$)

b) $\pi/F_x'(0) \leqq 2L < 2\pi/F_x'(0)$. In this case, another inhomogeneous stationary solution is added to the homogeneous solutions (Fig. 4.6a). It has one intersection with the line $\bar{x} = m_2$ (Fig. 4.6b). This solution is unstable[1]) with respect to shear perturbations ($\gamma_{01} > 0$) but is stable for any other perturbation modes ($\gamma_{i1} < 0$, $i > 0$). The homogeneous solution $\bar{x} = m_2$ is unstable with respect to two perturbation modes: a homogeneous perturbation ($\gamma_{0\infty} > 0$) and a perturbation with one point of intersection of the zero value of $\bar{x} = m_2$. For all other perturbation modes, the solution $\bar{x} = m_2$ is stable ($\gamma_{i\infty} < 0$, $i > 1$).

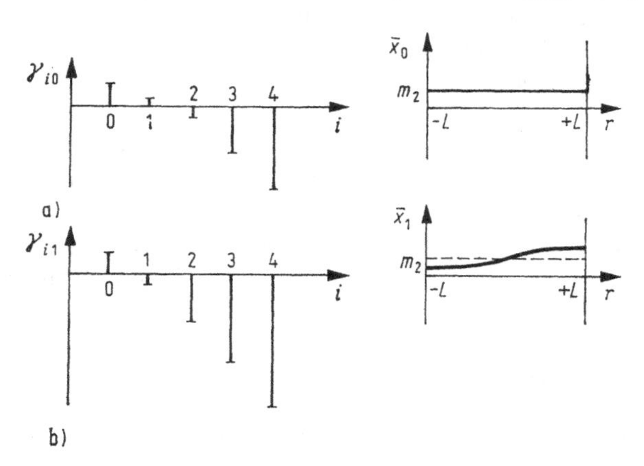

Fig. 4.6
Dependence of the instability increments on the perturbation number for two stationary solutions $x = m_2$ and $x = x(r)$ with one intersection of the value m_2 (the case $\frac{\pi}{F_x'(m_2)} \leqq 2L \leqq \frac{2\pi}{F_x'(m_2)}$) (the forms of the stationary solutions are shown on the right)

---

[1]) The data on the stability of inhomogeneous stationary solutions are obtained by a method given in Section 3.3.

c) $2\pi/F_x'(0) \leqq 2L < 3\pi/F_x'(0)$. In such a system, one homogeneous and two inhomogeneous stationary solutions can exist (Fig. 4.7). Their stability is characterized by the corresponding increments shown in this figure.

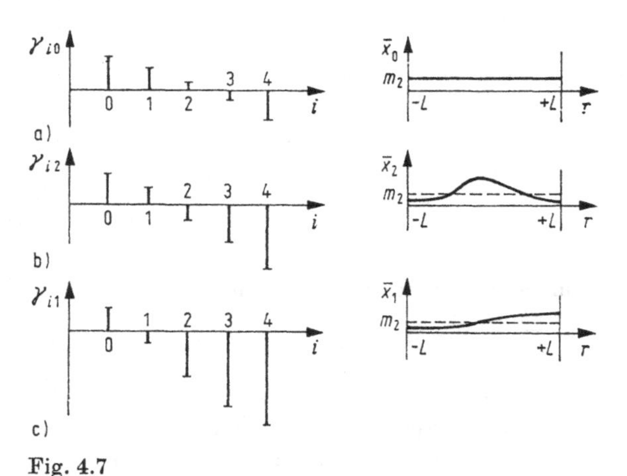

Fig. 4.7
Dependence of the instability increments on the perturbation mode number for three stationary solutions $\bar{x}_0 = m_2 = 0$, $x = \bar{x}_2(r)$ and $x = \bar{x}_1(r)$ (the forms of these solutions are shown on the right). The length of the segment ranges $\dfrac{2\pi}{F_x'(m_2)} \leqq 2L \leqq \dfrac{3\pi}{F_x'(m_2)}$

These examples show that inhomogeneous stationary solutions are stable with respect to such perturbation modes which have equal or larger number of zeroes than the stationary solution being investigated and are unstable for perturbations with smaller number of zeroes.

The next step in the qualitative analysis is to predict which space distribution the increasing perturbation modes will transform to. When considering instability schemes, it is natural to suppose that a perturbation mode with i zeroes will turn into a corresponding stationary solution with the same number of zeroes. This supposition was checked by numerical calculations. The initial condition was assigned in the form

$$x(r) = \sum_i \left(A_i \cos \frac{i\pi}{2L} r + B_i \sin \frac{i\pi}{2L} r\right).$$

By changing the initial amplitudes of different perturbation modes, we traced the evolution of a homogeneous solution $x = m_2 = 0$ to a stationary inhomogeneous one with the same number of zeroes (Fig. 4.8) and the transitions from one inhomogeneous solution to another (Fig. 4.9). Note that in the calculations it is necessary to choose an appropriate relation between the amplitudes of different modes since those perturbation modes have larger instability increments which have less zeroes. The numerical calculations support the assumption that an elementary perturbation transforms to a corresponding stationary solution with the same number of zeroes. However, there are certain deviations from the simple scheme we propose. These

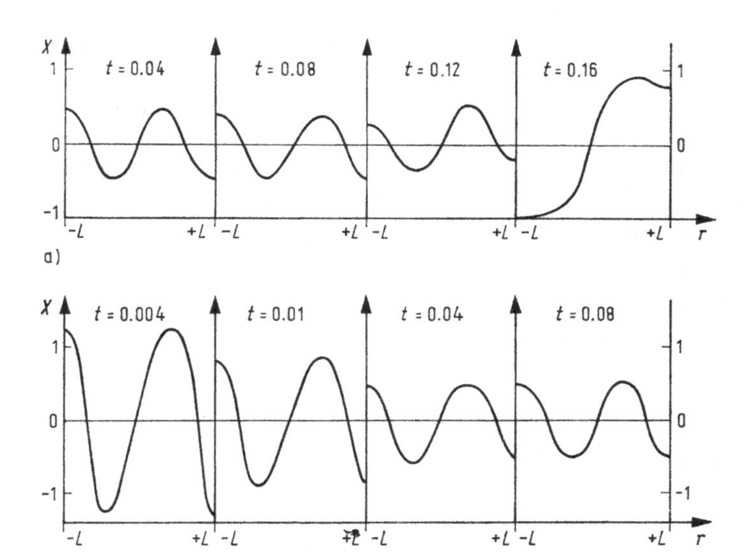

Fig. 4.8
Initial stages of the nonlinear processes described by Eq. (4.1) for different amplitudes of the initial perturbation $x_0(r) = A_3 \sin \dfrac{3\pi}{2L} r$: a) $A_3 = 0.5$; b) $A_3 = 3$

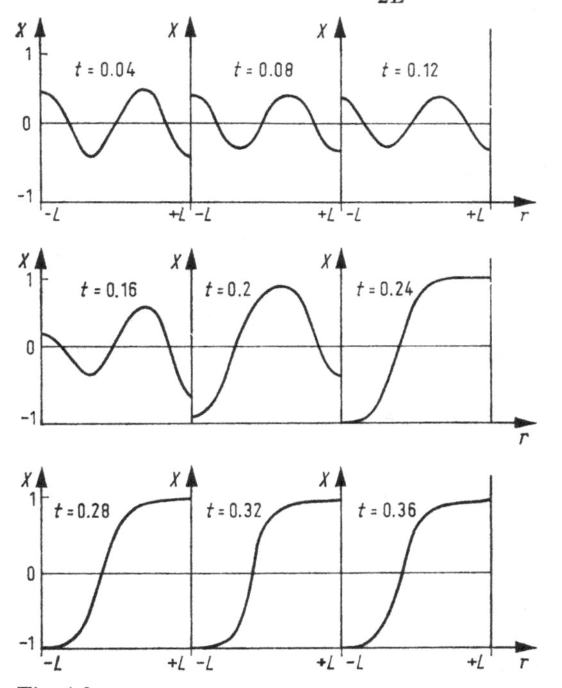

Fig. 4.9
A transient process at the initial perturbation $x_0(r) = 0.005 \cos \dfrac{\pi}{L} r + 2 \sin \dfrac{3\pi}{2L} r$, which leads to unstable stationary solutions with a lesser number of zeroes

deviations are caused by the unavoidable errors in the computer experiment when unstable structures are investigated (see Section 3.3).

On the whole, the above consideration supports the concept of a qualitative analysis of transient processes in distributed systems and permits one to distinguish its basic states. Namely, a transient process is determined by

a) the stationary solutions,

b) the character of their stability, and

c) laws for transition of unstable modes to known space–time structures.

4.3 The front velocity pulsations

We now consider models for two interesting phenomena associated with TF velocity pulsations: A. Self-excited oscillatory burning. B. Spin burning waves.

A. In a number of cases, a TF is described by two equations (see, e.g., Merzhanov 1980; Aldushin, Merzhanov et al. 1973):

$$\frac{\partial T}{\partial t} = D_1 \frac{\partial^2 T}{\partial r^2} + \gamma Q(T, n), \tag{4.5}$$

$$\frac{\partial n}{\partial t} = D_2 \frac{\partial^2 n}{\partial r^2} - Q(T, n). \tag{4.6}$$

Basically, these are the processes by which thermal waves propagate in chemical systems (Section 2.6).

If $D_1 \approx D_2$, then Eqs. (4.5) to (4.6) have an integral $H = \gamma T + n = \text{const}$ and reduce to one equation, which was investigated in Section 4.1. However, it is interesting to study the TF motions at $\mathcal{L} = D_1/D_2 \neq 1$ ($\mathcal{L}$ — is the Lewis number). Such solutions were investigated in sufficient detail in the theory of burning, because of the stable propagation velocity of thermal waves (Barenblatt, Istratov and Zeldovich 1962). The small perturbation method was used.

The analysis has shown that the stability of a stationary propagation of excitation waves is determined by two dimensionless parameters: $\mathcal{L}$ and θ_0, the dimensionless value of the TF amplitude. If $Q = n\left(\frac{k_0}{C}\right)\exp(-E/RT)$, where E is the activation energy, C is the specific thermal capacity, k_0 is the preexponential factor, R is the universal gas constant, then $\theta_0 = E\gamma/RC(T_0 + \gamma/C)^2$ (T_0 is the temperature in a cold medium). The part played by the parameters $\mathcal{L}$ and θ_0 has been investigated most thoroughly by Aldushin and Kasparyan (1978). Figure 4.10 shows a diagram which illustrates the results of such an analysis. It is seen that the stationary propagation of a TF is stable for $\mathcal{L} \simeq 1$ and small values of θ_0. A more complete analysis of the TF behaviour was made using numerical methods. Thus, it was shown by Shkadinskii, Khaikin and Merzhanov 1971 that for $\mathcal{L} \ll 1$ the instability shows itself as TF instantaneous velocity pulsations about the stationary value. The form of such pulsations is given in Fig. 4.11, which shows the dependence of the relative instantaneous velocity on the parameter θ_0. Harmonic oscillations are observed near the instability boundary, and relaxation oscillations far from the boundary. In the latter case the TF propagation velocity sharply increases and decreases alternatively. The space profiles of the variables T and n change strongly with time. When moving far from the instability boundary (temporary behaviour) the pulsations assume an increasingly complex structure. Evidently, the increased complexity of the pulsations can lead to chaotic motion of the front. However, the calculation of complex pulsations faces appreciable difficulties (Aldushin, Martyanova et al. 1973). The pulsational propagation of a burning front is generally referred to as self-excited oscillatory burning.

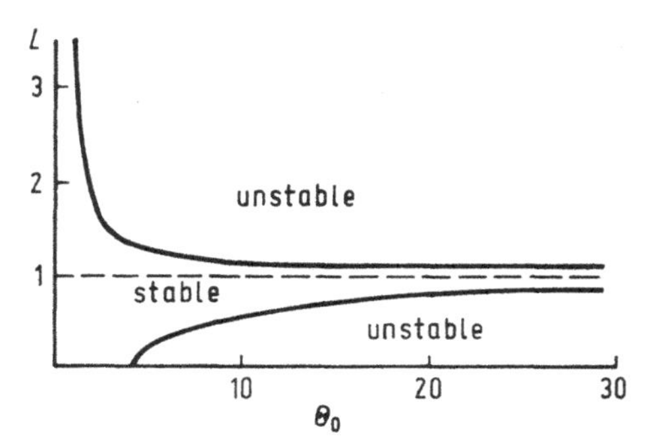

Fig. 4.10
Diagram of the stability of a plane front with respect to small perturbations of the spreading velocity (Aldushin and Kasparyan 1978)

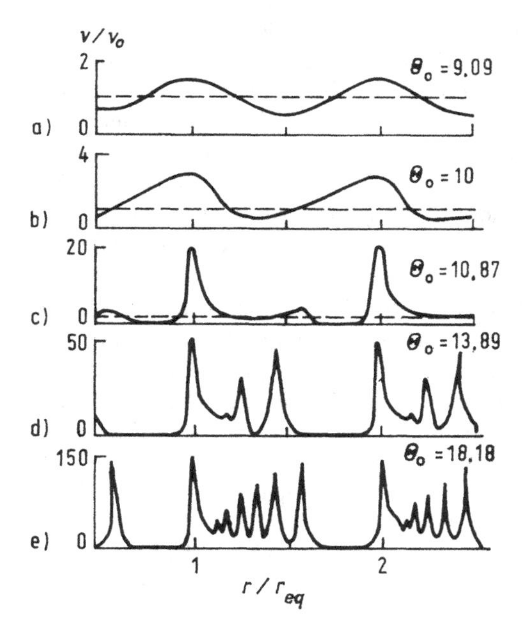

Fig. 4.11
Pulsations of TF velocity at different distances from the instability boundary. r_p is the width or space period of the TF pulsation (Shkadinskii, Khaikin and Merzhanov 1971).

B. Also, there are instabilities in the plane propagation of a TF in a two-dimensional medium which are generally called spin burning. The phenomenon consists of the fact that there is a localized source of increased burning temperature, which moves along the burning front. In a cylindrical sample, such a source spirals along the lateral face because of the propagation of the entire front (see Fig. 1.5).

A unified description of these phenomena was proposed by Shkadinskii, Ivleva and Merzhanov (1978). The authors considered a two-dimensional set of dimensionless burning equations

$$\frac{\partial \theta}{\partial t} = \frac{\partial^2 \theta}{\partial r^2} + \nu \frac{\partial^2 \theta}{\partial \varphi^2} - \alpha(\theta - \theta_{\mathrm{H}}) + (1 + \eta) \exp \frac{\theta}{1 + \beta\theta}, \tag{4.7}$$

$$\frac{\partial \eta}{\partial t} = \gamma(1 - \eta) \exp \frac{\theta}{1 + \beta\theta}. \tag{4.8}$$

The solution is written $r > 0$, $0 \leqq \varphi \leqq 1$, $t > 0$. The periodicity condition is defined by

$$\theta(r, \varphi, t) = \theta(r, \varphi + 1, t), \quad \eta(r, \varphi, t) = \eta(r, \varphi + 1, t),$$

the "end" condition by

$$\theta(0, \varphi, t) = \theta_{\mathrm{H}} + \gamma^{-1},$$

and the initial condition by

$$\theta(r, \varphi, 0) = \theta_0(r, \varphi), \quad \eta(r, \varphi, 0) = \eta_0(r, \varphi),$$

where θ and η are the dimensionless temperature and transformation depth.

These equations describe the burning of a cylindrical sample with end firing. It is assumed that θ does not propagate in the radial direction. There are two parameters in the problem: γ and ν. The γ parameter determines the instability limit, and is well known from burning theory; ν is a parameter which arises in a two-dimensional problem; it is inversely proportional to the sample's diameter squared. The computational results are listed in Fig. 4.12 and 4.13. Thus, Fig. 4.12 with r and φ coordinates shows lines of permanent transformation depth (η variable) at successive times. These lines demonstrate the position of a TF in a two-dimensional space. The calculations have been performed for one value of γ and three values of ν. All the TF lines are arranged in a periodic manner. This is an indication that the burning regimes are steady-state ones. Three burning regimes are possible. Figure 4.12a shows vertical straight lines. This is a one-dimensional self-oscillatory regime. Spin burnings with one and two sources of reaction are demonstrated in Fig. 4.12b. The dependence of the instantaneous velocity of burning on the r coordinate is presented in Fig. 4.12d. A two-dimensional temperature field for two-source spin burning is given in Fig. 4.13. The "heaps" in the field are the sources of reaction, which propagate across the basic wave.

The existence of other spin burning regimes is attested by a theoretical consideration that describes minor deviations from a stationary plane wave (Volpert and Mer-

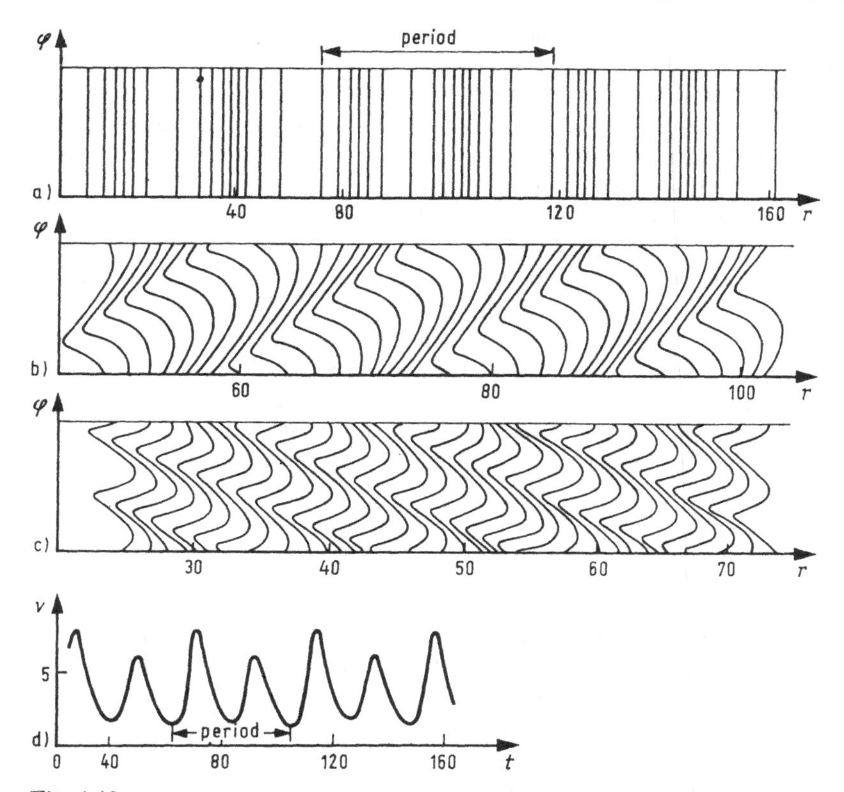

Fig. 4.12

Positions of a burning TF in two-dimensional space at successive times for $\gamma = 0.08$; $\beta = 0.08$; $\theta_H = -9.5$; $\alpha = 6 \times 10^{-4}$: a) $\nu = 1 \times 10^{-2}$, one-dimensional self-excited oscillatory regime; b) $\nu = 5 \times 10^{-4}$, spin burning with one active centre; c) $\nu = 2 \times 10^{-4}$, spin burning with two active centres; d) self-excited oscillations of the instantaneous velocity of burning in the one-dimensional regime (Ivleva, Merzhanov and Shkadinskii 1980)

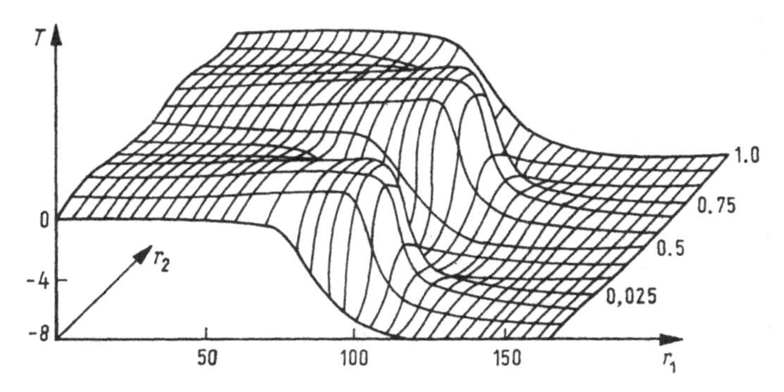

Fig. 4.13

A stationary field of burning temperatures in a plane sample with two active centres (Aldushin, Kasparyan 1978)

zhanov 1982). Such a consideration enables one to specify conditions for the onset of both self-oscillatory and spin regimes. Besides, regimes which are generally referred to as limiting ones were detected. In the simplest case, a burning centre moves first along the cylinder axis, reaches the surface and propagates along the sample in the form of a burning cycle and then goes back into the sample, and so on. Such a regime was observed experimentally by Merzhanov, Pak and Kuchin (1978). In complicated cases (at rather large radii), burning centres in the form of alternating cycles, symmetrical with respect to the cylinder axis, can propagate inside the sample.

Thus, a solitary TF, which is a basic autowave structure, exhibits diversified behaviour. In the stationary case, the mathematical image that corresponds to a TF is a saddle-to-saddle separatrix (trigger wave) or a node (phase wave). For the non-stationary propagation regimes, elementary mathematical images are not yet available. There is only a list of characteristic regimes: one-dimensional front pulsations, spin regime and limiting regimes.[1] Note that this important section of autowave theory is under intense study but has not been "constructed to completion" so far.

4.4 Stationary pulses

The second basic nonlinear structure is a travelling excitation pulse (TP). The behaviour of a TP can be analyzed on the basis of the motion of its two fronts. Two such coupled (or, to be more precise, joined by a smooth plateau) fronts form a localized structure, which exhibits autonomous behaviour. Hence, it is reasonable to confine a TP to a separate "image".

Let us consider a stationary TP using the model (3.1). We then pass over to an autowave variable (4.2) and obtain a set of ordinary equations. There are homoclinic trajectories (saddle-to-saddle trajectories) in the phase space of this system. Similarly to the TF case, these trajectories correspond to certain values of the TP velocity. Figure 4.14 gives a typical numerical calculation for the Fitz-Hugh-Nagumo system (see Section 2.6):

$$\varepsilon \frac{\partial x}{\partial t} = \frac{\partial^2 x}{\partial r^2} - y + f(x), \quad f(x) = x - \frac{1}{3} x^3, \tag{4.9}$$

$$\frac{\partial y}{\partial t} = -by - x - c, \tag{4.10}$$

which is an adequate basic model for many autowave processes. Two stationary pulse solutions can exist in the set (4.9)—(4.10). The first solution corresponds to a stable TP. The phase trajectory and the form of such a TP are denoted by "1" in Fig. 4.14a. Supposedly, for $\varepsilon \ll 1$, the phase trajectories corresponding to the leading edges of TP lie in the plane x, $\partial x/\partial t$, $y = \text{const}$. These planes are shown by hatching in Fig. 4.14a. The second solution (denoted by "2" in Fig. 4.14a, b) corresponds to

[1] Typical behaviour of thermal fronts in the peaking regimes will be considered in Section 7.7.

an unstable TP. Both solutions ("1" and "2") can exist only for values of ε which are smaller than a certain critical value ε_{cr} (see Fig. 4.14b). The unstable pulse is a sort of boundary which separates perturbations tending to a stable equilibrium state

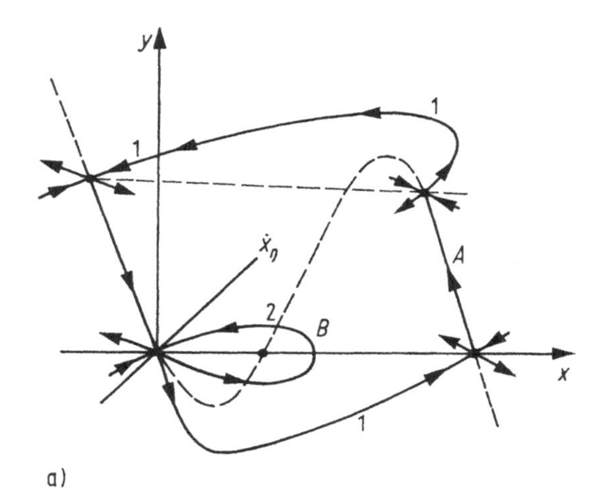

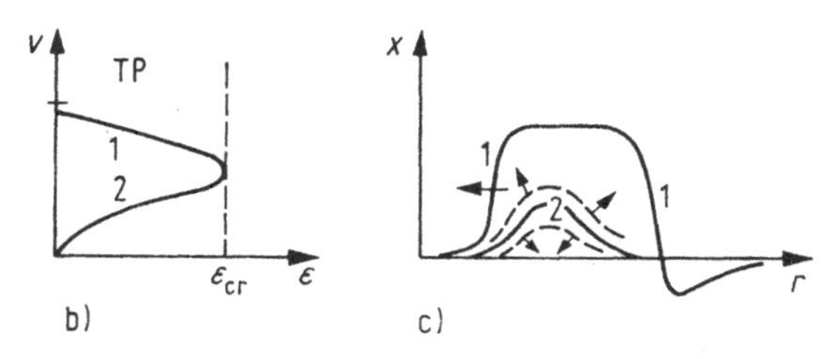

Fig. 4.14
A stationary TP in system (4.9)–(4.10): a) homoclinic trajectories in the phase space at $\varepsilon \to 0$; b) dependence of the velocity of stationary TP on ε; c) a stable (1) and an unstable (2) stationary TP. The dashed line denotes perturbations which relax either to an equilibrium state or to a stable TP.

and those leading to the formation of a stable TP. This holds only for a limited class of perturbations. This problem is still to be investigated. If the initial perturbation is slightly different from the form of an unstable pulse, then the results taken from Section 4.2 are useful. Integral inequalities for damped perturbations are also known (Scott 1975; Belintsev et al. 1978).

The TP velocities are the eigenvalues of the homoclinic trajectory problem. If $\varepsilon \ll 1$, then the velocities can be found analytically by the method of successive approximations (Casten et al. 1975). Figure 4.14a shows trajectories for a stable (1) and unstable (2) TP. The velocity of a stable TP can be determined from the velocity of its leading edge (from one equation (4.9) at $y = y_0$, as shown in Section 3.7). The correction to the TP velocity can be found by using an expansion

$$\begin{aligned} x &= x_0(\eta) + \varepsilon x_1(\eta) + \cdots, \\ y &= y_0(\eta) + \varepsilon y_1(\eta) + \cdots, \\ v &= v_0 + \varepsilon v_1{}^{\mathrm{A}} + \cdots. \end{aligned} \tag{4.11}$$

Substituting (4.11) into (4.9) and (4.10) and equating terms in ε of equal power we have:

$$\frac{\mathrm{d}^2 x_0}{\mathrm{d}\eta} + v_0 \frac{\mathrm{d}x_0}{\mathrm{d}\eta} - y + f(x_0) = 0, \tag{4.12}$$

$$\frac{\mathrm{d}y}{\mathrm{d}\eta} = 0 \quad \text{i.e.} \quad y_0 = \text{const}, \tag{4.13}$$

$$\frac{\mathrm{d}^2 x_1}{\mathrm{d}\eta^2} + v_0 \frac{\mathrm{d}x_1}{\mathrm{d}\eta} + f_x{}'(x_0)\, x_1 = y_1 - v_1 \frac{\mathrm{d}x_0}{\mathrm{d}\eta}, \tag{4.14}$$

$$v_0 \frac{\mathrm{d}y_1}{\mathrm{d}\eta} = -[b y_0 + x_0(\eta) + c]. \tag{4.15}$$

Using (4.12) and integrating by parts it is easy to show that the left-hand side of Eq. (4.14) is orthogonal to $\frac{\mathrm{d}x_0}{\mathrm{d}\eta} \exp(v_0 \eta)$. Thus, in view of (4.14) and (4.15) we have

$$v_1{}^{\mathrm{A}} = -\frac{1}{v_0} \frac{\int\limits_{-\infty}^{\infty} \left(\int\limits_{-\infty}^{\infty} [x_0(\tau) + b y_0 + c]\, \mathrm{d}\tau \right) \frac{\mathrm{d}x_0}{\mathrm{d}\eta} \exp(v_0 \eta)\, \mathrm{d}\eta}{\int\limits_{-\infty}^{\infty} \left(\frac{\mathrm{d}x_0}{\mathrm{d}\eta} \right)^2 \exp(v_0 \eta)\, \mathrm{d}\eta}. \tag{4.16}$$

The approximate formula $v \simeq v_0 + \varepsilon v_1{}^{\mathrm{A}}$ describes the upper branch of $v(\varepsilon)$ in Fig. 4.14b. The velocity of the trailing edge can be calculated from the same formulae if the value y_0 is substituted by a value y_* corresponding to the pulse decay. It is understood that the velocity of the trailing edge should be equal to the velocity of the leading edge. In other words, when taking such an approach, there are three parameters, which are the eigenvalues: v_{front}, v_{decay} and y_* (y_0 is treated as a fixed parameter). These three parameters are determined by three conditions, which are:

1. $v_{\mathrm{l.edge}}$ is determined from the homoclinicity condition at $y = y_0$.
2. $v_{\mathrm{tr.edge}}$ is determined from the homoclinicity condition at $y = y_*$.
3. y_* is determined from the condition $v_{\mathrm{l.edge}} = v_{\mathrm{tr.edge}}$.

The velocity of an unstable TP can be calculated by using an expansion

$$\begin{aligned} x &= x_0{}^{\mathrm{B}}(\eta) + \sqrt{\varepsilon}\, x_1{}^{\mathrm{B}}(\eta) + \cdots, \\ y &= y_0 + \sqrt{\varepsilon}\, y_1{}^{\mathrm{B}}(\eta) + \cdots, \\ v &= \sqrt{\varepsilon}\, v_1{}^{\mathrm{B}} + \cdots. \end{aligned} \tag{4.17}$$

This analogous procedure permits one to obtain a formula for the velocity of an unstable TP. The calculation results are shown in Fig. 4.14b.

In particular, such an analytical method was used in the analysis of the wave propagation in cardiac muscle tissues (Khramov and Krinskii 1977; Khramov 1978, 1971; Pertsov, Khramov and Panfilov 1981). The dependence of the TP velocity on the parameters of the system was obtained. Specifically, the dependence of the minimum wave length, $\lambda_{\min}$, in a cardiac fibre on the ion current parameters was investigated theoretically. It was demonstrated that in a physiologically normal membrane, $\lambda_{\min}$ increases with decreasing leading edge slope. This corresponds to the action of some cardiac antiarrythmic agents.

4.5 The formation of travelling pulses

The formation of a stable stationary TP can be easily traced in systems having strong relaxation properties (Ostrovskii and Yakhno 1975). To perform a qualitative analysis, it is sufficient to know the null isoclines of the system, $P(x, y) = 0$ and $Q(x, y) = 0$, and the dependence of the solitary front velocity on the slow variable at the position of the front $v = v(y)$ (see Section 3.7).

First we will consider the motion sequence in a simple case of TP formation and then pass over to complicated cases.

The analysis of the pulse formation in a system where each point is a "slave" self-oscillator is based on the use of a small parameter ε in Eq. (2.53; 3.35). In this case, the motions can be divided into "fast" and "slow" as is usually done for local oscillatory systems of a relaxation type and, for example, when describing the evolution of shock waves.

In the "fast" stage which takes a time of order ε, the y function has no opportunity to change; thus, in view of (2.53; 3.22) we have:

$$\frac{\partial x}{\partial \xi} = D_x \frac{\partial^2 x}{\partial r^2} + P(x, y), \tag{4.18}$$

$$y = y_0(r),$$

where $\xi = t/\varepsilon$ is the "fast" time.

For "slow" processes, discarding the term $\varepsilon\, \partial x/\partial t$ in (3.35) we have

$$D_x \frac{\partial^2 x}{\partial r^2} + P(x, y) = 0, \tag{4.19}$$

$$\frac{\partial y}{\partial t} = D_y \frac{\partial^2 y}{\partial r^2} + Q(x, y). \tag{4.20}$$

Let the P and Q functions be given by $P = -y + f(x)$ and $Q = -y - \gamma x$, where γ is a constant and the function $f(x)$ is N-shaped (Fig. 4.15). Such an approximation is sufficient for the description of the basic features of autowave processes and admits an analytical solution. Assume that the equilibrium equations $\gamma x = -y$ and $f(x) = y$ admit one solution x_{eq}, y_{eq} for a stable state.

First, let us discuss fast processes described by a nonlinear equation (4.18). The solutions for $y = \text{const}$ were considered above. At a given $y = y_0$, there is either one or, in more interesting cases, two or three equilibrium states corresponding to the zero values of the function $P(x, y)$. The stable[1]) states correspond to certain points on branches I and III in Fig. 4.15. The most important solutions to (4.18) are stationary TFs, which can be determined from the equation

$$D_x \frac{d^2 x_0}{d\eta^2} + v \frac{dx_0}{d\eta} - y_0 + f(x_0) = 0, \tag{4.21}$$

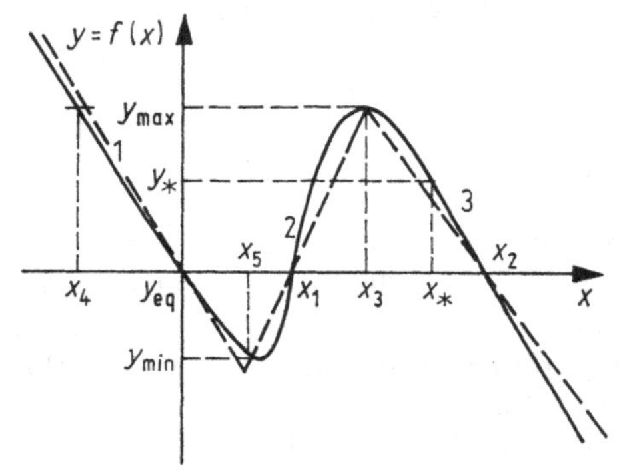

Fig. 4.15
View of the nonlinear function in Eq. (4.21). The dashed line denotes a piecewise-linear approximation:

$$f(x) = \begin{cases} -\alpha_1 x & x < x_5 \\ \alpha_2 x - (\alpha_1 + \alpha_2)\, x_5 & x_5 < x < x_3 \\ -\alpha_3 x - (\alpha_1 + \alpha_2)\, x_5 + (\alpha_3 + \alpha_2)\, x_3 & x_3 < x \end{cases}$$

implied by (4.18). Transition to a self-similarity variable $\eta = r - vt$, Eq. (4.21) leads to the TF motion problem investigated above (Section 4.1). The result is a dependence $v = v(y)$ shown in Fig. 4.16. These calculations were made for the piecewise-linear approximation of the function $f(x)$ shown by a dashed line in Fig. 4.15.

We now consider the evolution of a TP in a slave relaxation-type medium using a differentiation between "fast" and "slow" motions. Let the initial perturbation of the function $x(r, 0)$ be smooth enough and have a characteristic dimension greatly exceeding the diffusion scale $L_D \simeq \sqrt{D_x T}$. Assume for simplicity that at the initial instant of time y is a constant and equals the equilibrium value $y_{eq} < y_{cr}$ (see Section 3.7).

[1]) stable with respect to a fast motion

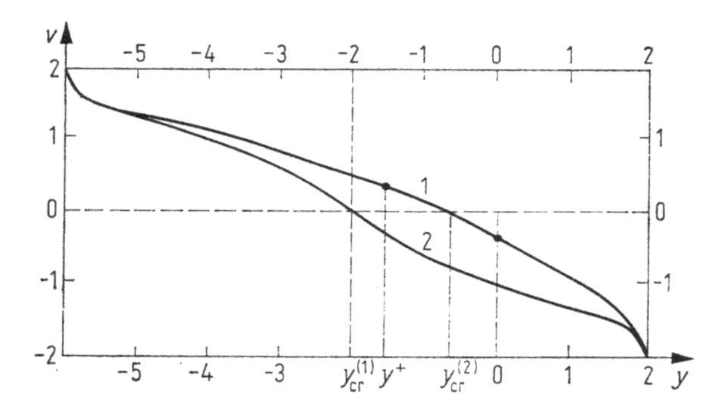

Fig. 4.16
Dependence of the TF velocity on the parameter y ($x_3 - x_5 = 2$; $1 - \alpha_1 = 1, \alpha_2 = 4, \alpha = 10$; $2 - \alpha_1 = 1, \alpha_4 = 4, \alpha_3 = 1$)

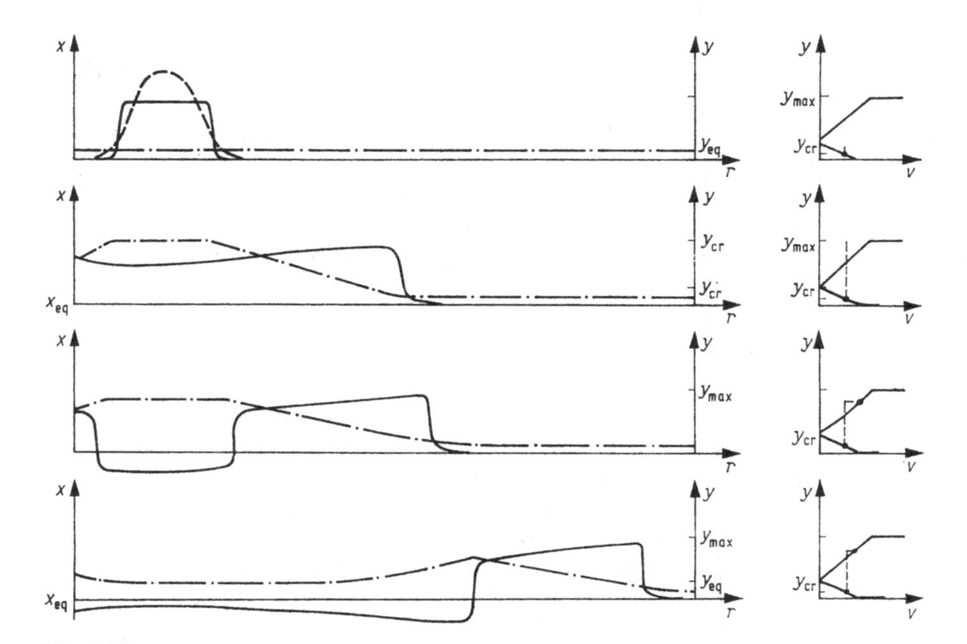

Fig. 4.17
The formation of a TP at $y_{eq} < y_{cr}$. The solid line is the spatial distribution in x. The dashed line is the spatial distribution in y. a)—d) The positions of a TP at successive times is shown on the left. The characteristics of a TP in the configurational space (v, y) are shown on the right (the dots on the dependence $v(y)$ correspond to a fast propagating front, and the dashed line to a front moving at the velocity v).

Thus, TP evolution can be divided into the following stages (Fig. 4.17):

a) The initial perturbation profile (dashed line in Fig. 4.17a) changes quickly (over a time $\Delta t \sim \varepsilon$) and assumes an almost rectangular form. The edges of a pulse formed here develop into stationary fronts and propagate in different directions.

The initial perturbation decays in a symmetric manner; thus, it is sufficient to trace the pulse evolution in one direction.

b) The pulse edge moves at a constant velocity $v(y_{eq})$. The crest of a TP is formed by slow motions independently at each point. The variations in x in the slow phase are given by

$$\tau = \int_{x_2}^{x} \frac{\mathrm{d}f(x)}{-\gamma x - f(x)} = \tau(x_2, x). \tag{4.22}$$

c) When x and y reach the values x_3 and y_{max}, there occurs a sudden fall to the refractory phase followed by the appearance of another fast stationary front, which is a trailing edge of the pulse (Fig. 4.17). This trailing edge, which corresponds to the pulse decay, begins to move at the velocity $v(y_{max})$ in the same direction as the leading edge, both forming a TP. The initial length of the pulse in space is equal to $l_{in} = v(y_{eq})\,\tau(x_2, x_3)$.

If the function $f(x)$ is relatively symmetric, then $v(y_{max})$ is always larger than $v(y_{eq})$ and the trailing edge shifts to a region with $y < y_{max}$ where its velocity is unambiguous. The length of the TP decreases monotonically. The velocity of the trailing edge declines and, finally, becomes equal to that of the leading edge. This occurs at a certain $y = y_* > y_{cr}$ (Fig. 4.15). Consequently, the whole TP assumes a stationary form and contracts asymptotically to a length $l_{st} = v(y_{eq})\,\tau(x_2, x_*)$, where x_* is found from the condition $f(x_*) = y_*$ (branch 3 in Fig. 4.15). The contraction time τ_0 can be estimated from the formula

$$\tau_0 = \frac{2\tau(x_*, x_3)}{[v(y_{max})/v(y_{eq}) - 1]}. \tag{4.23}$$

Using the piecewise-linear approximation for $f(x)$, which corresponds to the parameters $\alpha_1 = 1$, $\alpha_2 = 4$, $\alpha_3 = 10$, $x_3 = 4$, $x_5 = 2$ and $\gamma = 4$ (curve 1 in Fig. 4.16) it is easy to obtain the following pulse characteristics: $l_{in} = 0.172/\varepsilon$; $l_{st} = 0.0372/\varepsilon$; $\tau = 0.2$. The stationary leading and trailing edges have equal velocities: $v(y_{eq}) = v(y_*) = 0.4/\varepsilon$ (Fig. 4.16), where $y_{eq} = 0$ and $y_* = -1.6$.

If the function $f(x)$ is appreciably nonsymmetric, then the TP decay can occur at the values of y corresponding to the vertical sections of the graph $v(y)$ (Fig. 3.17 a). In this case, the formation of a stationary TP is complete when its trailing edge is formed.

d) The TP forms a refractory region behind which the medium points slowly return to the equilibrium state (Fig. 4.17 d).

The TP formation in a configurational space v, y is shown in the right-hand side of Fig. 4.17. The dots on the $v(y)$ dependence denote the leading and trailing edges of the pulse, and the dashed line the change of the y variable. The reader can easily compare the space distributions $x(r, t)$, $y(r, t)$ and their images in the v, y space.

Thus, the example shows how a nonstationary process can lead to the formation of a stationary TP. In one case this occurs asymptotically; in the second, the pulse becomes stationary immediately after a sharp decay.

This pattern of pulse formation takes place at $y_{eq} < y_{cr}$. But if $y_{eq} > y_{cr}$ and the initial perturbation corresponds to $x(0, r)$, then again a rectangular pulse forms in stage a) but the pulse edges now move towards the centre. The perturbation "collapses" and disappears over a finite time (Fig. 4.18).

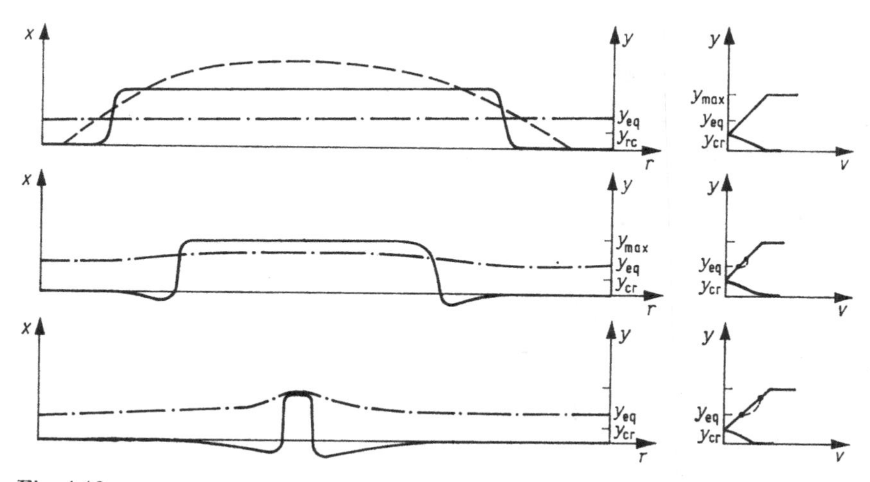

Fig. 4.18
Evolution of a perturbation for $y_{eq} > y_{cr}$. The collapsing of the initial pulse in the configurational space (y, v) is shown on the right.

Note that the change in the form of the functions $P(x, y)$ and $Q(x, y)$ or even an increased number of equations for slowly varying variables are insignificant for the qualitative characteristics of the processes until, e.g., the number or character of the equilibrium states of "fast" equations at $y = \text{const}$ change. This fact seems to be important since the initial equations such as (2.53) or (3.35) are usually based on more or less rough approximations and the interest is mainly in a qualitative study of the process.

The problem of the existence of periodic self-excited oscillatory processes is of particular interest. It should be readily apparent that for any pulsed perturbation with $y_{in} \neq y_{cr}$ the system returns to the equilibrium state. The situation can change only if $y_{in} = y_{cr}$ or if the initial perturbation $y_{in}(r)$ is inhomogeneous and passes through the values y_{cr} at some points. The formation of a stopped front is then possible, which is not described by the "fast" equation (4.18). This front can be unstable with respect to the slow motion phase (4.19), (4.20); thus, periodic waves diverging from the region where $y = y_{cr}$ can, in principle, arise (see Section 5.1).

Thus, Fig. 4.17 shows the formation of a TP, which is the simplest nonstationary process. It can be represented as a series of patterns $x(r, t)$ and $y(r, t)$ at successive times. This is an exact representation of the solutions, which, however, can be simplified by disregarding minor changes of the v, y variables in the excitation and rest regions. Let us denote the region of the excited state by hatching against the background of the nonexcited (blank) state (Fig. 4.19a) for the x variable. The region where $y > y_{cr}$ is shown by dots, and the region with $y < y_{cr}$ is left blank for the y variable. The meaning of such a differentiation consists in the qualitative difference of TF behaviour in the hatched and blank regions: the velocities of these fronts will be opposite in sign. Only two characteristics, I and III, are relevant to this solution.

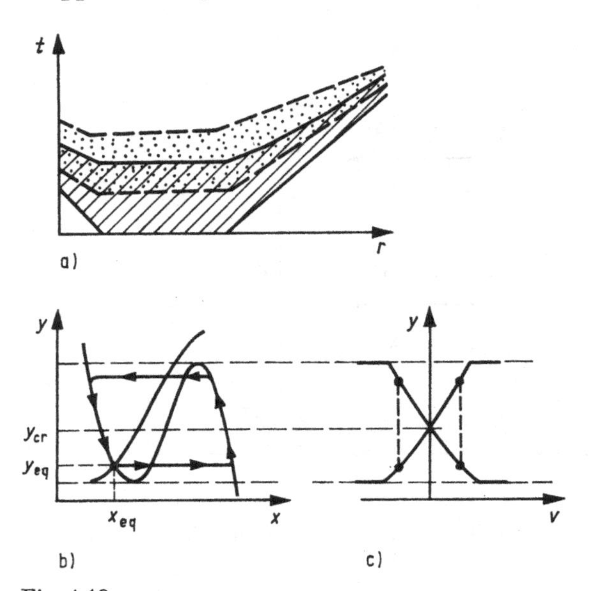

Fig. 4.19
The formation of a TP:
a) scheme of solution in the plane (r, t);
b) mathematical image of a TP in the phase space;
c) mathematical image of a TP in the configurational space

The TF image on characteristic III is a line corresponding to a transition between stable branches 1 and 3 on the null isocline $P(x, y) = 0$. The TF image on characteristic I is a point corresponding to the value y_{eq} (see Figs. 4.17 and 4.18). The TP image represents a closed trajectory on characteristic III, which includes two fronts (transitions) and two regions of slow motions (Fig. 4.19b).

One limitation of the qualitative method should be noted. This method does not permit one to describe the disappearance of a very short pulse. It is well known from the theory describing a stationary TP in the autowave medium (Section 4.4) that at low spreading velocities, $v \lesssim \sqrt{\varepsilon}$, a stationary TP cannot exist. In this instance, characteristics I and III are helpful. The point is that characteristic I was obtained for an isolated excitation front. But if the TFs converge rather close[1]) to each other (this occurs at low spreading velocities), then the interaction between the fronts becomes strong and the initial assumptions are no longer valid. A qualitative description of this case requires a condition to be added: If the TP length becomes

[1]) We mean to a distance of the order of or less than the excitation front length, l_{fr}.

smaller than $l_1 \simeq 3l_{\mathrm{fr}}$, then such a pulse can be treated as vanishing and the slow variable value can approximately be considered equal to y_{cr} at the time the pulse disappears (the true value of y differs from y_{cr} only by $\approx \varepsilon^{1/2}$). Such an additional condition permits one to describe the damped propagation regimes (see Fig. 4.20) and rhythmic transformation regimes.

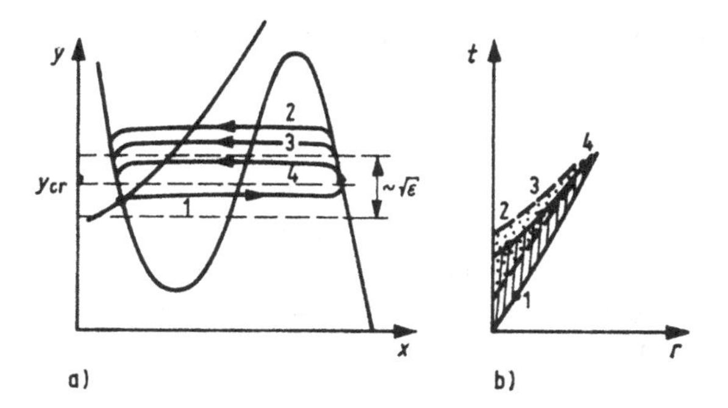

Fig. 4.20
The pulse behaviour for the parameters corresponding to damped propagation:
a) mathematical image of a TP in the phase space;
b) scheme of solution in the plane (r, t)

A qualitative approach to the study of the TP formation can be easily applied to the problems of propagation of pulse successions. The reader can consider independently the problems of equalization of intervals between TPs, determination of the velocity of a periodic succession of TPs (Scott 1975), and exclusion of some of the pulses with the maximum permissible stimulation frequency exceeded (Fomin and Berkinblit 1973).

The above consideration of a one-dimensional model can be applied to a two-dimensional case, given that the initial perturbation radius by far exceeds the stationary front lengths. Figure 4.17 illustrates the formation of a divergent ring pulse of finite length, which originates from a symmetrical initial perturbation.

4.6 The propagation of pulses in a medium with smooth inhomogeneities

It is well known that the medium inhomogeneities that change the refractory phase or pulse duration can lead to the rhythmic transformation of a succession of TPs. Such a process can be considered, for example, with the aid of axiomatic models for excitable media. However, the change in TP duration and repetition rate can be found in a most natural way from the basic model (2.53). Based on a simple analysis of these equations we will consider rearrangements arising in a succession of TPs as they pass through a smooth inhomogeneous medium.

Let the basic model be presented in the form of two equations

$$\varepsilon \frac{\partial x}{\partial t} = \frac{\partial^2 x}{\partial r^2} + P(x, y); \qquad \frac{\partial y}{\partial t} = Q(x, y, r). \tag{4.24}$$

The scale of t and r is chosen such that the coefficients on the variables are equal to unity. Since we do not consider in detail the form of a TP but only time variations in the position of its fronts, we will make use of the approximate method given above. Then the position of the n-th TF can be found from the equation

$$\frac{\partial r_n}{\partial t} = v[y_n(r, t)], \tag{4.25}$$

where $y_n(r, t)$ is the slow variable value before the front. The method of determining $v(y)$ from (4.24) (see Section 4.5) is valid if the characteristic scale of inhomogeneity by far exceeds the TF length. The change of the slow variable before the leading or trailing edges can be found from the equations

$$\frac{\partial y_n}{\partial t} = Q[y_n, x_1(y_n), r] = Q_1(y_n, r), \tag{4.26}$$

$$\frac{\partial y_n}{\partial t} = Q[y_n, x_3(y_n), r] = Q_3(y_n, r), \tag{4.27}$$

where $x_{1,3}(y)$ are the stable branches of the N-shaped null isocline. A combination of (4.25) with (4.26) and then (4.25) with (4.27) yields

$$\frac{\mathrm{d}t_{n,f}}{\mathrm{d}r} = \frac{1}{v(y_{n,f})}; \qquad \int\limits_{y_{n-1,l}}^{y_{n,f}} \mathrm{d}\eta / Q(\eta, r) = t_{n,f} - t_{n-1,l}; \tag{4.28}$$

for the leading edge of the n-th pulse, and

$$\frac{\mathrm{d}t_{n,l}}{\mathrm{d}r} = \frac{1}{v(y_{n,l})}; \qquad \int\limits_{y_{n,l}}^{y_{n,f}} \mathrm{d}\eta / Q_3(\eta, r) = t_{n,l} - t_{n,f} \tag{4.29}$$

for the trailing edge of the n-th pulse. The initial conditions at point $r = r_0$ represent the excitation times of the leading and trailing edges of the n-th pulse. The motion of n pulses can be traced by solving a system of $4 \cdot n$ (or $2n$ in simple cases) ordinary differential equations.

As an illustration, we analyzed the TP propagation in excitable tissue with a monotonic and periodic inhomogeneity. The results are given in Fig. 4.21. It is seen that this inhomogeneity leads to decreased velocities of the leading and trailing edges. This effect was noted, for example, in papers by Arshavskii, Berkinblit et al. 1964; Krinskii and Kholopov 1967; Kholopov 1968, where axiomatic models were used. The calculations presented here allow only a quantitative determination of the magnitude of the deceleration. These calculations give new information on the motion of a "cluster" of decelerated leading and trailing edges (the region of such a cluster is shown by a dashed line in Fig. 4.21). The features associated with the motion of such a cluster are as follows: First, the spreading velocity of the cluster is governed by the minimum velocity of the decelerated TF. Second, in the region the cluster has passed through, the tissue inhomogeneity is, so to say, averaged out and wears smooth. Hence the TPs change insignificantly in duration when propagating through such a region. These peculiarities of the pulse propagation can apparently be used for the diagnostics of the magnitude and location of inhomogeneities in nerve fibres and other excitable tissues.

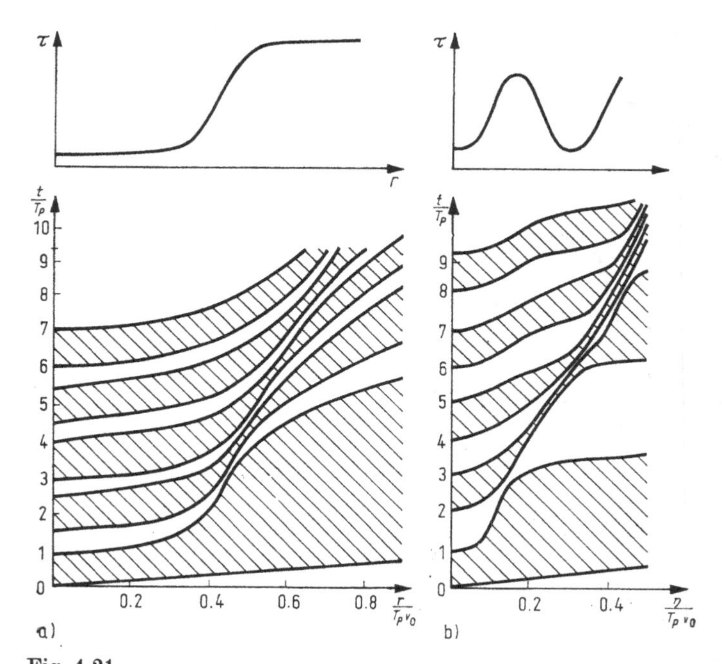

Fig. 4.21
Schemes of solutions describing pulse propagation through a smooth inhomogeneous medium:
a) monotonic variation of pulse duration in space;
b) periodic variation $\tau(r)$

4.7 Pulses in a medium with a nonmonotonic dependence $v = v(y)$

More diversified dynamic processes in TPs occur in autowave systems with a nonlinear function $P(x, y)$ which corresponds to a nonmonotonic dependence $v = v(y)$. Figure 4.22 shows characteristic forms of dependence of $v(y)$. They can be obtained, in particular, from the formulae given in Table 3.1 (points 1 and 2). Also, this nonmonotonic dependence of velocity on the slow variable occurs in systems where the autocatalytic variable diffusion coefficient decreases with increasing y. Such a nonmonotonicity mechanism can manifest itself, for example, in cardiac muscle tissues.

This case is associated with weakened electric coupling between cells because of increased concentration of Ca^{2+} ions in the cell.

The principal distinctive feature of systems with the characteristics shown in Fig. 4.22 is the existence of two stable and one unstable pulse. These pulses have the same spreading velocities (equal to $v(y_0)$) but differ in duration (Fig. 4.23c). The slow

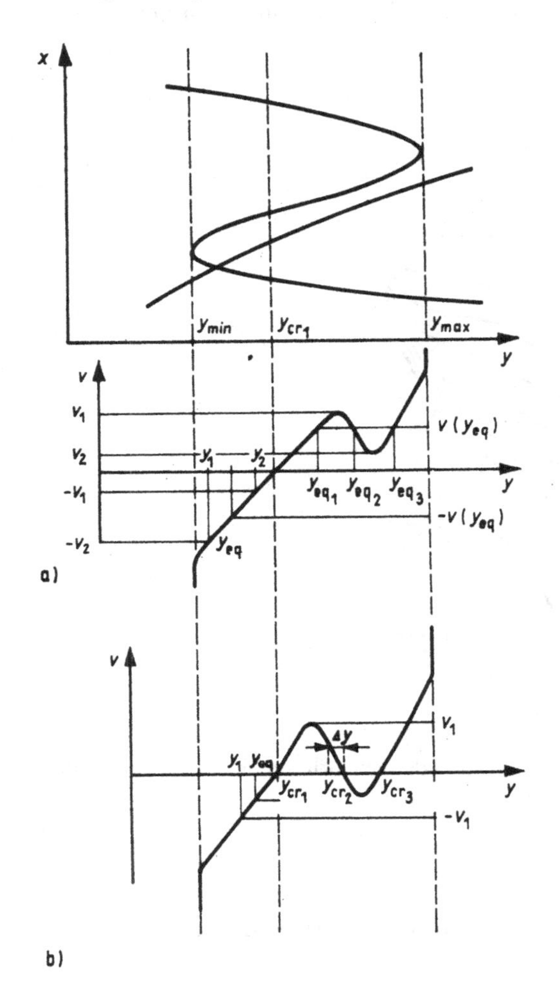

Fig. 4.22
Types of nonmonotonic characteristics $v(y)$: a) with one intersection of zero; b) with three intersections of zero

variable changes from y_{eq} to y_{eq_1} in the first stationary TP, from y to y_{eq_2} in the second, and from y to y_{eq_3} in the third (see Figs. 4.22a and 4.23c). Using the dependence it is easy to check the pulse durations for stability. The two stationary TPs (of lengths l_1 and l_2; Fig. 4.23c) are stable with respect to variations in length. The pulse of intermediate length is unstable with respect to the trailing edge shifts. This last pulse represents a "provisional" boundary for the initial conditions, which separates the initial perturbations leading to the formation of stable stationary TP.

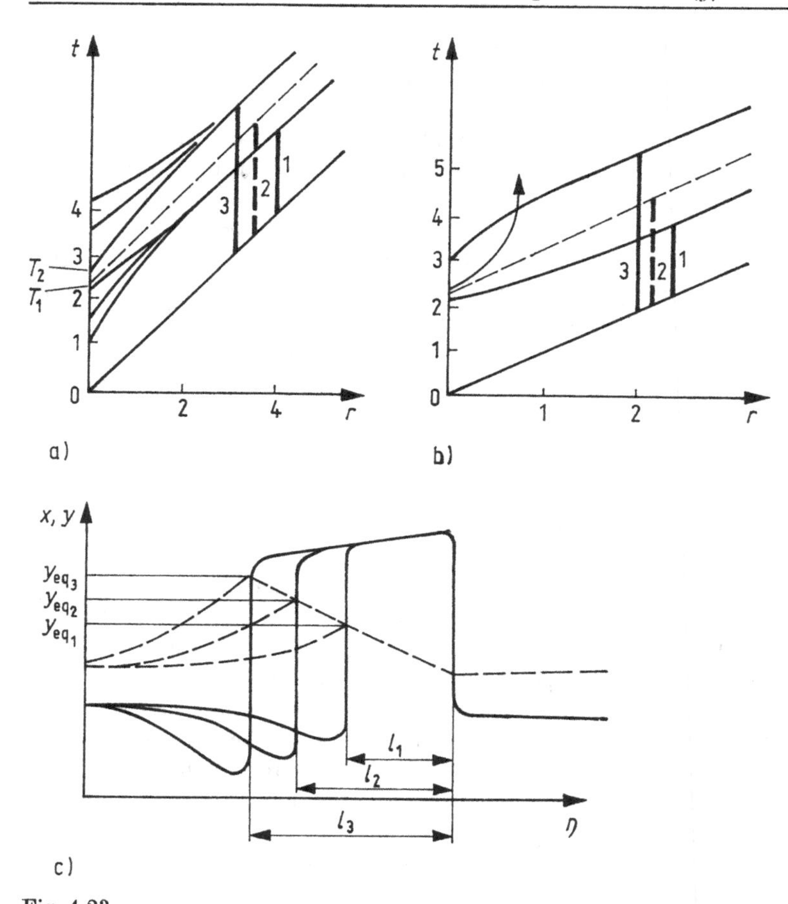

Fig. 4.23

Pulse propagation in a medium with a nonmonotonic dependence $v(y)$:

a) the formation of pulses of different durations depending on the duration of the initial perturbation (a nonmonotonic characteristic has one intersection of zero);

b) halt in decay of a TP in a medium with a nonmonotonic dependence $v = v(y)$ having three zeroes (see Fig. 4.22b). The formation of initial perturbations of a corresponding stationary pulse: (b) stationary pulses in a system with a nonmonotonic dependence $v = v(y)$: 1 and 3 — pulses, which are stable with respect to perturbations; 2 — an unstable pulse

A dynamic pattern of TP formation is given in Fig. 4.23a, b. The durations of two stable TPs are shown by solid lines, and of the unstable TP by a dashed line.

If $v(y)$ vanishes at three values of y_{cri} ($i = 1, 2, 3$) (Fig. 4.22b), then stopped excitation fronts can arise in the pulse formation. This occurs if the slow variable enters the interval Δy (see Fig. 4.22b) or the intervall $y_{cr_2} < y < y_{cr_3}$ in the place of pulse decay. In this case, the trailing edge is always slower than the leading edge. In other words, the trailing edge lags behind the leading edge, and the value of in the place of pulse decay increases, thus leading to a further decrease in velocity. A typical halt in decay of a TP is shown in Fig. 4.23b. The formation of TPs in systems with a nonmonotonic function $v = v(y)$ was considered in ample detail by Yakhno (1977, 1980).

Note that in real excitable systems, for example in self-oscillatory chemical reactions, cardiac muscle fibres, etc., direct measurement of the nonlinear characteristics faces appreciable difficulties. Thus, these characteristics are often determined by the system's response to an external perturbation. In the case of a complex response, attempts are made to explain the behaviour of the system by way of increasing the complexity of the initial model or, in particular, using more variables. It seems to be interesting to show how new modes of behaviour are revealed in a qualitative study of simple models. For this, it is only required that a special form of a qualitative characteristic such as $v = v(y)$ be chosen. Essentially, this dependence results from reducing the higher-order systems to a system of two equations.

4.8 Pulses in a trigger system

We now consider a trigger system with different qualitative characteristics: III for the null isocline and I for the dependence $v = v(y)$ (see Section 3.8, Fig. 3.17 in Chapter 3 and Fig. 4.24). Let the initial condition be the state of rest (y_{eq_1}, x_{eq_1}), which corresponds to the first stable point of the system. The second stable point denotes the state of activity (y_{eq_3}, x_{eq_3}). A pulsed perturbation is formed as a result of successive stimulation of transitions from the state of rest to the state of activity, and vice versa. Such transitions can be caused only by corresponding stimuli of finite amplitude. The instant the stimuli are switched on is shown by arrows in Fig. 4.24. A variety of pulsed perturbations are possible.

If $|v(y_{eq_1})| < |v(y_{eq_3})|$, then a stationary TP (see Fig. 4.34a) will form. The leading edge propagates at a constant velocity $v(y_{eq_1})$. Since the TF which causes the system to pass over from the active state to the state of rest moves faster than the first leading edgé, it overtakes the first one with time and forms a trailing edge. The pulse which arises here persists in form and amplitude. The pulsed perturbation of inverse polarity, which is the state of rest against the background of activity, cannot be stationary in such a system, and it will steadily increase in time.

In the case of the inverse relation of velocities $|v(y_{eq_1})| > |v(y_{eq_2})|$, a stationary TP of inverse polarity will form. The TP of direct polarity will increase in duration with time (see Fig. 4.24). It is apparent that if the velocities are equal, then stationary TP of both polarities with arbitrary duration can form.

Here again, like in the slave case, there is a minimum velocity of a stationary TP $(v_{min} \sim \sqrt{\varepsilon})$. Therefore for $v(y_{eq_1}) < v_{min}$ or $v(y_{eq_2}) < v_{min}$ a stationary pulse will not form. The leading and trailing edges converge during their propagation and the perturbation vanishes (Fig. 4.24). Also, there are trigger systems where pulsed perturbations cannot propagate. Such a system is shown in Fig. 4.24 (the case $y_{eq_1} > y_{cr} > y_{eq_3}$). These schemes show all the principal cases of TP propagation in trigger systems for $D_y = 0$.

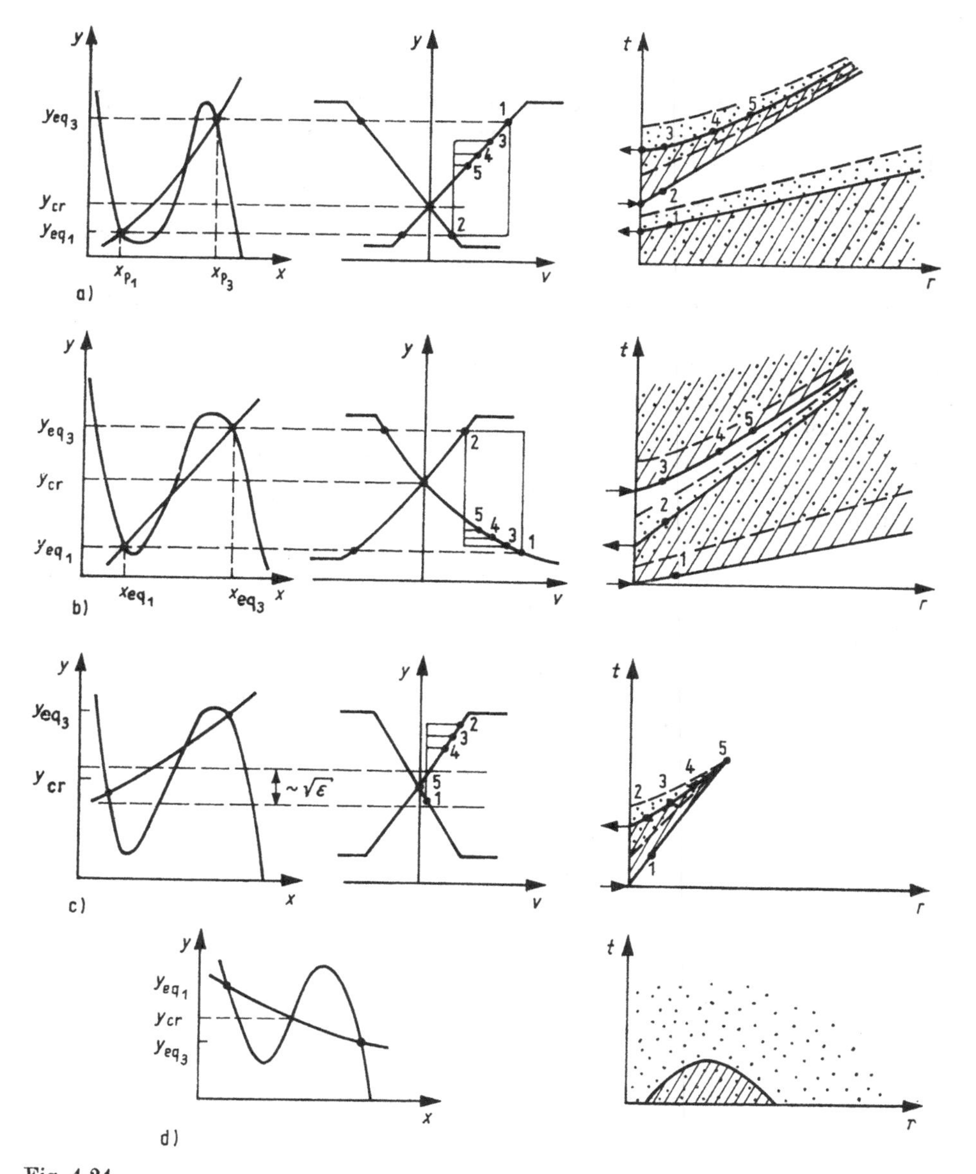

Fig. 4.24

The formation of initial perturbations in a trigger system. The null isoclines are shown on the left, and the schemes of solutions on the right. Corresponding mathematical images in the configurational space are shown in the centre:

a) the formation of a stationary TP in the nonexcited region;
b) the formation of a stationary TP in the excited region;
c) damped propagation;
d) a "collapsing" initial perturbation

4.9 Discussion

The cases of propagation of fronts and pulses considered above do not cover all modes of propagation of perturbations which occur in active kinetic systems. But we see characteristic features in describing such processes. Certain mathematical images have been obtained for pulsed perturbations:

a) limit cycles in the phase space y, x, $\mathrm{d}x/\mathrm{d}\eta$;

b) dots and lines between them in the dependence of $v(y)$ in the configurational space v, y (the dots correspond to stationary fronts).

Besides, the law of alternation of stable and unstable stationary waves holds for all solutions considered. Therefore stable stationary solutions are separated by an unstable stationary solution. Thus, it is possible to indicate a certain boundary or threshold for external perturbations where a transition from one stable nonlinear structure to another can occur. Also, it is seen that in highly relaxational systems a qualitative analysis of space–time processes can be performed with the aid of two characteristics: a null isocline and a dependence $v = v(y)$. Such a qualitative consideration should be developed in future. Specifically, when studying the dynamics of pulsed perturbations in the plane, one must take into account the dependence of the velocity on the K-curvature of the wave front: $v = v(y, k)$. Scientific work in this direction is now being carried out (Krinskii and Mikhailov 1982; Zykov 1981). Basically, the efforts are aimed at describing the excitation reverberation mechanism (see Section 5.6). It should be confessed, however, that the investigation of characteristic nonlinear structures in two-dimensional systems is time-consuming and that there are many problems in this field which are still to be investigated.

Chapter 5
Autonomous wave sources

Autonomous wave sources (AWS) hold a prominent place among the characteristic nonlinear structures in a homogeneous autowave medium. They are regions of finite dimensions where nonstationary motions lead to the generation of TPs. It is well known that there are so-called pacemakers in an inhomogeneous space which are localized regions of a self-sustained oscillatory medium, where the oscillation frequency is higher than anywhere in the surrounding space. The formation of a TP by such sources occurs under any initial conditions.

In this chapter, primary attention will be paid to those autowave sources which can generate TPs in a medium with homogeneous properties and are formed there due to special initial conditions. Such sources in a homogeneous space can be called self- or autopacemakers. A description of a source located at a border between two homogeneous media (see Section 5.2) is also given, although, strictly speaking, this source is not a self-pacemaker.

Self-pacemaker sources in a one-dimensional space can be also of the "echo" and "fissioning front" types, which are regimes caused by the retriggering of two neighbouring small regions of active medium (see Section 5.1). Such regimes occur in two-component media and are not stable in a rigorous sense. Stable self-pacemaker sources can exist in a three-component medium (see Sections 5.3, 5.4). These are so-called leading centres (LC) and standing waves.[1]) Also, there are regimes intermediate between standing waves and LCs — so-called alternating sources of TPs, which can be treated as migratory LCs.

Sources such as rotating spiral waves or reverberators, which arise in two-dimensional and three-dimensional media will be considered in Section 5.5.

All the regimes mentioned above occur in distributed chemical reactors, certain types of living cells and colonies of microorganisms, and are especially important in governing pathological regimes in the activity of a cardiac muscle.

---

[1]) We sometimes mean by "leading centres" both proper stable LCs and sources such as echoes and fissioning fronts.

5.1 Sources of echo and fissioning front types

Can periodic solutions exist in a model for homogeneous medium because of the interaction of its neighbouring active elements which are in the slave or trigger regimes? Figure 5.1 demonstrates mechanisms of the formation of such sources by means of an axiomatic description. It is seen that an echo (Fig. 5.1 a) can exist only

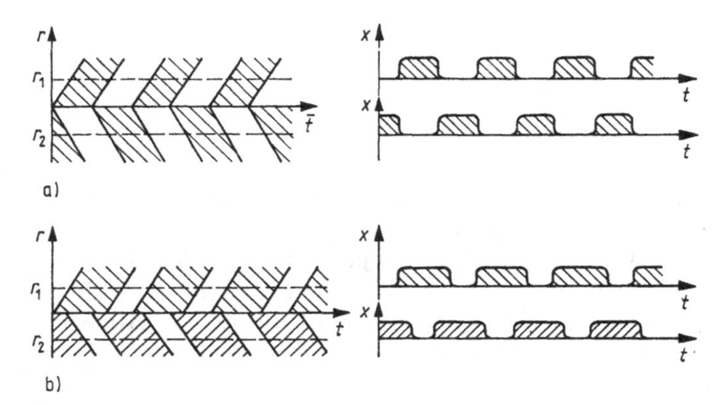

Fig. 5.1
Schematics of TP sources:
a) "fissioning front" in a non-auto-oscillatory medium. The solutions are constructed in the plane (r, t). The arrows show directions of wave propagation. Variations in time of the states of two elements of the medium at points r_1 and r_2 are shown under the schemes (Krinskii and Zhabotinskii 1981).
b) "echo", i.e. mutual retriggering of neighbouring elements of the medium.

if the duration τ of a TP is related to the refractory time R by the inequality $\tau/R > 1/2$. The source consists of two elements and has "zero" dimension (Krinsky 1968). Similar conditions were obtained in a model composed of two relaxation-type slave oscillators with diffusion coupling (Krinsky, Reshetilov, Pertsov 1972). A source which is a fissioning front is a particular case of an echo and occurs at $\tau = R/2$ (Fig. 5.1 a). Both echo and fissioning front regimes can be investigated by means of a basic model with two variables provided $\varepsilon \ll 1$.

All possible solutions can be analyzed by using the qualitative characteristics I, II and III (see Section 3.7). Using the characteristic $v = v(y)$ it has been ascertained that a TP source will be produced only if the initial conditions are given in the form of a stopped excitation front (see Section 4.5). Any other initial conditions lead to the formation of one or two divergent TPs. The subsequent evolution of the stopped excitation front can take either of two paths. If the velocities of the slow motions on the opposite sides of a stopped front are chosen appropriately (when the characteristics $F = 0$ and $\varphi = 0$ in (2.53) are symmetrical the velocities of the slow motions are equal, Rosenblum, Starobinets, Yakhno 1981), the front will split up (see Fig. 5.2 b). In a configurational space, the first evolution of a stopped TF is associated with the motion of a representative point which passes through the instability region 1 (on

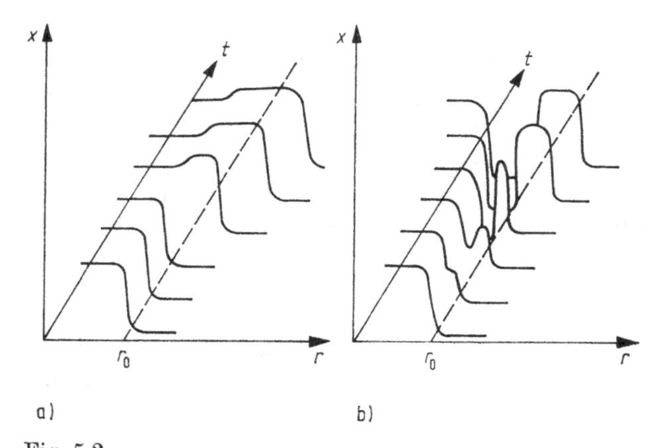

Fig. 5.2
Evolution of a stopped front of excitation pulses: a) departure of the front from the place of stopping (Goltsova, Zhislin and Yakhno 1976) b) fission of a front in the case that the slow motions on both sides of the front are matched;

characteristic II) from the initial position O_1 (where $y_3 = y_1 = y_{cr}$) to the boundary $r - r'$. As soon as the representative point reaches the boundary, the front begins splitting up (see Fig. 5.3). As was mentioned in Section 3.7, this point can reach the boundary $r - r'$ only by moving along the separating trajectories. A shear perturbation mode cannot arise here, in accord with the definition of these trajectories.

An alternative case is that in which the slow motion on opposite sides of the front are not matched. In this case, the front will move off in the direction in which the changes in the slow variable are faster (Fig. 5.2a). In a configurational space, the representative point in region 1 (see Figs. 5.3 and 3.12) escapes from the separating

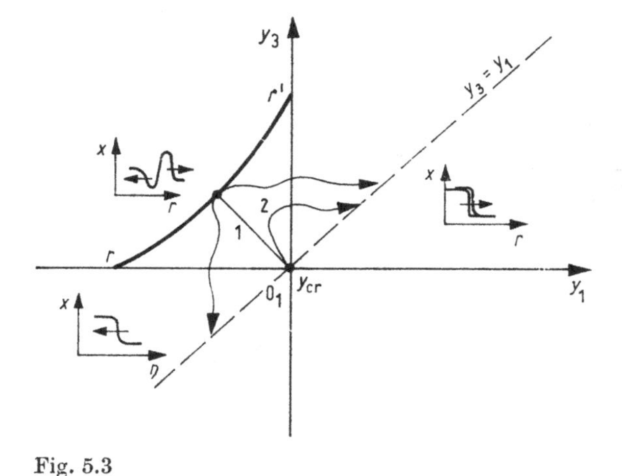

Fig. 5.3
Typical trajectories in the configurational space y_1, y_3 leading to the front fission (shear perturbations do not increase when moving along the separating trajectory "1") and to the front departure from the point of stopping (shear perturbations increase along the trajectory "2")

trajectory. This corresponds to the appearance of increasing shear perturbations, which cause the front to escape from the unstable stopping point.

Thus, it is seen that both ways of evolving are associated with the instabilities of an immobile front with respect to different perturbation modes: An antisymmetric mode having one intersection with zero increases when the front is split, and a shear mode when the front escapes from the stopping point. In both cases the parameters can be chosen such that a strictly periodic autowave process will arise in the system. This can be illustrated by using the results given in Section 3.7.

Consider first a periodic fission regime. Let the initial condition be given in the form of an immobile front with positive slope at $y = y_{cr}$. This point $y_1 = y_3 = y_{cr}$ is denoted by O_1 in Fig. 5.4. Slow motions in the place where the front is located lead to a discontinuity in the slow variable. The representative point moves in the unstable region 1 to the boundary point D_1 along the separating trajectories. This means that the velocities of slow variations of y on different sides of the front are consistent and that the amplitude of the unstable shear mode is always equal to zero. As soon as the representative point reaches the boundary region 1 the front begins splitting

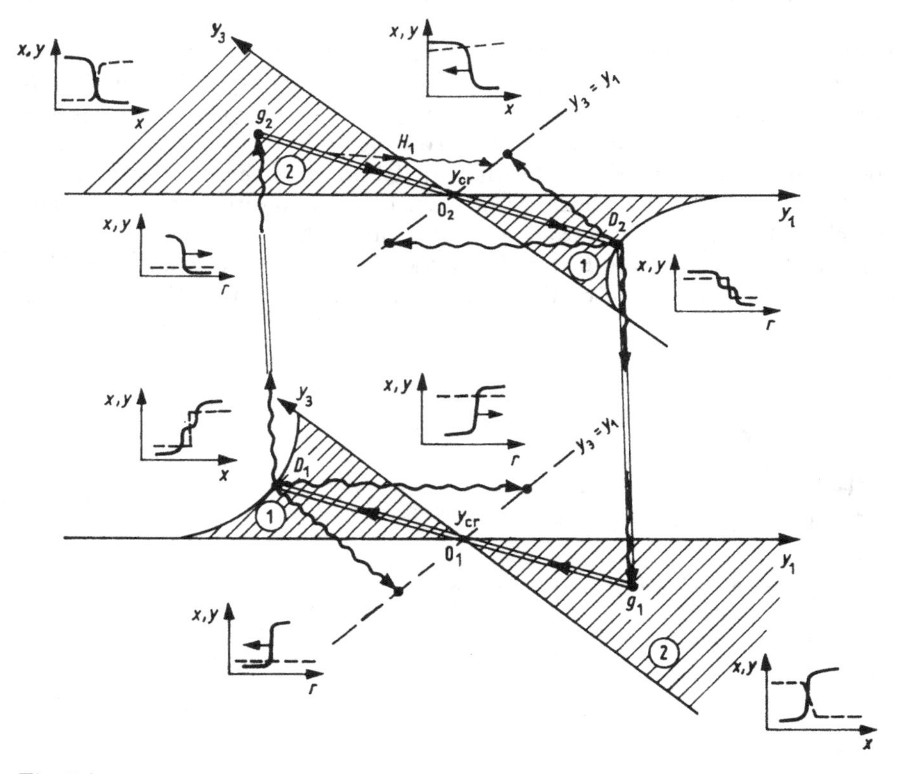

Fig. 5.4
Scheme of operation of a periodic TP source — "fissioning front". The closed trajectory $g_1D_1g_2D_2$ passes through two planes y_1, y_2 and y_1, y_3, which correspond to fixed fronts with different slopes (Yakhno 1981).

up thereby leading to the formation of a new immobile front (with opposite slope) and two divergent fronts.

We now pass over to the plane y_1, y_3, which corresponds to an immobile front with negative derivative dx/dr. The discontinuity in the slow variable decreases, i.e. the front moves from the point g_2 to the point O_2. Another condition for periodicity is consistency in the variations of y on the different sides of the front such that these variations reach the values y_{cr} simultaneously (the trajectory from g_2 leads exactly to O_2). The motion from O_2 to O_1 through g_2 and g_1 is quite the same as that from O_2 to O_1 through D_2 and g_1. Thus, in a configurational space which consists of two planes, the periodic process is represented by a closed trajectory $O_1D_1g_2O_2D_2g_1O_1$. This corresponds to the case $\tau/R = 1/2$ in the axiomatic model.

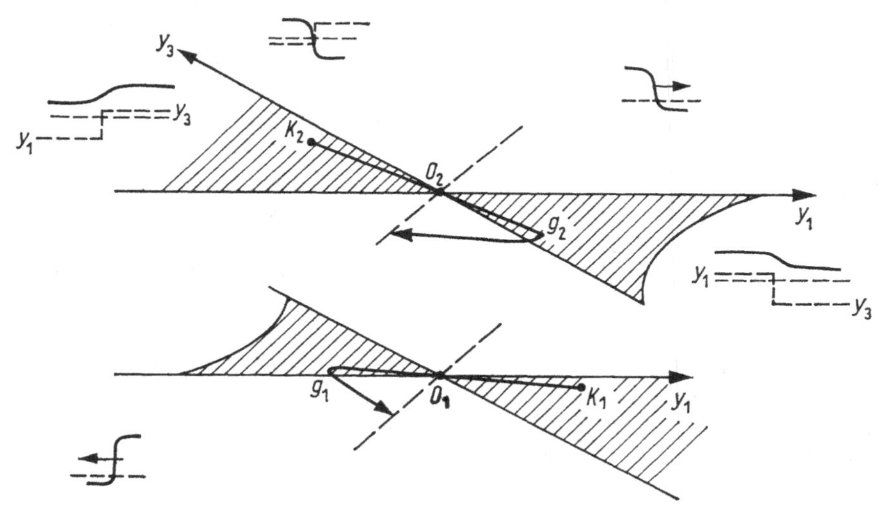

Fig. 5.5
Scheme of operation of a TP source of "echo" type in the configurational space y_1, y_3

The functioning of an "echo" source in the distributed system can be explained by analyzing a recurrent process, which is a departure of TFs from the stopping place followed by the appearance of a new front. The emergence of another immobile front is caused by the inhomogeneous distribution of the slow variable, which was formed by a preceding immobile front. Conditions for periodic recurrence of such motions can also be specified by a qualitative analysis. Let the initial condition be again given in the form of an immobile front at $y = y_{cr}$. Then the periodic motion will have a trajectory as shown in Fig. 5.5. The slow variable on both sides of the immovable front changes so that an increasing shear perturbation mode arises, which leads to the departure of the front from the place where it stops. In a configurational space, this state is represented by point g_1 (see Fig. 5.5). The TF moves to a region in which the system is in the state of rest and, therefore, the stopping region transforms into a perturbed state with a new inhomogeneity in the distribution of the slow variable. This inhomogeneity leads, after a certain time, to the formation of a new immobile front (see point k_2 in Fig. 5.5), which again moves away to a region in which the system is in the state of rest, and so on.

The "echo" regime is periodic if the slow motions are matched so that the values y on both sides of the immobile front achieve the values y_{cr} simultaneously. In Fig. 5.5, this corresponds to a trajectory composed of two pieces, $O_1g_1k_2O_2$ and $O_2g_2k_1O_2$. Thus, the qualitative analysis shows that both regimes, periodic "front fission" and "echo", are possible only if there is a coordination between the velocities of the slow motions on the different branches of the N-shaped null isoclines. Otherwise the front will move away from the stopping place to the region of the excited state and the whole system will revert to the state of rest (Goltsova, Zhislin, Yakhno 1976). Note also that for $\tau < R/2$ autonomous sources of TPs cannot arise.

Another important problem is to ascertain the characteristic size of self-pacemakers. Computationally, the "echo" and "fissioning front" sources have a size of the order of the length of the excitation front. The source will break up if the system decreases to the characteristic size of the source.

Thus, the following information has been obtained by means of a qualitative description of the dynamical model. First, it is shown that the "echo" source can exist both in the auto-oscillatory and slave systems, while the "fissioning front" pacemaker can also operate in a trigger medium. Second, periodic generation can be obtained only with appropriate forms of the functions $F(x, y)$ and $\varphi(x, y)$. Therefore, a periodic regime is not so rough with respect to the parameter variations in a two-component system. In this relation, it is interesting to ascertain possible modifications of a two-component system which could ensure structural stability of the solutions. This problem will be considered in Section 5.3 and 5.4. Third, the characteristic size of a one-dimensional source has been determined to be equal to the length of the front.

Numerical calculations for one-dimensional sources in systems with two variables are now available although it is still difficult to account for the particular functioning of these sources by means of qualitative methods (see, e.g., Zhabotinsky, Zaikin 1973).

5.2 Generation of a TP at a border between "slave" and "trigger" media

In a system with two variables, TPs can also arise from the medium inhomogeneity, which corresponds to a transition from slave to trigger regime in a local model (Zaikin, Kokoz 1977). A complete description of the mechanism for such a source is still not available. However, this stage of evolution to front fission due to the inhomogeneity can be considered using a qualitative approach. Figure 5.6 shows that the excitation pulse encounters the inhomogeneity at $r = r_1$. A situation can arise in which the leading edge of the TP passes into a region with trigger properties, while the trailing edge stops at the entrance. In this case three regimes are possible:

a) The trailing edge, after stopping at the entrance, nevertheless goes into the trigger region (Fig. 5.6a);

b) the stopped trailing edge generates several TPs and then passes into the trigger region (Fig. 5.6 b);

c) the trailing edge stops at the entrance and continuously produces TPs (Fig. 5.6 c).

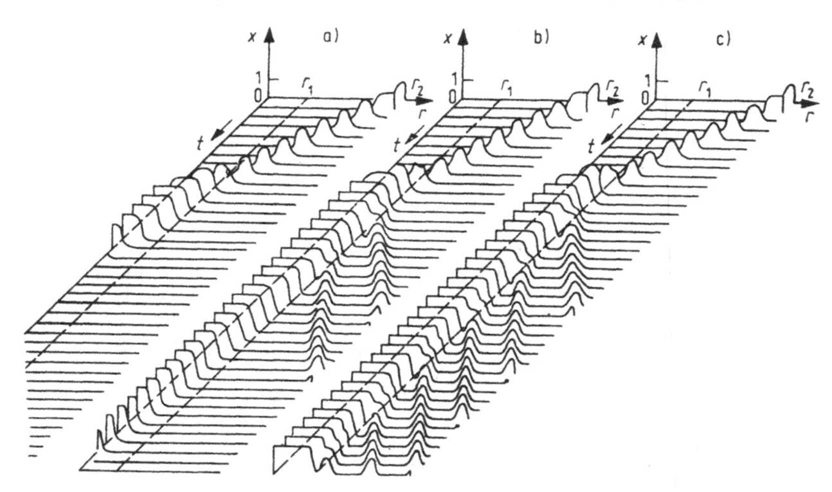

Fig. 5.6
A TP source at a border of media with slave and trigger properties (Zaikin, Morozova 1979)

These processes were considered using a model (2.53) with the nonlinear function given by

$$F(x, y) = \frac{A}{1 + \exp[(1 - x - y)/\delta]} - x, \tag{5.1}$$

$$\varphi(x, y) = \frac{B}{1 + \exp[(1 - x - y)/\delta]} - y. \tag{5.2}$$

Possible variants of null isoclines in different regions of the system under study are shown in Fig. 5.7. Conditions under which the trailing edge stops can be specified by

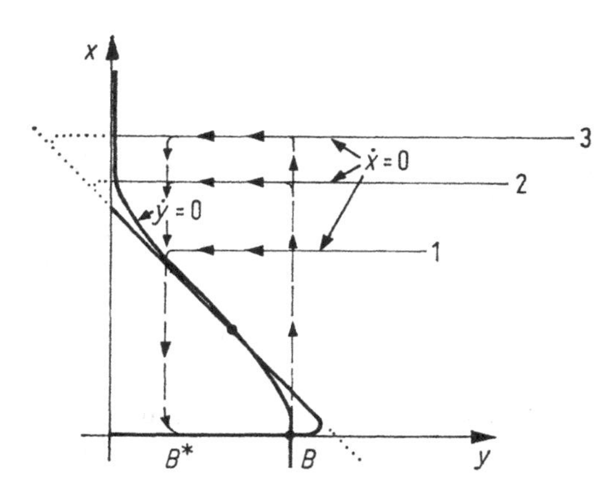

Fig. 5.7
Null isoclines corresponding to slave ("1") and trigger ("2" and "3") regimes of system (5.1)—(5.2) (Zaikin and Kokoz 1977)

analyzing the null isoclines. In the slave regime, $A = A_* < 1$ and the x null isocline is shown by curve 1. The TP corresponds to a trajectory shown by a dashed line. At the trailing edge of such a pulse the slow variable equals $B^* = 2 - A_* - B$ with $y_{cr} = 1 + A/2$. If the parameters of the trigger region are such that $y_{cr,tr} > B^*$, i.e., $1 - A_{tr}/2 > 2 - A - B$, then the trailing edge running against the boundary will, although decelerated, move without stopping. But if $0 < y_{cr,tr} = 1 - A_{tr}/2$

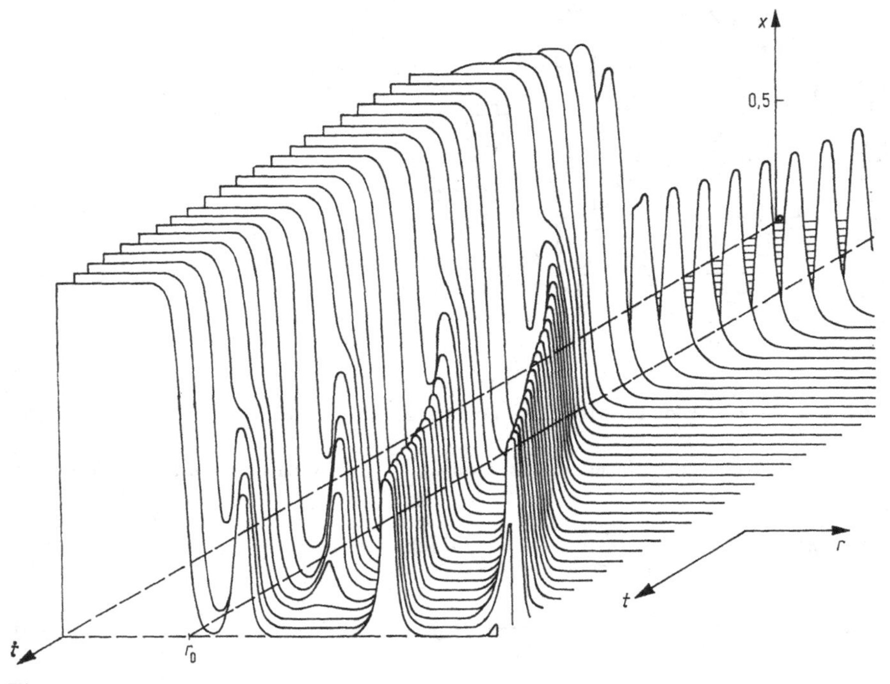

Fig. 5.8
Transformation of the rhythm of TP emitted by the interface of trigger and slave media (Zaikin and Morozova 1979)

$< 2 - A - B$, then the trailing edge will necessarily stop. Further evolution of the front will depend on the relation between the velocities of the slow motions on different sides of a stopped trailing edge, in a manner similar to that in the "echo" type source. For a system that moves in the fashion shown in Figs. 5.6b and 5.6c, the condition $y_{cr,tr} < B^*$ leads to the splitting of a stopped trailing edge. It is apparent that the processes by which TPs are formed or periodically generated cannot be adequately described by means of the qualitative approach taken here. The point is that the formation of a TP is a weak relaxation process where the motions at distances of the order of the front length should be taken into account. This is seen not only from the form of the TP (Fig. 5.6) but also from the transient regime (Fig. 5.8). The generated frequency of the TP even exceeds their maximum repetition rate in such a medium. Figure 5.8 demonstrates that the third pulse drops out. As was noted in Section 3.7, the TP having frequencies near to cut off cannot be described by means of an approximate method.[1])

5.3 Stable leading centres

In spatially homogeneous systems with two variables considered above, wave generation is possible only if the parameters of the system are chosen appropriately. In this section, we will explore whether the LC operation can be stabilized taking into account the third component. However, the stabilization of "echo" type sources has not yet been investigated. Nevertheless three-component models for LCs having characteristics qualitatively consistent with experimental data have been constructed (Vasilev, Zaikin 1976).

Thus, experimental records of the distributions at successive times for the LC regime of the Belousov-Zhabotinsky reaction were shown in Chapter 1 (Fig. 1.2b). It is seen that the amplitude, slope of the fronts and other oscillation parameters change in the generation region but these are quantitative changes rather than qualitative ones as in the "echo" case. The situation is similar to the theory of pacemakers in inhomogeneous two-component systems where the wave generation is governed by the altered parameters in the region of increased natural frequency. The LCs, which are regions of self-pacemakers, can form in three-component systems. The third component records the difference in the TP parameters at the point of the wave generation and far away from it, keeps them in memory for the time of slow motions and uses them for the acceleration of these slow motions. Consequently, the oscillation frequency in the wave generation region increases and a stable LC is formed.

Two classes of three-component systems with stable LCs will be discussed below. Rather simple models for both classes have been chosen. Lumped systems contain an auto-oscillatory subsystem (x and y variables) and the third variable determines the additional delayed feedback, which strongly affects the oscillation parameters. In such distributed models, a TP is produced by the auto-oscillatory subsystem, while the changes in its form arising in the course of the propagation depend on the spatial distribution of the third variable. The types of the models considered here differ in the relations between the third component and the auto-oscillatory subsystem. In some of the models, the stability of the LC is due to the process by which the third component is transferred in the slow motion phase, while in other models, the third component "memorizes" the differences in the pulses at the wave generation point, as well as at other points of the medium. These models are simple enough to be used as basic models.

---

[1]) Under real conditions, stationary distributions of x and y variables at the boundary will not be discontinuous. This circumstance should be taken into account when determining the generation frequency of a TP.

Solutions such as stationary LCs can be obtained by using a self-similarity variable

$$\eta = t + \alpha(r, \varepsilon), \tag{5.3}$$

where ε is, as usual, a small parameter. The LC problem reduces to determining the dependence of the oscillation phase α on the coordinate. The exact solution has not been obtained but a simple scheme of the method of successive approximations is applicable here: One need only substitute series expansions in ε of the functions $\alpha = \sum \alpha_n(r)\, \varepsilon^n$, $x = \sum x_n(r, \eta)\, \varepsilon^n$ and $\omega \simeq \omega_0 + \sum \omega_n \varepsilon^n$ in the initial system, where ω and ω_0 are the oscillation frequencies in the LC regime and in the local system. Application of this method requires knowledge of the form of the limiting cycle for $\varepsilon = 0$, e.g. of a local system. Then equations for the subsequent approximation can be obtained by means of averaging with respect to the trajectories obtained in the previous approximation.

Obtaining an analytical solution for such a scheme encounters technical difficulties. The scheme was rigorously reproduced only for the quasiharmonic case of two-component systems, e.g. those with Van-der-Pol kinetics, brusselators. We will not give intermediate calculations and note only that the basic result of the first approximation is the positive increase in frequency in the LC regime and the following form of the phase distribution equation: div grad $\{\exp(\alpha)\} = \exp(\alpha)$. The phase vs. coordinate dependence for a one-dimensional case is shown in Fig. 5.9. This dependence corresponds to the generation of plane waves observed experimentally.

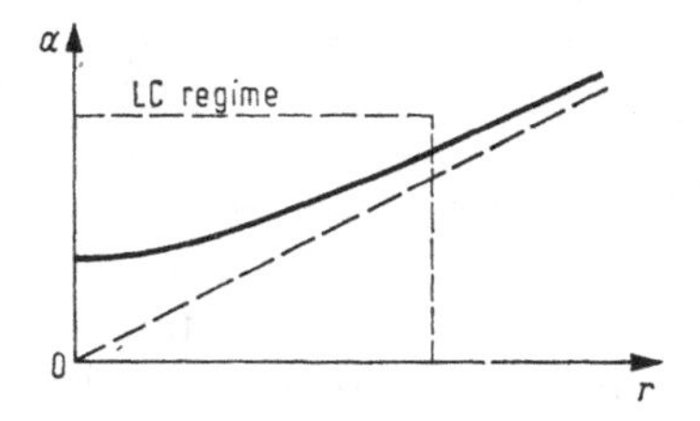

Fig. 5.9
Graph of the dependence of the oscillation phase of space points on the distance to a LC; the dependence for plane travelling waves is shown by a dashed line.

The next problem concerns the stability of the solutions obtained. It should be readily apparent that such solutions are unstable in quasi-harmonic systems. They transform into in-phase oscillations (systems having two or more components) or into standing waves (systems having three or more components) (see Section 5.4). That is why we will not discuss the properties of the above solutions. LCs can be stable in relaxation-type systems. Application of analytical methods is here very difficult. However, mechanisms responsible for the stabilization of LCs in the models mentioned above can be qualitatively analyzed in a rather simple way. Such an analysis will be supplemented with numerical experimental data.

As was mentioned, stable LCs are possible in systems having three components, one of which serves to "memorize" the TP generation region and to accelerate the slow motion stage, i.e. to reduce the oscillation period. This can be illustrated by the following model

$$\frac{\partial x}{\partial t} = y(1-x) - (\gamma_0 + z)\,x + D_x \frac{\partial^2 x}{\partial r^2},$$
$$\frac{\partial y}{\partial t} = \beta\, y(1-x) - \frac{(py - \varkappa)\,x}{qy + \tau} - cy^2 + D_y \frac{\partial^2 y}{\partial r^2}, \quad (5.4)$$

$$\frac{\partial z}{\partial t} = W + ky^2 - Rz + D_z \frac{\partial^2 z}{\partial r^2}, \quad (5.5)$$

where $D_y \gg D_x$.

For $D_x = D_y = 0$ the x and y variables form a relaxation-type auto-oscillatory subsystem (5.2), the phase plane of which is shown in Fig. 5.10. This subsystem is a reduced version of a model for the Belousov-Zhabotinsky reaction, a so-called "oregonator" (Section 2.6). Here, y plays the part of a "fast" autocatalytic variable. If the null isoclines are arranged in the manner shown in Fig. 5.10, then the motion of the representative point along the upper part of the y-null isocline does not depend significantly on z but this point is accelerated at high values of z when moving along the lower part. Note that the dependence of the velocity of the representative point on z is monotonic, i.e. even instantaneous additions in the lower part reduce the oscillation period. Moreover, it is namely this feature of a complete system which ensures the existence of a parameter region where only "hard" excitation of auto-oscillations occurs. In the latter case, the projection of the singularity lies on the lower branch of the y-null isocline.

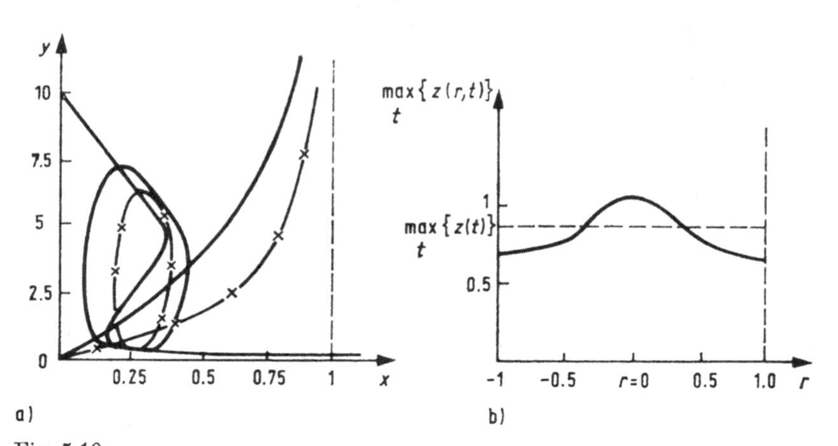

Fig. 5.10
The LC in system (5.4)–(5.5):

a) null isoclines of subsystem (5.4) and projections of integral curves (limiting cycles) onto the phase plane of the subsystem from the space points $r_0 = 0$ and $r = 1$;
b) stationary distributions of $\max\{z(r,t)\}$ (results of numerical integration of Eqs. (5.6), (5.5) on a computer ($k = 1$; $\varepsilon = 1/B = 10^{-7}$; $p = 5$; $\varkappa = 10^{-3}$; $q = 4$; $\tau = 1$; $c = 0$; $\nu = 0.02$; $W = 0$; $R = 0.1$; $D_x = D_y = 0$; $D_y = 0.2$; $\beta = 1$)

In a distributed system, the projection onto the phase plane of the trajectory of a representative point, which is a space element in the LC region, is a cycle that includes all the projections of the phase trajectories corresponding to the space elements, those along which TPs generated by an LC propagate. This is due to the fact that the jump of the representative point from the upper part of the y-null isocline to the lower part (and back) at the centre of a region where TPs form can only slacken. The point is that the y-component is transferred from here diffusionally thus accelerating the jump in the neighbouring parts and decelerating the jump at the centre.

The third equation of the system describes the relaxation of the variable to its current equilibrium value. The relaxation time is the most important parameter governing the stability of LCs. In the case of instantaneous relaxation (at relatively high values of R), the regimes of the system considered are the same as those of two-component systems. Characteristic of large relaxation times is complex dynamics, e.g. pitch generation occurs in a local system. The LC regime arises at relaxation times of the order of the oscillation period, with $\max_{t,r} \{z(r, t)\}$ in the wave initiation region, as is shown in Fig. 5.10. Such a distribution of $\max_{t} \{z(r, t)\}$ ensures a sharp decrease of the time of slow motions (see the first equation of system (5.4)) and therefore a decrease of the oscillation period and stabilization of LCs.

We should also explore the possibility that alternative regimes can occur, for example a migratory LC. Actually, $\max_{0<t<T_0} \{z(r, t)\}$ is always in the TP initiation region. This maximum is larger than $\max_{t} \{z(t)\}$ that corresponds to a local model (see Fig. 5.11). Thus, in a distributed system, the oscillation frequency increases compared to ω_0 and, therefore, subsequent TPs are generated. However, if the distribution of $\max_{t} \{z(r, t)\}$ is a mildly sloping one, then the value $y_{\min}$ (point of fall in the phase plane) can be first achieved by the representative point of the spatial domain far away from the wave initiation region. Hence the next TP will be initiated by this new region, and so on. If the distribution of $\max_{t} \{z(r, t)\}$ is too smooth or the diffusion of the z component is rather strong, then the gradients of the y variable decrease before subsequent TP are initiated. Consequently, the distribution $\max_{t} \{z(r, t)\}$

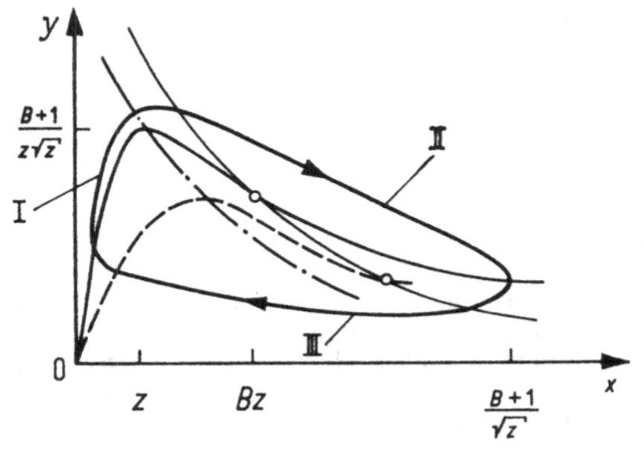

Fig. 5.11
Structure of the phase plane of the auto-oscillatory subsystem of the model (5.7)—(5.8)

becomes increasingly smooth and the system returns to the homogeneous state after several TPs are generated.

Thus, it was easy to make preliminary estimates for the parameters which allow the excitation of the regime of a stable LC. Then our concept of such a regime was confirmed computationally (Fig. 5.11). Note that the profiles of the x component at successive times coincide qualitatively with the Belousov-Zhabotinskii reaction (see Fig. 1.2b). The numerical experiments have also demonstrated that solutions such as migratory LCs can exist and auto-oscillations can be excited in systems the local kinetics of which has no limiting cycles.

We emphasize that in our qualitative analysis of the performance of subsystem (5.2), we assumed that the null isocline was of a relaxation type and N shaped, as is usually done for active systems. The main role is played by the coupling between the subsystem and the third component. However, Eq. (5.5) can have diversified structures. For example, the coupling with subsystem (5.4) in (5.5) can be replaced by the equation

$$\frac{\partial z}{\partial t} = \theta(x - x_0) - Rz + D_z \frac{\partial^2 z}{\partial r^2}, \tag{5.6}$$

where $\theta(x)$ is a step function ($\theta = 0$ at $x \leqq 0$ and $\theta = 1$ at $x > 0$). Such a form of the source should be treated as a limiting case of the usual kinetics, e.g., in oxidation–reduction reactions considered by A. N. Zaikin (1975). The nonanalytical form of the functions $\theta(x)$ should not confuse for this permits one to investigate dynamic systems in terms of the theory of finite automata.

The models (5.4) and (5.6) have an interesting feature: the oscillation frequency in the LC regime, with parameters and R chosen appropriately, can be too high to permit stable propagation of a sequence of TPs (see Chapter 4). The numerical experiments show that in the latter case the region of TP generation, i.e. the LC, changes its position from period to period and the regime of a single LC becomes unstable. Such a regime can be stabilized by decreasing the frequency of the LC. This can be done, e.g. by increasing the value of D_z.

Of course, other types of sources are possible in (5.3)—(5.4), e.g. $k(1 - x)/(k_0 + x + y)$, kx^2, etc., which ensure a stabilization of the LC in the manner discussed above. An ample description of the whole class of such sources (and systems) is rather problematic and is apparently not so useful. The mere understanding of the LC stabilization mechanism seems to be an equivalent for describing the general structure of the basic models of this type.

An alternative mechanism for LC stabilization can be illustrated by the following model (Vasilev, Zaikin 1976):

$$\begin{aligned} \frac{\partial x}{\partial t} &= zy + x^2 y - (B + 1)\,x + D_x \frac{\partial^2 x}{\partial r^2}, \\ \frac{\partial y}{\partial t} &= Bx - x^2 y + D_y \frac{\partial^2 y}{\partial r^2}, \end{aligned} \tag{5.7}$$

$$\frac{\partial z}{\partial t} = W + kx - Rz + D_z \frac{\partial^2 z}{\partial r^2}. \tag{5.8}$$

The autowave two-component subsystem (5.7) is a simple modification of the widely used "brussellator" model (Section 2.6). Figure 5.12 shows the null isoclines and limiting cycle of a corresponding local system for a fixed value of z. The velocity of the representative point is a function of z: it increases at large z in the vertical section I and decreases in the horizontal section III. In the relaxation case, fast motions weakly dependent on z occur in the slope section II. In a complete three-component

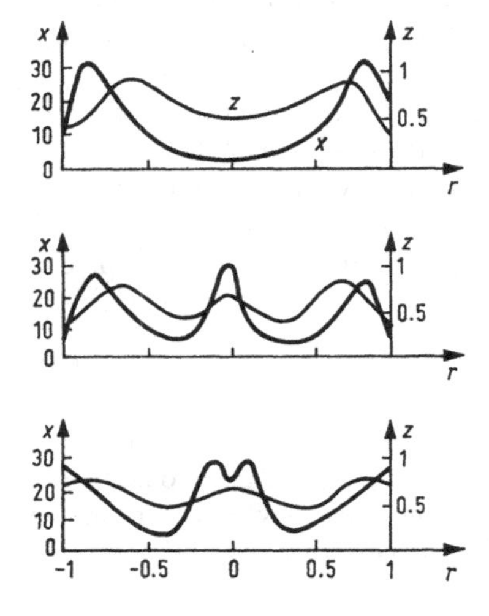

Fig. 5.12
Profiles of waves initiated by the LC of the model (5.7)—(5.8) at successive times (results of numerical integration on a computer ($B = 36$; $W = 0.36$; $k = 0.05$; $R = 1$; $D_x = 2.5 \times 10^{-4}$; $D_y = 0$; $D_z = 2.5 \times 10^{-3}$; $T_{LC} \simeq 3.3$; $T_0 \simeq 3.6$)

system (5.7), (5.8) the z component increases if the projection of the representative point is in the lower section III and decreases in the vertical section. The position of the dropping points also depends on z (see Fig. 5.12): the drop from the vertical section is accelerated with increasing z. This data on the local kinetics seems to be enough for a qualitative investigation of the LC stabilization mechanism.

In order to stabilize the LCs, it is necessary to decrease the oscillation period in the wave initiation region. For this, it is sufficient that the current values of z increase in section I and decrease in section III.

Consider the wave profiles of the z component shown in Fig. 5.12. The z component tends to a homogeneous distribution in space due to diffusion. Consequently,

a) if the projection of the wave initiation region is in the horizontal section, then the z component is transformed from there diffusionally and therefore the current values of z in this section slightly decrease;

b) if this projection lies in the vertical section, then the z component is transported into the LC region and the current values of z increase. Thus, diffusion of the z component leads to the acceleration of the slow motion in the initiation region. Therefore the oscillation period is reduced and the LC is stabilized.

Typical numerical integration for a system with a stable LC is shown in Fig. 5.12.

Let us enumerate some of the properties of the LC mechanism. The oscillation period is a function of D_z: it decreases for high values of D_z. However, if a certain critical value of D_z ($= D_{cr}$) is exceeded ($D_z > D_{cr}$), then the regime of a migratory wave initiation region arises. The point is that the phase of the decreasing x variable is delayed in the wave initiation region and another region overtakes it with respect to the phase of oscillation. The LC is unstable if D_y (i.e., the TP velocity is high) exceeds a critical value. In this case, in-phase oscillations are established. Note also that the oscillation period depends on the distance between the initiation region and the boundary of the system if this distance is less than the wavelength.

We emphasize that mutual diffusion is vitally important here: it can stabilize the LC even in those systems which behave as two-component ones in the absence of mutual diffusion. This can be illustrated by the model of Vasilev and Polyakova, 1975

$$\frac{\partial x}{\partial t} = z + x^2 y - (B+1)\,x + D_{xx}\frac{\partial^2 x}{\partial r^2},$$
$$\frac{\partial y}{\partial t} = Bx - x^2 y + D_{yy}\frac{\partial^2 y}{\partial r^2}, \tag{5.9}$$

$$\frac{\partial z}{\partial t} = W - Rz + \frac{\partial}{\partial r}\left[D_{zz}\frac{\partial z}{\partial r} + D_{zx}(z)\frac{\partial x}{\partial r} + D_{zy}(z)\frac{\partial y}{\partial r}\right], \tag{5.10}$$

where the autowave subsystem (5.9) is a Brussellator, and the source in (5.10) does not depend on the x and y variables. The mechanism for LC stabilization in this model is the same as that considered above but the z component is transported in the horizontal and vertical parts of the cycle (c.f. Fig. 5.12) due to the x and y gradients on the travelling wave profiles. It is apparent that the most favourable conditions for the appearance of LCs hold in systems with $D_{zx} > 0$ and $D_{zy} < 0$. The quantities D_{yz} and D_{xz} are not essential to the autowave process. Numerical experiments (Vasilev, Polyakova 1975) show that the quantities D_{zx} and D_{zy} needed for the stabilization of LCs should be somewhat less than D_{zz} and should meet the requirements of the thermodynamics of irreversible processes (see Section 2.3).

Note also that in the earlier models, the autowave excitation required localized perturbations of high relative level, while in the model with mutual diffusion, the autowave excitation can be soft.

Let us compare the two mechanisms for LC stabilization. The basis for the first mechanism (5.4), (5.5) is adapting the form of the TP, which depends on the distance to the initiation region, and memorizing this form by the third variable. Central to the second one is redistribution of the third component at the slow motion stage. It is easy to see that the second mechanism works only in systems having the characteristic scale of diffusion of the z component comparable to the length of the travelling wave, e.g. in systems with relatively high oscillation frequency and low velocity of the TP. The second mechanism places opposite demands on the parameters of a system.

The analysis of such models can be divided into several basic stages: a qualitative analysis of the dynamics in the determining phase plane, an estimate of the parameter region where LC are possible, and a numerical experiment which is intended to refine the estimates obtained at the earlier stages. We emphasize that a numerical experiment is the determining stage in the LC problem in which theoretical predictions are substantiated. In distributed relaxation-type multicomponent systems, a numerical experiment requires much effort and machine time which will be appreciably reduced if regimes of a corresponding simple discrete model are considered first (see Section 3.6). Actually, if a large, overthreshold perturbation of a homogeneous state is assigned to two diffusion-coupled "boxes", then the following solutions are possible:

a) damping inhomogeneous perturbations, i.e. in-phase oscillations;

b) auto-oscillations with one of the elements advancing in phase;

c) auto-oscillations with elements advancing in phase alternately;

d) antiphase auto-oscillations;

e) a stationary state with different values of variables in the elements.

Only these qualitative types of solutions occur in a simple discrete model. In a distributed model, a stable LC is possible only if the solution b) of a discrete model is realized, a migratory LC corresponds to solution c), and standing waves to solution d) (see Section 5.4). Thus, the investigation of a discrete model helps to ascertain the possibility of the existence of certain types of AW solutions in corresponding distributed models and to estimate the range of parameters for which these solutions are realized. Note that if the model has a parameter region where the regime of a stable LC occurs, then there are regions of standing waves and an intermediate region of migratory LCs. However, the presence of these latter regions does not necessarily mean that a region of stable LCs must be present.

Let us compare the experimental data on LCs with the simulation results. In the Belousov-Zhabotinsky reaction (Section 2.6) x is a slow variable. Figure 12b shows that the duration of a TP is about one eighth of the oscillation period. Since the "echo" regime requires $\tau_p/\tau_{\text{refr}} > 0.5$, such a mechanism for auto-oscillations will not work. Essential features of this phenomenon can be adequately described by three-component basic models for LCs. Taking the space-time characteristics T, λ of experimental LC into account, the mechanism of the model in (5.4), (5.5) is clearly preferrable for explaining the experimental evidence.

5.4 Standing waves

On a segment with boundary conditions of the second kind, any solution can be represented in the form of a superposition of normal modes, i.e. standing waves. Near the bifurcation point of oscillation type (see Section 3.3), when the real parts of the roots of the dispersion equation are small, there are quasi-harmonic single-mode solutions in the form of standing waves. There are many systems, however, where the self-excitation condition is met not only for a mode with $k \neq 0$ but also for a zero mode. It is namely this mode which causes the most "dangerous" perturbation of standing waves. Therefore the problem of stability of standing waves is, above all, the problem of their interaction with the zero mode (Vasilev, Romanovskii 1975; 1976; Vasilev 1976).

As was noted above (see Section 3.6), the problem of interaction between two harmonic modes is equivalent to that of the regimes of a simple, "two-chamber" model. We will first consider this model as it gives a physically relevant and illustrative interpretation of the interactions being investigated. Primary attention will be focused on the qualitative aspect. Then we will expound a rigorous method for quasi-harmonic analysis of multicomponent distributed systems.

The influence of diffusion coupling between generators of auto-oscillations is different in two- and three-component systems. Thus, stable out-of-phase quasi-harmonic auto-oscillations are possible only in three- (or more) component systems, while these oscillations are unstable in two-component systems. We stress that this assertion holds only for quasi-harmonic oscillations, and is, generally, realized asymptotically if there is a simultaneous coupling in x and y (see Chapter 6).

When studying such systems it is reasonable to use normal coordinates $\boldsymbol{S}$ and $\boldsymbol{\Delta}$ (see Section 3.6). A substitution of the variables yields:

$$\begin{aligned} \frac{\mathrm{d}S_i}{\mathrm{d}t} &= \sum_{j=1}^{n} M_{ij} S_j + \Phi_{si}(\boldsymbol{\Delta}, \boldsymbol{S}), \\ \frac{\mathrm{d}\Delta_i}{\mathrm{d}t} &= \sum_{j=1}^{n} (M_{ij} - \delta_{ij} d_{ii})\, \Delta_j + \Phi_{\Delta i}(\boldsymbol{\Delta}, \boldsymbol{S}), \end{aligned} \tag{5.11}$$

where $i = 1, 2, \ldots, n$, d_{ii} and M_{ij} are the constant coupling coefficient and matrix elements of a linearized local system (see Section 3.3). The functions $\Phi_{is}(\boldsymbol{\Delta}, \boldsymbol{S})$ and $\Phi_{i\Delta}(\boldsymbol{\Delta}, \boldsymbol{S})$ describe the nonlinear properties of $\boldsymbol{S}$ and $\boldsymbol{\Delta}$ generators, as well as their interaction. We emphasize that the form of Eqs. (5.11) and of the functions $\Phi_{i\Delta}$ and Φ_{is} remain the same for a two-mode approximation; thus, $S_i \to A_{0i}(t)$ and $\Delta_i \to A_{ik}$, where A_{0i} and A_{ki} are the time factors of the 0-th and K-th modes.

The nonlinear terms can be grouped in the following fashion:

$$\begin{aligned} \Phi_{si} &= \Phi_{si(\mathrm{ex})}(\Delta) + \Phi_{si(\mathrm{param})}(\boldsymbol{\Delta}, \boldsymbol{S}) + \Phi_{si(\mathrm{in})}(S), \\ \Phi_{\Delta i} &= 0 + \Phi_{\Delta i(\mathrm{param})}(\boldsymbol{\Delta}, \boldsymbol{S}) + \Phi_{\Delta i(\mathrm{in})}(S). \end{aligned} \tag{5.12}$$

We specify the functions from the right-hand side of (5.12) for the "Brussellator" model (see Section 2.6):

$$\begin{aligned} \Phi_{sx(\mathrm{ex})} &= -\Phi_{sy(\mathrm{ex})} = (B/A)\, \Delta_x^2 + 2A\Delta_x\Delta_y, \\ \Phi_{\Delta x(\mathrm{param})} &= -\Phi_{\Delta y(\mathrm{param})} \\ &= (B/A)\, S_x\Delta_x + 2A(S_y\Delta_x + S_x\Delta_y) + 2S_xS_y\Delta_x + S_x^2\Delta_y, \\ \Phi_{sx(\mathrm{param})} &= -\Phi_{sy(\mathrm{param})} = (\Delta_x^2)\, S_y + 2(\Delta_x\Delta_y)\, S_x, \\ \Phi_{\Delta x(\mathrm{in})} &= -\Phi_{\Delta y(\mathrm{in})} = \Delta_x^2\Delta_y, \\ \Phi_{sx(\mathrm{in})} &= -\Phi_{sy(\mathrm{in})} = (B/A)\, S_x^2 + 2AS_xS_y + S_x^2S_y. \end{aligned} \tag{5.13}$$

From these equations it is seen that the inequality $\Phi_{is(\mathrm{ex})}(\Delta) \neq 0$ is satisfied only for systems where even powers of the variables are present in the right-hand sides of the "reduced" local equations. On the contrary, $\Phi_{is}(\mathrm{ex}) = 0$ for systems with odd powers (e.g. Van-der-Pol generators). There is a well-known theorem which suggests that only in-phase auto-oscillations can occur in such two-component systems (Romanovskii, Stepanova, Chernavskii 1975). It is assumed that $\left(\mathrm{Re}\,(P_s), \mathrm{Re}\,(P_\Delta)\right) \ll \left(|\mathrm{Im}\,(P_s)|^{-1}, |\mathrm{Im}\,(P_\Delta)|^{-1}\right)$ are the roots of the characteristic equations of S and Δ generators). An analogous statement is valid for "Brussellator" type generators. Essentially, the proofs are as follows:

Let in-phase auto-oscillations ($\Delta = 0$; $S(t) \neq 0$) be established in a system ($\mathrm{Re}\,(P_s) > 0$ and $\mathrm{Re}\,(P_\Delta) > 0$). Consider the condition for self-excitation of a Δ generator driven by "an external force" $S(t)$. Assuming that $\mathrm{Im}\,(P_s)$ and $\mathrm{Im}\,(P_\Delta)$ are incommensurable and applying the averaging method to the equations of the Δ generator we have: $\mathrm{Re}\left\{\left\langle P_\Delta\left(S(t)\right)\right\rangle\right\} \lesseqgtr -(d_x + d_y)$. We stress that the decrease of $\mathrm{Re}\,\{P_\Delta\}$ and therefore the stability of in-phase auto-oscillations are due to the presence of the two last terms of $\varphi_{i\Delta(\mathrm{param})}$ in (5.3) (cubic nonlinearity in the kinetic equations). In other words, such a stability is caused by *asynchronous quenching* of the oscillations of a Δ generator.

Out-of-phase quasi-harmonic oscillations are unstable as the self-excitation conditions for an S generator are still satisfied.

In systems where the kinetic equations do not contain cubic or higher odd powers of the variables, the asynchronous quenching of a Δ generator cannot occur since $\langle \Phi_{i\Delta(\mathrm{param})} \rangle = 0$ in the truncated equations. Therefore if the coupling coefficients d_{xx} and d_{yy} are such that $\mathrm{Re}\,\{P_\Delta\} = \mathrm{Re}\,\{P_s\} - (d_{xx} + d_{yy}) > 0$, then the in-phase regime is unstable. In order to determine the amplitudes of auto-oscillations, one has to use the second and higher approximations. Note that, as a rule, the single-frequency regime is stable, while the beating regime ($\mathrm{Im}\,(P_s)$ and $\mathrm{Im}\,(P_\Delta)$) is unstable. An illustration of such properties can be a model for auto-oscillations in the dark photosynthesis reactions, which was proposed by D. S. Chernavskii and N. M. Chernavskaya back in 1960. Two such generators were investigated by means of an analogue computer by Lishutina and Romanovskii (1967).

Things are somewhat different for three-component systems. This can be illustrated by a completed "Brussellator" model:

$$F_x = z + x^2 y - (B + 1)\,x, \quad F_y = Bx - x^2 y, \quad F_z = W + kx - Rz. \tag{5.14}$$

Equations (5.14) for a two-component subsystem are exactly the same as those for the "Brussellator" but z is a variable in the latter case. The equation for z can be interpreted as feedback superimposed on the "brusselator". The conditions for self-excitation of in-phase and antiphase regimes can be easily found by making use of Eqs. (3.9) to (3.11). We omit simple calculations and report only the final result (the case $d_{xx} = d_{yy} = 0$ has been chosen for illustrative purposes)

$$\begin{aligned} C_s &\approx \mathrm{sign}\,\mathrm{Re}\,(P_s) \approx \mathrm{sign}\,\{R(k - g^2) + g(R^2 + z^2) - k(z^2 + g)\}, \\ C_\Delta &\approx \mathrm{sign}\,\mathrm{Re}\,(P_\Delta) \approx \mathrm{sign}\,\{C_s + 2d_z[k + gR + g(R - g + 2d_z)]\}, \end{aligned} \tag{5.15}$$

where $g = B - 1 - \bar{z}^2$. The inequality $g > 0$ is the condition for self-excitation of a "Brussellator". From (5.15) it follows that if $g > 0$ then at rather high values of d_{zz} the self-excitation conditions for a generator will be satisfied even for $C_s < 0$.

Out-of-phase auto-oscillations can also occur if $C_s > 0$. Let $B = 5$; $W = 2.2$; $R = 4.1$; $k = 3$; $d_{xx} = 0.075$; $d_{yy} = 0$; and $d_{zz} = 1$. Then the characteristic equations have the roots $P_s = 0.085 \pm j\,1.03$ and $P_\Delta = 0.019 \pm j\,1.6$. The analysis and numerical integration of such a system on a computer (see Fig. 5.13) show that the out-of-phase regime is stable. Note that if $C_s > 0$ the stability of out-of-phase

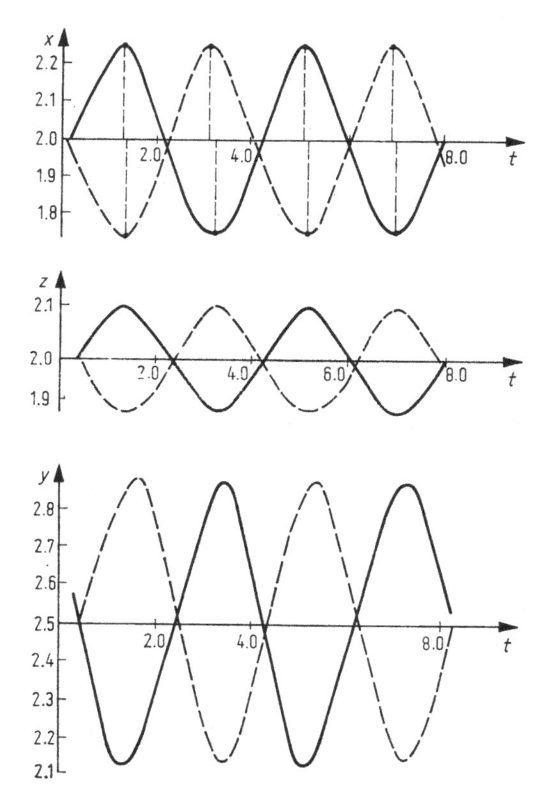

Fig. 5.13
Stable regime of out-of-phase oscillations of a three-component model (5.14) (the distance between thin dashed lines is the deviation from an ideal antiphase regime) (Vasilev and Romanovskii 1975)

oscillations depends on the relations between the eigenfrequencies of the S and Δ generators ($\mathrm{Im}\,(P_s)$ and $\mathrm{Im}\,(P_\Delta)$). For example, if $d_{xx} = 0$ and $d_{yy} = 0.075$ (other parameters have the previous values), then $\mathrm{Im}\,(P_\Delta) = 1.01 < \mathrm{Im}\,(P_s) = 1.03$ and only the regime of in-phase auto-oscillations is stable although $\mathrm{Re}\,(P_\Delta) = 0.9 > \mathrm{Re}\,(P_s)$.

If $C_s > 0$ in-phase auto-oscillations are *always* stable. Therefore if out-of-phase auto-oscillations are stable too, as in an example given above, then such a system can be treated as a frequency trigger. The differences between the oscillation frequencies in these regimes can be substantial but, as a rule, $\omega_\Delta > \omega_s$.

If $|\mathrm{Re}\,(P_\Delta)| \ll |\mathrm{Im}\,(P_\Delta)|$, then, as is seen from Fig. 5.13, the oscillations are nearly harmonic. Since $\Phi_{is(\mathrm{ex})} \neq 0$ (5.13), a pure antiphase regime is impossible. The oscillation spectrum of a Δ generator consists of even harmonics of the fundamental mode but the amplitudes of these harmonics are relatively small. In calculating the oscillation amplitudes, it seems reasonable to use only the first approximation. The scheme for constructing truncated equations is essentially the same as that for a distributed model.

Solutions which satisfy the boundary condition of the second kind can be represented as an infinite series (see Section 3.4) of the spatial harmonics $\cos(\pi kr/L)$ with time-dependent coefficients A_{ik}. Substituting this series in the equations being studied and expanding the nonlinear functions in Fourier series we choose terms having the same spatial factors. Finally, we obtain an equation for the coefficients of these series A_{ik}, which reads

$$\frac{\mathrm{d}A_{ik}}{\mathrm{d}t} = \sum_{j=1}^{n} [M_{ij} - \pi^2 k^2 D_{ij}/L^2] + \Phi_i(A_{0k}, A_{1k}, \ldots), \tag{5.16}$$

$$i = 1, 2, \ldots, n; \quad k = 0, 1, 2, 3, 4, \ldots$$

The infinite-dimensional set of equations (5.16) is fully equivalent to a distributed system, and is, of course, not a simpler one. Nevertheless if all $\mathrm{Re}\,\{P_k\} < 0$ corresponding to a mode with the period $2L/k$ and $|\mathrm{Re}\,\{P_k\}|^{-1}$ are substantially smaller than the characteristic times of the process under study, then such modes are clearly not essential for the process and can therefore be ignored. Moreover, even interactions between three or more modes in weakly nonlinear systems are negligibly small. Hence it stands to reason to construct solutions on the basis of single-mode approximations and to investigate their stability using two-mode approximations.

If a zero (in-phase) mode is included in a two-mode approximation, then Eqs. (5.16) take the form of (5.11): $S_i \to A_{0i}$ and $\Delta_i \to A_{ki}$. This implies that the qualitative data for two coupled oscillators are useful here. Specifically, it can be inferred that in distributed two-component systems with local kinetics (such a Brussellator or Van-der-Pol kinetics), only the in-phase regime is possible, while quasi-harmonic standing waves do not occur. However, the in-phase regime is unstable in systems with a "purely quadratic" nonlinearity if the condition $\mathrm{Re}\,\{P_k\} > 0$ $(k \neq 0)$ is satisfied for one mode. In this case, not the standing wave regime but more composite ones, for example a ripple regime (Vasilev, Zaikin 1976), are established.

For $n = 3$ equations such as (5.11) can have stable solutions with $A_{ik} \neq 0$. Assuming that there are no resonances between these solutions we obtain a set of truncated equations. These can be easily reduced to standard equations by means of the averaging method. Such a transformation can be illustrated by using only one cell corresponding to a certain value of k, since the matrix of the system is cell-diagonal.

A set of eigenvectors $\boldsymbol{b}$ corresponding to the eigenvalues of the matrix $\{M_{ij} - \pi^2 k^2 D_{ij}/L^2\}$ forms a matrix $\hat{B}_k$. Let us introduce complex variables $\tilde{A}_{jk}$

$$A_{ik} = B_{ijk} \tilde{A}_{jk}. \tag{5.17}$$

In view of the fact that the matrix $\{M_{ij} - \pi^2 k^2 D_{ij}/L^2\}$ is diagonal in its representation, the equations for the complex variables A_{jk} have the form

$$\frac{\mathrm{d}A_{jk}}{\mathrm{d}t} = P_{jk} \tilde{A}_{jk} + \Phi_{jk}(B_{1jk}, A_{jk}, \ldots). \tag{5.18}$$

The averaging method of Bogolyubov-Mitropolskii (1956) is, of course, practicable here. However, it is more suitable to use first real variables introduced by ($n = 3$, $\omega_k = \text{Im}\,\{P_k\}$)

$$\tilde{A}_{1k} = u_k \exp\left\{j\left(\omega_k t + \varphi_k(t)\right)\right\},$$
$$\tilde{A}_{2k} = u_k \exp\left\{-j\left(\omega_k t + \varphi_k(t)\right)\right\}, \quad A_{0k} = \omega_k. \tag{5.19}$$

When the averaging method is applied to the truncated equations of the first approximation in the equations for μ_k, φ_k and w_k, one must drop the vibration terms. Thus, we arrive at the following equations for the mode amplitudes and phases

$$\frac{\mathrm{d}u_k}{\mathrm{d}t} = (\text{Re}\, P_k)\, u_k + f_k(u_k, w_k, u_{k_1}, w_{k_1}, \ldots),$$
$$u_k \frac{\mathrm{d}\varphi_k}{\mathrm{d}t} = g_k(u_k, w_k, u_{k_1}, w_{k_1}, \ldots), \tag{5.20}$$
$$\frac{\mathrm{d}w_k}{\mathrm{d}t} = P_{0k} w_k + f_{0k}(u_k, w_k, u_{k_1}, w_{k_1}, \ldots),$$

where P_{0k} is the real root of the third-order characteristic equation (3.11). This root is negative in the absence of the DS instability.

Using the conditions for "physical relevance" of a local system (absence of infinitely increasing solutions) it can be proved that if $P_{0k} < 0$ and $\text{Re}\,\{P_k\} > 0$, then Eqs. (5.20) have a stationary solution ($\overline{u}_k \neq 0$). We will not give the proof itself but note only that the procedure is similar to the proof of sufficient conditions for the existence of a DS (Chapter 7). The quantity $\overline{u}_k$ is the stationary oscillation amplitude (5.19). Therefore the relations $P_{0k} < 0$ and $\text{Re}\,\{P_k\} > 0$ are sufficient conditions for the existence of a limit cycle. This helps in understanding the meaning of the use (see Section 3.3) of these relations in determining the oscillation instability. If the conditions for oscillation instability are met for different values of k, then the question arises whether or not the limit cycle corresponding to a certain wave length ($2L/k$) is stable. A complete solution of the problem depends on many factors. Thus, it is reasonable to restrict oneself to studying some typical examples.

Consider a modification of the "Oregonator" model (Section 2.6)

$$\frac{\partial x}{\partial t} = y(1 - x) - \gamma x + D_{xx} \frac{\partial^2 x}{\partial r^2},$$
$$\frac{\partial y}{\partial t} = \beta y(1 - x) - pyz - cy^2 + \varepsilon z + D_{yy} \frac{\partial^2 y}{\partial r^2}, \tag{5.21}$$
$$\frac{\partial z}{\partial t} = x - qyz - \tau z + D_{zz} \frac{\partial^2 z}{\partial r^2}.$$

A characteristic feature of the point kinetics in (5.21) is that all the elements $M_{ii} < 0$ ($i = 1, 2, 3, \ldots$) holds for any parameter value in model (3.8) and that no oscillation subsystem can be distinguished. Nevertheless there is a parameter region where a local three-component system is an auto-oscillatory. Self-excited standing waves are possible even if the stationary state of the local system is stable.

For the parameter values listed in Fig. 5.14 we have:

$P_0 = -0.14 \pm \mathrm{j}\,1.4$ ($P_{00} = -2.8$); $P_8 = 0.08 \pm \mathrm{j}\,2.3$; $P_9 = 0.06 \pm \mathrm{j}\,1.9$. For $k < 6$ and $k \geq 13$ ($L = 1$) the inequality $\mathrm{Re}\,\{P_k\} < 0$ is satisfied, i.e., both an upper and a lower limit are placed on the spectrum of the excited modes. Truncated equations for the modes with $k = 8$ and $k = 9$ are given by

$$\frac{\mathrm{d}u_8}{\mathrm{d}t} = [\mathrm{Re}\,P_8 + a_8 W_0]\,u_8, \quad \frac{\mathrm{d}u_9}{\mathrm{d}t} = [\mathrm{Re}\,P_9 + a_9 W_0]\,u_9,$$

$$\frac{\mathrm{d}W_8}{\mathrm{d}t} = P_{08}W_8 + W_0(b_8 W_8 + C_8 W_9), \quad \frac{\mathrm{d}W_9}{\mathrm{d}t} = P_{09}W_9 + W_0(b_5 W_9 + C_9 W_8), \tag{5.22}$$

$$\frac{\mathrm{d}u_0}{\mathrm{d}t} = [\mathrm{Re}\,P_0 + \varkappa_8 W_8 + \varkappa_9 W_9 + \varkappa_0 W_0]\,u_0,$$

$$\frac{\mathrm{d}W_0}{\mathrm{d}t} = P_{00}W_0 + \sum_{i=0,8,9}\Big(\alpha_i u_i^2 + \sum_{j=0,8,9} \beta_{ij} W_i W_j\Big).$$

Equations for W_0, u_0 should necessarily be included in the set of truncated equations, since the nonlinear terms in (5.22) are all quadratic; thus, if the interaction between the mode $k = 8$ (or $k = 9$) and the mode $k = 0$ is neglected, then there is no stationary value of the oscillation amplitude of the mode $k = 8$ (or $k = 9$). The modes $k = 8$ and $k = 9$ interact only via the mode $k = 0$. There are two types of stationary solutions: $u_8 \neq 0$; $u_9 = 0$ and $u_8 = 0$; $u_9 \neq 0$. In the first case, $\mathrm{Re}\,\{P_8\} = a_8 W_8$. Here, the inequality $\mathrm{Re}\,\{P_9\} - a_9 W_0 < 0$ is satisfied ($u_8 = 0.2$; $W_0 = 0.05$; $W_8 \approx W_9 \approx 0.01$) and asynchronous quenching of the mode $k = 9$ occurs. The solution $k \neq 0$, $k = 0$ is unstable as in this case $\mathrm{Re}\,\{P_8\} - a_8 W_0 > 0$. A substitution of any other mode for the mode $k = 9$ in (5.22) will not alter the result obtained. Thus, only a standing wave of length 2/8 ($L = 1$) can be stable. Such an analysis cannot be considered rigorously. That is why Eqs. (5.21) were numerically integrated on a computer. The results are given in Fig. 5.14. Note that the oscillation amplitudes obtained here

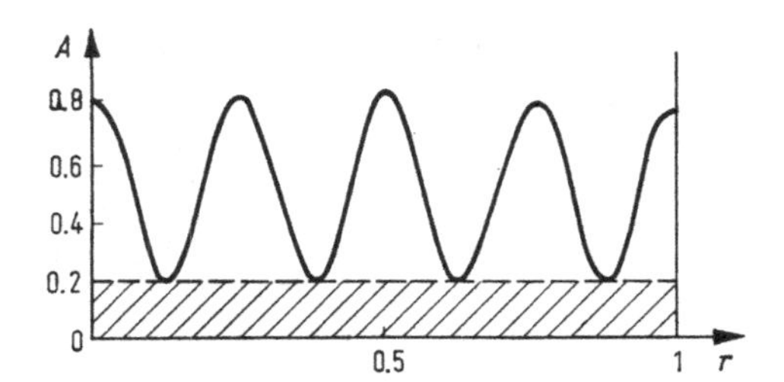

Fig. 5.14
Distribution of oscillation amplitudes in space (results of numerical integration of system (5.21) on a computer ($\gamma = 0.02$; $B = 30$; $P = 30$; $\varepsilon = 0.45$; $q = 1$; $\tau = 0.2$; $c = 0.06$; $D_{xx} = D_{yy} = 0$; $D_{zz} = 0.05$)). The height of the dashed region indicates the level of deviation from an "ideal" standing wave (Vasilev and Romanovskii 1976)

are almost twice as large as the predicted ones. It is apparent that a good qualitative agreement is not to be expected, since only truncated equations of the first approximation were obtained for a system with quadratic nonlinearity. However, the basic qualitative result is fully confirmed: all modes (standing waves) with $k \neq 8$ are unstable.

We now report some data concerning a completed "Brussellator" model (5.14) with cubic nonlinearity. Figure 5.15 is a computation of the dependence of oscillation amplitude on r. In this case Re $\{P_0\} < 0$ and for $k \geqq 3$, Re $\{P_k\} > 0$. The self-excitation conditions are met for an infinite number of modes. This is characteristic of systems containing auto-oscillatory subsystems. Equations for $\dot{u}_k$ $(k \neq 0)$ consist of the following terms $(m \neq k)$: $\dot{u}_k \sim [\text{Re}\,\{P_k\} + W_k + W_m + W_k^2 + W_m^2 + u_k + u_0 + u_k^2 + u_m^2]\, u_k$ (terms of the form $u_m^2 u_k \delta(2m - k)$ are neglected), etc.

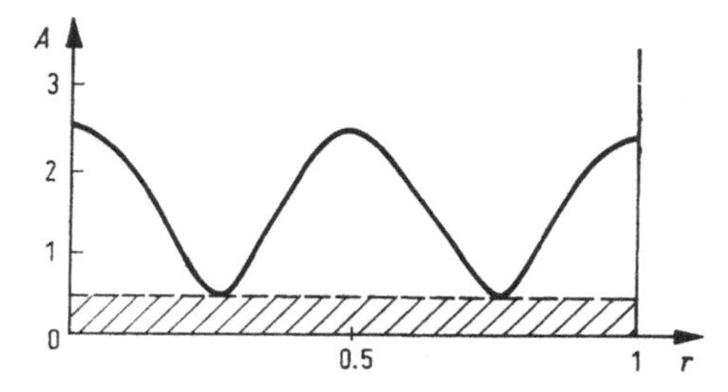

Fig. 5.15
Spatial distribution of the amplitudes of auto-oscillations (results of numerical integration of the model (5.14); $B = 3.9$; $W = 0.342$; $k = 1.8$; $R = 2$; $D_{xx} = D_{yy} = 0$; $D_{zz} = 0.03$). The height of the dashed region characterizes the deviation from the regime of an "ideal" standing wave.

Note that the values for u_k found from the single-mode set of truncated equations and those obtained by means of a numerical integration of the initial equations coincide within $\approx 20\%$. This is a much better agreement than that in (5.21). The analysis of a two-mode set of truncated equations shows that direct mode interaction ensures *asynchronous quenching* of this or that mode depending on initial conditions.

If the parameter values are such that Re $\{P_0\} > 0$, then the mode $k = 0$ is stable. In this case, modes with $k \neq 0$ can be stable only if Im $\{P_k\} >$ Im $\{P_0\}$. These systems can be treated as distributed frequency triggers, since the oscillation periods of modes with different space periods do not coincide.

In conclusion, if the self-excitation conditions are fulfilled for multiple modes (e.g., for $k = 1$, $k = 2$), then the interaction between these modes is essential to the dynamics of the systems. For example, transitions between multiple modes occur in a jump-like way as the length of the system changes. Here, exactly as in the previous case of a stationary DS, hysteresis phenomena take place (hysteresis loops in a diagram u_k, L). Note also that the auto-oscillations are far from being harmonic away from the bifurcation point of the oscillation type. Different regimes of self-sustained oscillatory activity occur in such systems.

5.5 Reverberators: a qualitative description

So far, we have considered mechanisms of TP sources in one-dimensional systems. Qualitative models for such mechanisms have been constructed and analytical methods developed for investigating them. Things are different with the study of radiation from AW sources in two- and three-dimensional systems which is mainly based on numerical calculations. Therefore it seems useful to summarize in describing the associated qualitative effects. This problem was discussed in ample detail in a number of review papers and monographs (see Ivanitskii, Krinskii, Selkov 1978; Polak, Mikhailov 1983; Zykov 1984).

The natural and numerical experiments show that reverberators, which are two-dimensional wave sources, represent spiral structures stable with respect to external perturbations. An important feature of reverberators is that these AW sources have the highest frequency. By generating pulses at a maximum frequency such sources force out self-pacemakers of any other types from the medium. As a result, the medium is filled with only spiral waves. This fact is essential to autowave processes in many biological systems. The existence of spiral waves in living objects was mentioned back at the beginning of this century. However, a mathematical description of such waves in terms of axiomatic models was initiated by Wiener and Rosenbluth in 1946, half a century later. Balakhovsky (1965) simulated the first spiral wave in a continuous single-coupled medium. Krinsky (1965) studied qualitative regimes in inhomogeneous media, specifically the appearance and disappearance of reverberators due to inhomogeneities and the multiplication of reverberators. The existence of such sources in chemical reactions was first demonstrated by Zaikin and Zhabotinsky (1970, 1971). Their experiments provide support for the predicted mechanisms for the appearance of reverberators due to inhomogeneities and due to breaks of the wave front. Reverberators in three-dimensional space were obtained by Winfree (1973). Rotating waves were detected in cardiac muscle, cortex and other living objects (see Chapter 1).

The basic parameters of a reverberator are the frequency of generated waves, the critical dimension and characteristics of the structure of the kernel, and the velocity of the kernel in space. Commonly, there are two characteristic regions in a reverberator: a kernel having an excitation "tongue" entering it and spiral waves generated by the kernel. The kernel can have the form of a circle or a ring (a so-called "anomalous" kernel). If the source drifts, the form of the kernel is involved and can hardly be determined unambiguously. A stationary reverberator can be described mathematically by means of the Archimedian spiral $\varphi(\varrho) = \int\limits_{\varrho}^{\varphi} \sqrt{R^{-2} - r^{-2}}\, \mathrm{d}r$, where ϱ and φ are the polar coordinates. In a formal sense such a wave is a spiral which rotates about the "hole" in an active medium.

Typically, the emergence of a reverberator is associated with discontinuities in the TF. Such a discontinuity can originate from initial conditions or loss of excitation in a certain region of the medium, or escape of the excited region from an obstacle being enveloped, etc. The edge of a plane wave begins to twist, thus forming a closed path of excitation (Fig. 5.16). There are lines in the excitation region, which corre-

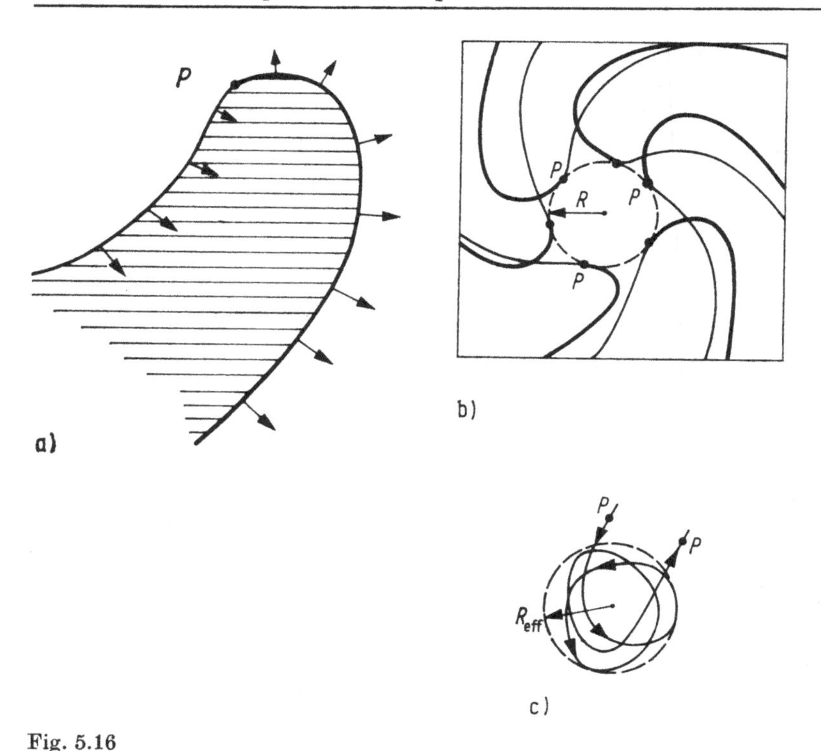

Fig. 5.16
Basic elements of a reverberator:
a) region of excited state, P is the point of transition of the leading edge (heavy line) into a trailing edge (thin line);
b) a periodically rotating reverberator, R is the radius as an estimate of the reverberator kernel;
c) the line of motion of point "P"-phase variation in a nonstationary reverberator, R_{eff} is the radius as an estimate of the kernel of such a reverberator (Zykov 1984)

spond to the leading edge (heavy line in Fig. 5.16a, b) and the trailing edge of the TP. The point in the plane at which the leading edge transforms into a trailing edge is generally referred to as the point of phase variation (point P in Fig. 5.16a, b). This point can either be at rest or move along a circle or move nonperiodically in the plane, for example in the form of a "rosette" depending on the parameters of the medium (Fig. 5.16b). Qualitatively, the possibility that such a rosette can exist was demonstrated by Zykov (1984) who analyzed the dependence of the wave propagation velocity on the wave period.

The calculations show that the size of the kernel and the period of rotation of a reverberator have a U-shaped dependence on the parameters of the medium (see Fig. 5.17). A qualitative picture of the reverberator properties can be obtained by analyzing the influence of the wave front curvature in a two-dimensional space on the TP propagation velocity. It is well known that the higher the curvature of the excitation front, the lower is its velocity. Thus, at a certain critical curvature $K = K_{cr}$

(or critical radius of curvature $r = r_{cr} = K_{cr}^{-1}$), the excitation propagation velocity will be equal to zero. If $K > K_{cr}$ the excitation front moves in the opposite direction. Typical dependence of the velocity of a stationary TP on K is shown in Fig. 5.18a. The physical meaning of such a phenomenon is that the degree of curvature of the

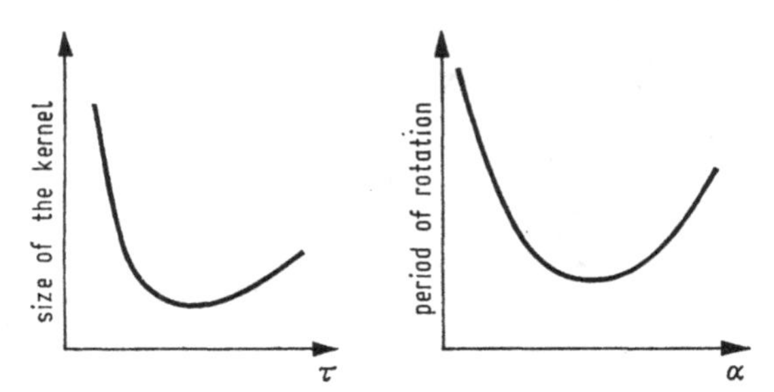

Fig. 5.17
Characteristic dependences of the size of the reverberator kernel and period of its rotation on the active medium parameters: τ is the TP duration in the medium; g_F is the slope of the incident part in the nonlinear function of the source (Zykov 1984)

front changes its ability to "ignite" or cause the space region at rest to transform into an excited state. Thus, the same diffusion flow will be transported to a larger distance with increased curvature of the front. Consequently, the excitation threshold of the medium will be exceeded at a smaller distance from the front and, therefore, the wave decelerates. The same process determines the critical size of the initial perturbation region. Such a region shown by a dashed line in Fig. 5.18b is an unstable structure of radius $R = r_{cr}$. If the initial radius $R_0 > r_{cr}$, then the perturbed region will increase in size, and the front velocity will increase monotonically with increasing

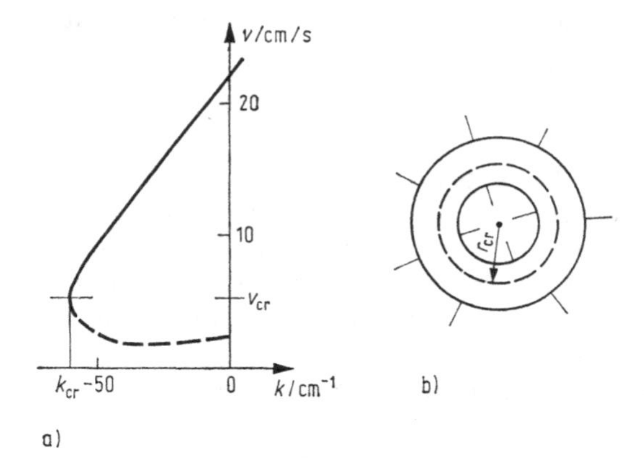

Fig. 5.18
Influence of the front curvature on the velocity of the excitation region;
a) of a stable ("1") and unstable ("2") curved TP with a front of constant curvature K by Zykov and Morozova (see Zykov, 1984);
b) directions of motion of the edge of the excitation region in a two-dimensional medium for the radius above and below the critical value

radius up to a value equal to the velocity of a plane wave. If $R_0 < r_{cr}$ the excited region will collapse.

We now explore the question of what is the influence the curvature of the excited region exerts on the propagation of the wave discontinuities. The associated processes recently considered by Panfilov and Pertsov (1981, 1983) are shown in Fig. 5.19. In the first case, the excitation wave envelopes an obstacle and does not escape from

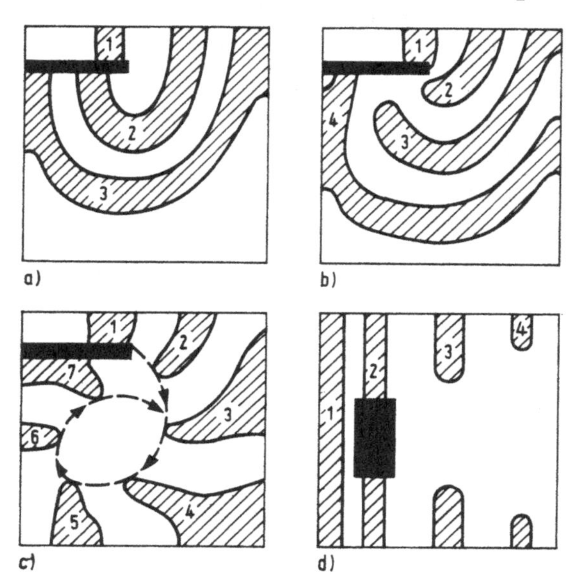

Fig. 5.19

Motions of the edge of the excitation region when enveloping an obstacle:

a) enveloping with escape;
b) enveloping with partial escape from the obstacle in a medium with decreased excitation;
c) formation of a reverberator when the excitation region escapes from the obstacle (Panfilov, Pertsov 1982);
d) increase of wave breaking in the case of strongly decreased excitation of the medium (Pertsov, Panfilov, Medvedev 1983). 1,2,3, etc. are successive positions of the excitation region.

it. Such behaviour can be well described in terms of an axiomatic theory (see Fig. 5.19a). More involved behaviour is observed if the wave escapes from the obstacle (Fig. 5.19b, c). As a result, a region forms behind the obstacle into which the excitation wave cannot enter at all. This regime is only possible in media with reduced excitation where the TP length decreases down to a value at which the curvature of the excitation front begins to influence the discontinuity. In the third case, the TP length decreases so that the edges of a discontinuity produced by the obstacle can only diverge (Fig. 5.19d). The second case is especially interesting. It is seen that conditions can be chosen such that the enveloping of the obstacle by the edge of the excited region (a sort of diffraction) will lead to the formation of a reverberator (Fig. 5.19b). Thus, the calculations of Panfilov and Pertsov demonstrate a new means of reverberator formation.

In a medium the properties of which do not change with time, a reverberator can be excited by only one plane wave. Study of such a reverberator was initiated earlier. It was shown that its kernel has the form of a ring. That is why it was called a reverberator with an "anomalous" kernel (Fig. 5.20a). Pertsov and Panfilov (1981) showed that there is a region inside the kernel which is in the state of rest although it borders the excited regions. The presence of such a region can be explained by the fact that the excitation front (in the form of a "tongue" as shown in Fig. 5.20a) in places of contact with this region is curved so that it cannot propagate towards the kernel.

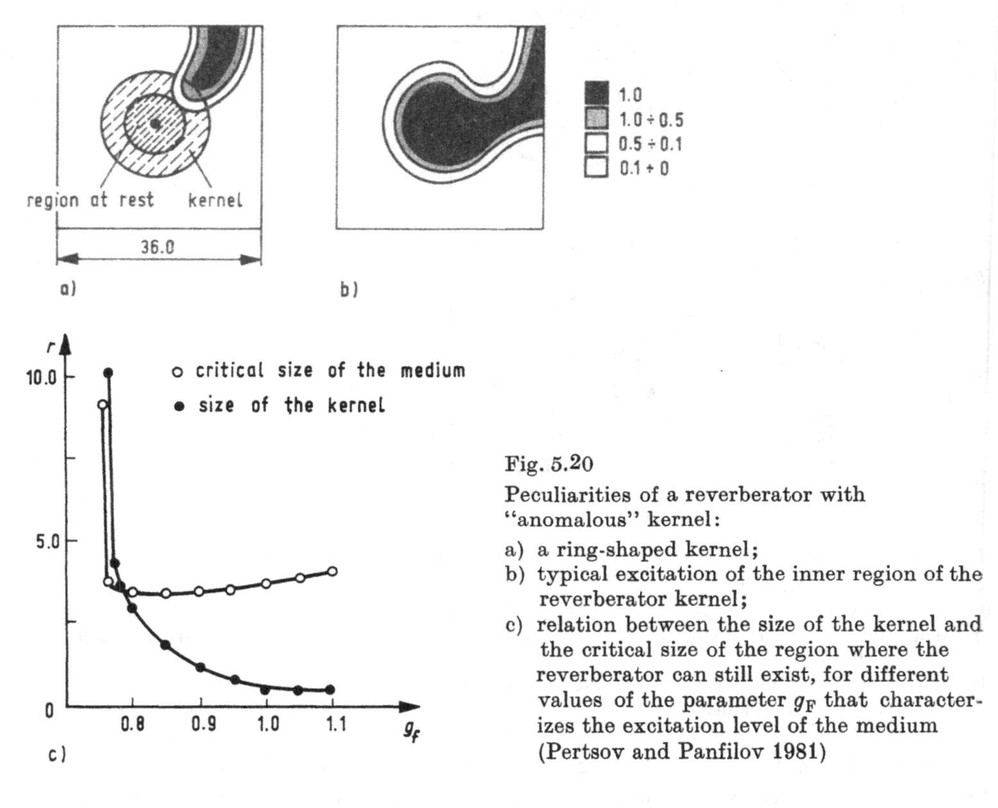

Fig. 5.20
Peculiarities of a reverberator with "anomalous" kernel:
a) a ring-shaped kernel;
b) typical excitation of the inner region of the reverberator kernel;
c) relation between the size of the kernel and the critical size of the region where the reverberator can still exist, for different values of the parameter g_F that characterizes the excitation level of the medium (Pertsov and Panfilov 1981)

The calculations of Pertsov and Panfilov suggest that the radius of the centre and the characteristic size of a limited active medium in which a reverberator can rotate are not always proportional to each other. If the characteristic size of the medium is less than a certain critical value, the reverberator will first deflect from a circular trajectory and then move to the boundary of the medium and break up. Figure 5.20c shows the dependence of the size of the kernel and critical size of a region where the reverberator can exist on the coefficient g_F, which characterizes the slope of the incident part of the function $F(x, y)$ (see basic model (2.53)). It is seen that these characteristics differ strongly. Thus, the critical size of a reverberator should not be estimated directly from the radius of its centre, such estimates appear occasionally in the literature and should not be taken literally. The influence of the boundaries of the medium is less for a smaller TP leading edge length. The order of magnitude of this length coincides with that of the size of the kernel.

In recent experiments with the Belousov-Zhabotinsky chemical reaction, it was demonstrated that other types of reverberators consisting of several spirals (so-called "multiarm reverberators") can exist (Agladze, Krinskii 1982). Such structures are also referred to as reverberators with topological charges $N > 1$ (see Fig. 5.21).[1]

Fig. 5.21
Multispiral reverberators in a chemically active medium:
a) a reverberator consisting of two spirals; the interval between frames is 4 min;
b) a reverberator having three spirals (Krinskii, Agladze, 1982)

The possibility that such structures can exist was assumed in theoretical papers (see, e.g., Koga 1980). The experimental evidence of Krinskii and Agladze is the first proof of the existence of multiarm reverberators. The theoretical predictions were exceeded. It was shown that usually multispiral reverberators exist while a chemical reaction proceeds (for more than 30 min). This fact attests to the stability of such structures with respect to perturbations, which are usually present in the reactions.

[1]) For topological approaches to studying complex reverberators in two- and three-dimensional media see Winfree 1980; Gurija, Livshits 1983.

Rare cases of the breaking of the AW sources are evidently caused by a certain stability threshold of reverberators consisting of several spirals being exceeded. A two-spiral source breaks most easily. Such a source changes, after a certain time, into two sources with one spiral. The beginning of the process is shown in Fig. 5.21 a. Cases of breaking of three-arm reverberators were much more rare. Breaking of a four-spiral source was not observed at all. As a three-arm source rotates, the spirals are locked and unlocked alternately (see Fig. 5.21 b). The causes of such an interaction between spirals, as well as mechanisms responsible for the threshold stability of multielement nonlinear structures have not yet been explained.

Chapter 6
The synchronization of auto-oscillations in space as a self-organization factor

Mutual synchronization (self-synchronization) is a remarkable, powerful factor of self-organization in distributed auto-oscillatory systems or in networks of discrete auto-oscillatory systems of collective activity, which are most diverse in nature. By synchronization we mean a spontaneous adjustment to a unique synchronous frequency and the establishment of certain phase relations between oscillations, which are stable with respect to perturbations, in some parts of a distributed system or in partially discrete auto-oscillators.

A tendency for mutual synchronization is the opposite of a tendency for chaos. Depending on internal relations, a tendency for self-organization prevails or, on the contrary, quasi-stochastic regimes are established in the same complex system.

The problem of synchronization has been considered in many monographs and review articles (e.g. Romanovskii, Stepanova, Chernavskii 1975, 1984; Blekhman 1981; Landa 1980). In this book, we shall consider synchronization as a condition vital to the behaviour of autowave systems. Primary attention will be paid to those processes which can be described by the basic model (1.1) and its discrete analogues. It is important to note that synchronization occurs not only in spatially homogeneous but also in inhomogeneous systems. A two-component model for an inhomogeneous system can be written

$$\begin{aligned} \frac{\partial x}{\partial t} &= P(x, y, r) - D_x \frac{\partial^2 x}{\partial r^2}, \\ \frac{\partial y}{\partial t} &= Q(x, y, r) + D_y \frac{\partial^2 y}{\partial r^2}. \end{aligned} \tag{6.1}$$

If the "reactor" is divided into N sections, then the discrete analogues of such a system are given by Eqs. (1.3), where $F_1 = P$ and $F_2 = Q$ depend on the parameter r.

The properties of an autowave system are determined by those of the corresponding local system. A detailed analysis of such models is given in monographs mentioned above. We thus restrict ourselves to some problems of synchronization theory including the important problem of the stochastization of a set of coupled oscillators. The examples given below refer to the problem of peristaltic motion considered in Chapter 9. Also, some problems that arise in designing synchronous networks of oscillators in electronics are briefly discussed.

6.1 Synchronization in homogeneous systems

Let the functions P and Q in a model (6.1) be independent of the coordinate r. The distributed system can then be represented as a continuum of identical partially diffusion-coupled self-excited oscillatory systems.[1])

At a first glance, it seems that synchronous self-excited oscillations are always established in such homogeneous systems (of course, if the internal and external noises are neglected). Nevertheless this is not necessarily true. In two-component active media with weak coupling, in-phase auto-oscillations with a unique synchronous frequency are indeed the only stable processes, while in three-component media several stable processes are possible (see Section 4, Chapter 5).

It should be emphasized that the assertion on unique steady operation in second-order systems is nontrivial. The point is that in a number of papers (Polyakova, Romanovskii, Sidorova 1968; Romanovskii, Stepanova, Chernavskii 1975) it is shown that apart from in-phase auto-oscillations, there are solutions of the form

$$y(r, t) = A(r) \cos [\omega t + \varphi(r)], \tag{6.2}$$

where the amplitude and phase depend on the coordinate r. However, all these solutions, except for $A = \text{const}$ and $\varphi = \text{const}$, are unstable. As an illustration let us consider a nontrivial autowave system constructed of a localy, almost harmonic oscillator with hard excitation (Polyakova, Romanovskii 1971; Romanovskii, Stepanova, Chernavskii 1975). In this case, the P and Q functions in a model (6.1) take

---

[1]) The meaning of the coupling coefficients d_x and d_y in electronics can be illustrated by classical Van-der-Pol oscillators. Let these coefficients be not too small, i.e., rigorous inequalities $d_x \ll \delta$ and $d_y \ll \delta$ are not satisfied (but $\omega_i \gg \delta$!). The circuits of the i-th and $(i + 1)$-th oscillators will then be coupled via the common R_c and C_c (see Fig. 6.1). The coupling capacitance, $C_c = (2d_x d_y \mathcal{L})^{-1} - (2d_x\delta)^{-1}$, can be either positive or negative, depending on the increment δ. For weak coupling $R_c = (d_x + d_y)\, 2\mathcal{L} \gg (C_c \nu_i)^{-1}$, i.e., the coupling is purely galvanic.

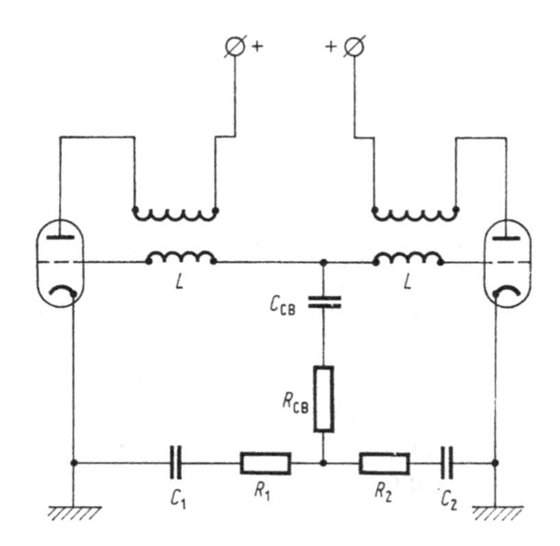

Fig. 6.1
Two Van-der-Pol generators with diffusion coupling

the form

$$P(x, y) = y,$$
$$Q(x, y) = -\omega_0{}^2 x - 2(\delta_0 - \delta_2 x^2 + \delta_4 x^4)\, y. \tag{6.3}$$

Let us investigate the behaviour of the discrete analogue of such a system. First, two simplest regimes are possible: 1. $A_1 = 0$: All the oscillators are not excited and are in stable equilibrium. 2. $A_2 = \text{const} > 0$: All the oscillators are excited and have amplitudes equal to the stable limit cycle of a point oscillator. Second, it is possible to excite only some of the oscillators, for instance those located in the left-hand part of the chain. Stimulated oscillations near a stable equilibrium state will then arise in the nonexcited oscillators in the right-hand part of the chain. For weak coupling the amplitude of stimulated oscillations is smaller than the amplitude of the unstable limiting cycle. The amplitude distribution along the chain $A_3(r)$ will finally become a step function, stable with respect to small perturbations. In accordance with (3.6) and for constant diffusion coefficients the coupling coefficients d_x and d_y as well as the amplitudes of stimulated oscillations should increase with an increasing number N of oscillators in the chain. It is expected that in transition from a discrete chain to a continuous analogue, the boundary between excited and non-excited oscillators will eventually disappear. In other words, the step distribution $A_3(r)$ will become unstable. Any perturbation imposed will cause the step to shift to the right or to the left.

Polyakova and Romanovsky obtained the distributions $A(r)$ directly for models in partial derivatives. For stationary amplitude and phase the truncated second-order equations have the form

$$\frac{\mathrm{d}^2 A}{\mathrm{d}r^2} + A\left(\frac{\mathrm{d}\varphi}{\mathrm{d}r}\right)^2 = \varepsilon f(A) + \dots,$$
$$\frac{\mathrm{d}}{\mathrm{d}r}\left(A^2 \frac{\mathrm{d}\varphi}{\mathrm{d}r}\right) = \varepsilon^2 A^2 \Phi(A) + \dots \tag{6.4}$$

Here $\varepsilon \sim \delta/(D_x + D_y)$ and the boundary conditions, assuming impermeable boundaries, are given by

$$\left.\frac{\partial \varphi}{\partial r}\right|_{r=0} = \left.\frac{\partial \varphi}{\partial r}\right|_{r=L} = \left.\frac{\partial A}{\partial r}\right|_{r=0} = \left.\frac{\partial A}{\partial r}\right|_{r=L} = 0. \tag{6.5}$$

To a first approximation, $\varphi = \text{const}$ and the amplitude equation is simple. In the hard excitation case (see (6.3)) we have

$$\frac{\mathrm{d}^2 A}{\mathrm{d}r^2} = \frac{2}{D_x + D_y} A\left(\delta_0 - \frac{1}{4}\,\delta_2 A^2 + \frac{1}{8}\,\delta_4 A^4\right). \tag{6.6}$$

A qualitative analysis of the solutions of this equation (which, in principle, can be expressed in terms of elliptical integrals) and a computer simulation show that all these solutions are unstable, except for trivial ones: $A_1 \equiv 0$ and $A_2 \equiv \text{const}$. By analyzing the second-order equations we can obtain solutions $A_3(r)$ nearly in the form of steps (see Fig. 6.2) but they are also unstable. Mathematically, this problem was considered by Dvoryaninov (1980), Butuzov and Vasileva (1983).

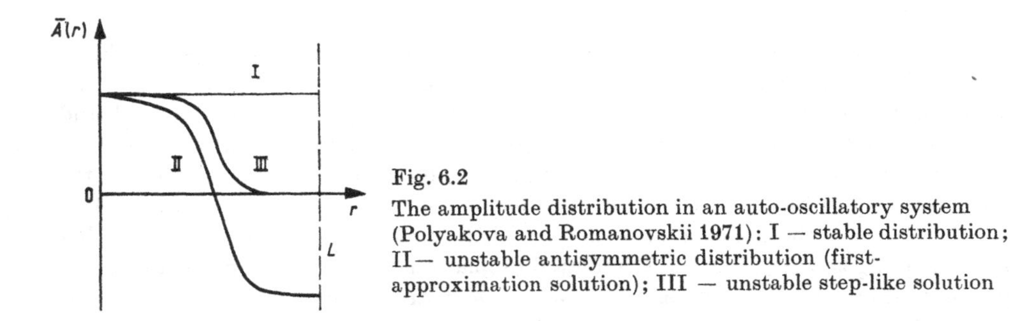

Fig. 6.2
The amplitude distribution in an auto-oscillatory system (Polyakova and Romanovskii 1971): I — stable distribution; II — unstable antisymmetric distribution (first-approximation solution); III — unstable step-like solution

6.2 Synchronization in inhomogeneous systems. The equidistant detuning case

Any real autowave system and its discrete analogues, which are networks of coupled oscillators, have parameter scattering leading to the appearance of different partial frequencies of self-excited oscillations. In such a system, there are sources of internal (natural) and external noise. The nature of such noise occurring in kinetic systems is considered in detail in Chapter 8. Noise arising in networks of electronic oscillators is also studied in ample detail. If the coupling between oscillators is weak, then the synchronous operation is violated by the inhomogeneity of the system and by the noise. Alternatively, the synchronous regime is the more stable, the stronger the coupling between oscillators and the larger the dimension of the oscillator network. To be more precise, the synchronous frequency fluctuations decrease with increasing coupling, while the synchronization band broadens.

If the partial auto-oscillatory systems are quasi-harmonic, $A_1 = \text{const}$, and the coupling coefficients, d_x and d_y, are small, then the synchronous frequency ω_s, synchronization band Δ_s and stationary phase differences $\theta_i = \varphi_{i+1} - \varphi_i$ in a chain composed of N oscillators are determined from the formulae (Malafeev, Polyakova, Romanovskii, 1970; Romanovskii, Stepanova and Chernavskii 1975, 1984):

$$\omega_s = \frac{1}{n} \sum_{i=1}^{N} \omega_i^2, \qquad \Delta_s = (d_x + d_y)\, f_1(\omega_i),$$

$$\sin\theta_i = \frac{1}{d_x + d_y} f_2(\omega_i - \omega_s) \tag{6.7}$$

$$(i = 1, 2, \ldots, N-1).$$

Here ω_i are the frequencies of self-excited oscillations in partial systems, and the functions $f_1(\omega_i)$ and $f_2(\omega_i)$ are determined from the distribution of these frequencies.

As the increment of partial auto-oscillatory systems increases, the oscillations assume a relaxation form, and the coupling coefficients, d_x and d_y, are no longer equally important. Let the degree of relaxation be characterized by a parameter ε_r. The expression for the synchronization band can then be written

$$\Delta_s \simeq (d_x/\varepsilon_r + \varepsilon_r d_y) f_1(\omega_i). \tag{6.8}$$

Here d_x determines the coupling in the "slow" variable that has no disruptions, and d_y, the coupling in the "fast" variable that undergoes a disruption. From Eq. (6.8) it follows that the synchronization band Δ_s broadens by a factor of ε_r^{-1} for coupling in the slow variable and, on the contrary, narrows for coupling in the fast variable only. In the next paragraph, we shall consider a situation in which a desynchronization of self-oscillations in space occurs in a relaxation-type autowave system for $d_x \equiv 0$ and $d_y \neq 0$ (with $\Delta_s \to 0$, if $\varepsilon_r \ll 1$).

Typical synchronization in networks of diffusion-coupled auto-oscillators is considered in monographs mentioned above. In this paper, we shall deal only with the important case of equidistant distribution of partial frequencies in the chain of auto-oscillators. Assume for simplicity that the amplitudes of all the oscillators are equal and that the partial frequencies are given in the form

$$\begin{aligned} &\omega_i = \omega_0 + \Delta(i - 1) \\ &(i = 1, 2, \ldots, N), \end{aligned} \tag{6.9}$$

where ω_0 is the frequency of the first oscillator and Δ is the frequency difference between two neighbouring oscillators. We also assume that the oscillators are Van-der-Pol systems with soft excitation. This means that the P and Q functions in (6.1) are given by

$$P = y_i, \qquad Q = 2\delta(1 - \delta_2 y_i^2)\, y_i - \omega_i^2 x_i. \tag{6.10}$$

As is shown in the book by Romanovskii, Stepanova and Chernavskii 1975, such a form of P and Q does not restrict significantly the generality of the results.

We try a solution in the form

$$\begin{aligned} x_i &= A_i(t) \cos [\omega_s t + \varphi_i(t)], \\ y_i &= -\omega_s A_i(t) \sin [\omega_s t + \varphi_i(t)], \end{aligned} \tag{6.11}$$

where $A_i(t)$ and $\varphi_i(t)$ are slowly varying functions of time. At $A_i =$ const for the phase difference θ_i and φ_1 the following set of truncated equations is obtained:

$$\begin{aligned} \frac{d\varphi_1}{dt} &= \frac{1}{2} (d_x + d_y) \sin \theta_1 + (\omega_s - \omega_0), \\ \frac{d\theta_1}{dt} &= \frac{1}{2} (d_x + d_y) [-2 \sin \theta_1 + \sin \theta_2] + \Delta, \\ \frac{d\theta_k}{dt} &= \frac{1}{2} (d_x + d_y) [\sin \theta_{k-1} - 2 \sin \theta_k + \sin \theta_{k+1}] + \Delta, \\ &\cdots\cdots\cdots\cdots\cdots\cdots \\ \frac{d\theta_{N-1}}{dt} &= \frac{1}{2} (d_x + d_y) [\sin \theta_{N-2} - 2 \sin \theta_{N-1}] + \Delta. \end{aligned} \tag{6.12}$$

We then find the stationary phase differences $\bar{\Theta}_i$, synchronous frequency ω_s and synchronization band Δ_s, i.e., we specify expression (6.7) for the equidistant frequencies case. Putting $d\theta_i/dt \equiv 0$ we have:

$$\left\|\begin{matrix} -2 & 1 & 0\ldots0 & \ldots\ldots & 0\ldots0 \\ \cdots & & & & \\ 0\ldots0 & +1 & -2 & +1 & 0\ldots0 \\ \cdots & & & & \\ 0 & \ldots & & +1 \;.\;. & -2 \end{matrix}\right\| \left\|\begin{matrix} \sin\bar{\theta}_1 \\ \cdots \\ \sin\bar{\theta}_k \\ \cdots \\ \sin\bar{\theta}_{N-1} \end{matrix}\right\| = -\frac{2\Delta}{d_x + d_y}. \tag{6.13}$$

Thus,

$$\sin\bar{\theta}_k = -\frac{\Delta}{d_x + d_y}(k^2 - Nk) \tag{6.14}$$

$$(k = 1, 2, \ldots, N - 1).$$

The maximum phase shift is observed at the centre of the chain at $k = N/2$. The phase shift between extreme oscillators at both ends of the chain is the same as that between two diffusion-coupled oscillators. The phase wave is supposed to run along the chain once a period, so that the maximum values of x_i will be achieved alternately. The wave will be faster at the edges and slower at the centre of the chain.

The synchronization band Δ_s, or the maximum detuning at which the phase difference distribution is stable, can be found from the condition $|\sin\bar{\theta}_k| = 1$. Thus,

$$\Delta_s = \max_{k\in[1,N-1]}\{\Delta_k\} = \max_{k\in[1,N-1]}\left\{\frac{d_x + d_y}{k^2 - Nk}\right\} = \frac{4(d_x + d_y)}{N^2}. \tag{6.15}$$

Note that if $N = 2$ (two coupled oscillators), then $\Delta_s \sim d_x + d_y$. The synchronization band Δ_s is the smaller the larger N is in the chain. The middle oscillators are the "weakest" units. The synchronous operation will be violated by fluctuations at the centre of the chain. Probably, the chain will be divided into two synchronous subclusters.

The synchronous frequency is given by

$$\omega_s = \frac{1}{\sqrt{N}}\left[\sqrt{\sum_{k=0}^{N-1}(\omega_0 + k\Delta)^2}\right]. \tag{6.16}$$

At $\Delta \cdot N \ll \omega_0$

$$\omega_s \simeq \omega_0\left[1 + \frac{\Delta}{2\omega_0}(N - 1)\right], \tag{6.17}$$

i.e., ω_s is equal to the arithmetic mean of the partial frequencies.

In deriving relations (6.15) to (6.17), which determine the synchronization region, two basic assumptions were made:

1) The amplitude of the oscillations is constant throughout the chain.
2) The nonlinearity of the system is small (quasiharmonic oscillations are assumed).

Thus, the relations here obtained have restricted accuracy. Numerical experiments were performed to check them for applicability in different situations. Equations

(6.1) were computed by the finite difference method. Before we report our results we recall that an autowave system with a smooth eigenfrequency gradient is a model for the so-called slow-wave activity of peristaltic organs. Phenomenological aspects of the autowave processes here arising will be discussed in Chapter 9. We emphasize that our interest is in the onset of nonsynchronous regimes. We shall first consider these regimes and then discuss the applicability of Eqs. (6.15)—(6.17).

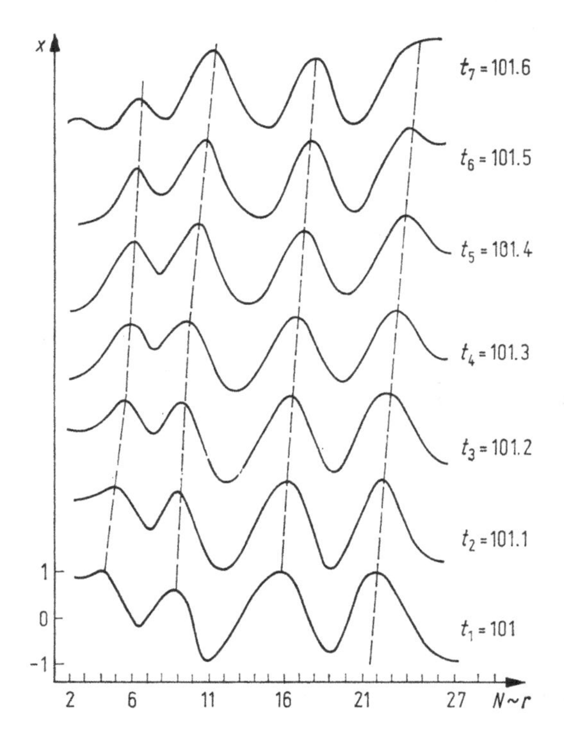

Fig. 6.3
Distribution profiles of the x variable at successive times $(\Delta t \ll T = \frac{2\pi}{\omega_{\max}}$, $\omega(r) = 3.14 + 3.14(1 - r)$; $\delta = 3$; $\delta_2 = 1$; $D_{xx} = D_{yy} = 0.001$

The distribution profiles of the variable x at successive times are shown in Fig. 6.3. The transient processes can be considered to be complete by the instant t_1. It is seen that running waves propagate in the system, with a distribution directed opposite to the eigenfrequency gradient of the local oscillators. The amplitude distribution is approximately homogeneous. The wave lengths increase in the region of smaller frequencies. It should be readily apparent that the process under investigation is a nonsynchronous one (note that $\omega_0 - \omega_N < \Delta_s$). However, the desynchronization regime is more clearly demonstrated by the spectral processing.

The amplitude distributions of the harmonic components in the energy spectra of oscillations along the system are given in Fig. 6.4 (Fig. 6.4a corresponds to Fig. 6.3). Pronounced clustering is the main feature of these distributions. Oscillations occur only at some frequencies characteristic of a particular cluster, not at every frequency that is possible in the system. The sum spectrum, which is continuous in the absence of coupling ($d_x = d_y = 0$), converts into a discrete one. As the coupling coefficients

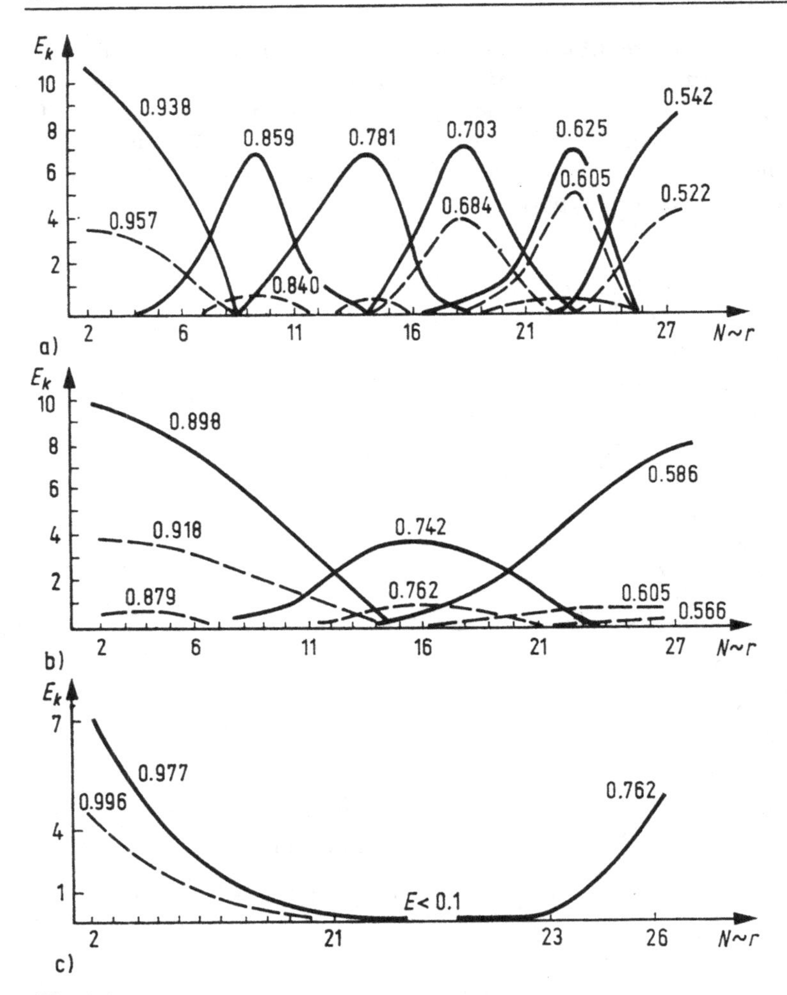

Fig. 6.4
The amplitude distribution of the harmonic components of oscillation spectra with a constant eigenfrequency gradient. Observation time $\approx 100 \frac{2\pi}{\omega_{\max}}$;
a) $\delta = 3$, $\delta_2 = 1$, $\omega(r) = 3.14 + 3.14(1 - r)$, $D_{xx} = D_{yy} = 0.001$;
b) same except for $D_{xx} = D_{yy} = 0.01$;
c) $\delta = 0.3$; $\delta_2 = 1$, $\omega(r) = 3.14 + \frac{3.14}{2}(1 - r)$, $D_{xx} = D_{yy} = 0.001$.
The digits in the ovals denote a cyclic harmonic frequency.

increase, the number of clusters decreases (Fig. 6.4a, b) tending to unity (synchronization region).

It is important to note that the behaviour of the system depends appreciably from the degree of relaxation (see Fig. 6.4a, b). In the weakly nonlinear case (Fig. 6.4c), regions in which oscillations are practically absent are formed. Note that this phenomenon resembles an asynchronous suppression of generation, which is known from the theory of oscillations (Strelkov 1972). In the strongly nonlinear case such phenomena take place at much higher eigenfrequency gradients. In any case, when such effects are present, the use of the "constant amplitude" approximation leads to qualitative errors. Specifically, at any given δ and δ_2 there are critical values of the difference $\omega_0 - \omega_N$ at which synchronous regimes are impossible. This last fact was ignored in the theory considered above.

Note also that as the nonlinearity of the system increases, the synchronization region increases compared to the value calculated from (6.15). A more accurate value can be obtained from (6.8). However, ω_s determined from (6.17) is in good agreement with the numerical experiment. Moreover, ω_s for clusters can also be determined from (6.17), where $\Delta = \Delta_k = [\omega_0 - \omega_N]/k$ (k is the number of clusters). The number of clusters can also be estimated from the relation $k \simeq [\omega_0 - \omega_N]/\Delta_s$ (6.15).

Thus, the synchronization theory complemented with a numerical experiment explains the basic features of autowave processes in the inhomogeneous systems. When analyzing such systems, one should first construct the synchronization regions and the characteristics of point systems driven by a periodic external force. A combination of these characteristics will permit one to qualitatively correctly predict the behaviour of autowave systems.

6.3 Complex autowave regimes arising when synchronization is violated

In this section, we shall consider regimes close to randomness. Such a behaviour of distributed systems has not been studied sufficiently so far. Therefore we shall dwell upon some of the aspects of the problem. First, we shall demonstrate that a certain structure of the phase distribution of auto-oscillations can arise in a distributed auto-oscillatory system with the coupling in the fast variable. Second, we shall consider one of the possible stochastization mechanisms of auto-oscillations of two coupled auto-oscillatory systems. Complex regimes also arise in three-component systems. This problem of "moving" leading centers (LCs) was briefly discussed in Chapter 5.

1. *Consideration of the desynchronization of auto-oscillations in a homogeneous system (Yakhno, Goltsova, Zhislin 1976)*

The analysis will be made for a relaxation-type system (at $\varepsilon \ll 1$):

$$\frac{\partial}{\partial t} x = y + \gamma x - \beta x^3, \tag{6.18}$$

$$\varepsilon \frac{\partial}{\partial t} y = -x + y - y^3 + \frac{\partial^2}{\partial r^2} y \tag{6.19}$$

assuming that its length $2L$ satisfies the conditions $l_{\mathrm{fr}} < 2L \ll v_0 T_0$, where l_{fr} is the characteristic length of the excitation front, T_0 is the period of natural auto-oscillations, v_0 is the characteristic velocity of the excitation front in the system. The last inequality means that the time of travel of an excitation along the system is negligible compared with the time of subsequent slow motions. In this case, slow motions will prevail, which can be described by ordinary differential equations (see (3.7)). The parameters will be chosen such that the system persists in the auto-oscillatory regime.

Let a homogeneous spatial distribution of the variables x and y be subject to a perturbation at the initial moment of time. We shall keep watch on the change of the variable x at two points spaced far enough from each other. We designate these points by c and d. Let initially $x_c = x_{\max}$ and $x_d = x_{\max} - \Delta x_0$ (Fig. 6.5), where Δx_0

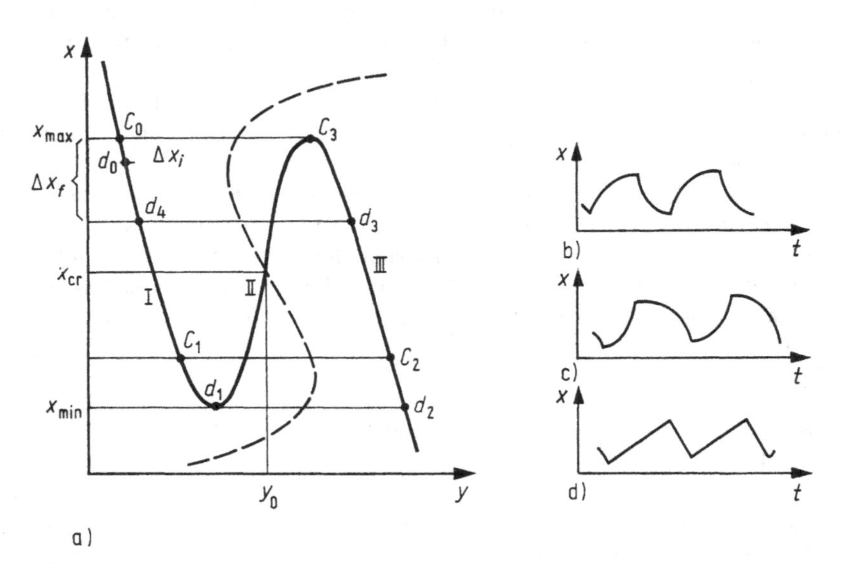

Fig. 6.5

a) The phase plane of the system under study. The isocline $P(x, y) = 0$ is shown by a solid line, and $Q(x, y) = 0$ by a dashed line. Time variation of x in different cases of stability of a homogeneous distribution of variables:

b) a system, which is stable with respect to space perturbations;

c) an unstable system;

d) a system where the perturbation does not change with time

is the amplitude of the initial inhomogeneity. Depending on whether the values of the variable x at points c and d converge or diverge with the passage of time, the homogeneous phase distribution of auto-oscillations will be stable or unstable. First, the points will move on branch I of a nonlinear function $P(x, y) = 0$ (see Fig. 6.5). Over a time

$$t = \int\limits_{x_{\max}-\Delta x_0}^{x_{\min}} \frac{\mathrm{d}x}{Q_1(x)} = \int\limits_{x_{\max}}^{x_{\mathrm{med}}} \frac{\mathrm{d}x}{Q_1(x)}, \tag{6.20}$$

where $Q_1(x) = y_{\mathrm{I}}(x) + \gamma x - \beta x^3$ is the rate of change of the slow variable on branch I, achieves a minimum value, $x_{\min}$, at point d and an intermediate value, x_{med}, at point c. The values of y on branch I will transfer to the corresponding values on branch III.

The variable x will increase with time; therefore point c will be the first to achieve a state with $x = x_{\max}$. Then the values of y will return from branch III to branch I. The time the points reside on branch III is given by

$$t_2 = \int_{x_{\mathrm{med}}}^{x_{\max}} \frac{\mathrm{d}x}{Q_3(x)} = \int_{x_{\min}}^{x_{\max}-\Delta x_k} \frac{\mathrm{d}x}{Q_3(x)}, \tag{6.21}$$

where $Q_3(x) = y_{\mathrm{III}}(x) + \gamma x - \beta x^3$ is the rate of change of the slow variable on branch III, $\Delta x_k = x_c - x_d$ is equal to the difference between the values of x at points c and d after one oscillation period at point c. The stability criterion can be determined by a parameter

$$\delta = \Delta x_k / \Delta x_0 . \tag{6.22}$$

Then for $\delta < 1$, the homogeneous distribution of variables will be stable since the desynchronization decreases for a time equal to the oscillation period. But if $\delta > 1$, the perturbations will increase and the homogeneous phase distribution will be violated.

In the case of small initial perturbations, $\Delta x_0 \sim \Delta x_k \ll (x_{\max} - x_{\min})$, the value δ will not depend on Δx_0 and can therefore be represented in a simple form.

Small perturbations Δx_k can be found from Eq. (6.21): $\Delta x_k = Q_3(x_{\max}) \cdot \Delta x_{\mathrm{med}} / Q_3(x_{\min})$, and Δx_{med}, from Eq. (6.20): $\Delta x_{\mathrm{med}} = Q_1(x_{\min}) \cdot \Delta x_0 / Q_1(x_{\max})$. Thus, the parameter δ takes the form

$$\delta = \frac{\Delta x_k}{\Delta x_0} = \frac{Q_1(x_{\min}) \cdot Q_3(x_{\max})}{Q_1(x_{\max}) \cdot Q_3(x_{\min})}. \tag{6.23}$$

It is seen that the desynchronization in a distributed system depends on the relation between the velocities of slow motions on the stable branches of the N-shaped characteristic.

The simplest cases of the variation of x with time for stable and unstable homogeneous phase distributions are shown in Figs. 6.5 b and 6.5 c, respectively. The change of $x(t)$ shown in Fig. 6.5 d corresponds to the intermediate case. Some features of such a behaviour will be reported at length below.

Using the formulae (6.20) to (6.23) obtained here one can only differentiate between the cases of stable ($\delta < 1$) and unstable ($\delta > 1$) homogeneous distributions. Further evolution of the system should be studied by numerical methods. Calculations were performed for Eqs. (6.18)—(6.19). The velocities of slow motions in different parts of the stable branches of the N-shaped characteristic $P(x, y) = 0$ were chosen with the aid of a special S-shaped load characteristic $Q(x, y) = 0$ (see Figs. 6.5 and 6.7). This means that this system possesses active autocatalytic properties both for the fast and the slow variable.

A computer experiment has confirmed that for $\delta < 1$ the initial phase perturbation decreases as oscillations proceed and that the variables x and y tend to a homogeneous phase distribution of oscillations in space (Fig. 6.7 a). More complicated behaviour was exhibited for $\delta > 1$ where a smooth perturbation transforms to a strongly inhomogeneous distribution which can be called a "cellular" structure (see Figs. 6.6 and 6.7 b, c).

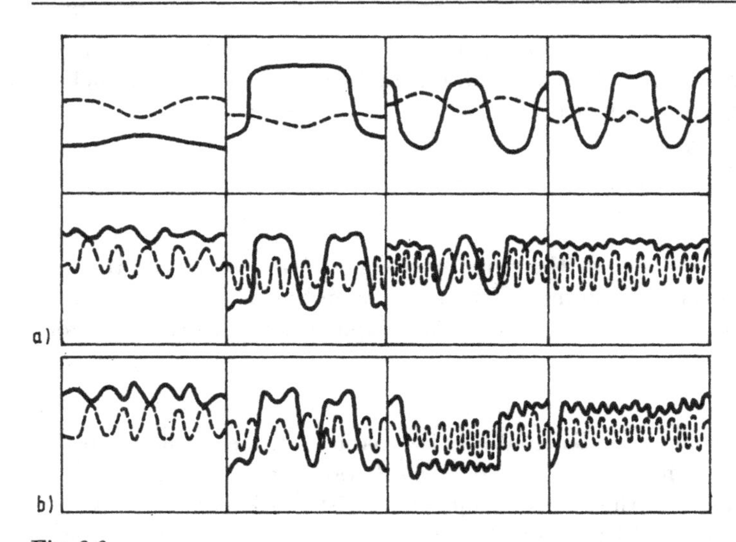

Fig. 6.6
Evolution of the initial perturbation (x is a solid line, and y is a dashed line) in system (6.18) to (6.19) with unstable homogeneous distribution. The initial conditions in b) are obtained from the initial conditions in a) by a shift of $0.01L$ in space. Beginning at $t = 6$, the solutions are very different. The initial conditions in case a) are $x_0(r) = -0.4$; $y_0(r) = 0.9 - 0.3/(1 + 0.025r^2)$. The eigenperiod in a point model $T_0 \simeq 2.8$. The parameters are: $\gamma = 3.5$; $\beta = 2.7$; $L = 20$; $\varepsilon = 0.01$ (Yakhno, Goltsova and Zhislin 1976).

It is important to note that the transient processes of division into "cells" strongly depend on the initial conditions (Fig. 6.6). Test calculations have shown that the spatial distribution profile changes appreciably in several oscillation periods. This means that small fluctuations in the form of the nonlinear functions or other parameters will produce a strong effect on the behaviour of the system and that conventionally, the transient process can be called a quasistochastic motion.

If the counting time is large enough, then two cases of final phase patterns are possible.[1]) In the first case, an inhomogeneous cellular distribution of the slow variable $x(r)$ arises, which, as a whole performs minor oscillations. The variable y performs synchronous auto-oscillations. The values of y in space represent a slightly wavy line, in accord with the arrangement of the cells in y. Such a regime can be explained as follows. From Eqs. (6.20) and (6.21) it follows that the parameter δ is a function $\delta = \delta(\Delta x_0)$. If the perturbation amplitude Δx_0 is small, then $\delta > 1$. However, at a finite amplitude Δx_{0cr}, the value δ tends to unity; this corresponds to auto-oscillations with a settled amplitude of inhomogeneities in the slow variable.

In the second case, the auto-oscillations are violated and the system transforms into an inhomogeneous stationary distribution with cellular structure in both variables. These oscillations illustrate the formation of contrasting dissipative structures (Fig. 6.7c). This means that homogeneous auto-oscillations are violated by the instability

[1]) The computing experiment was made by L. P. Chernysheva.

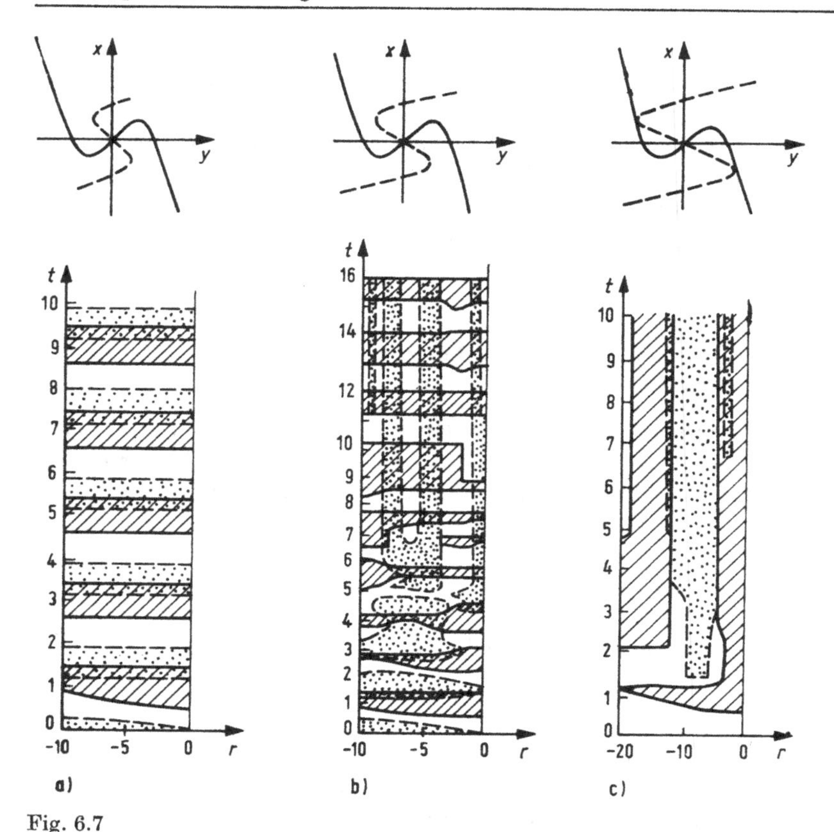

Fig. 6.7
Transient processes in a homogeneous auto-oscillatory system:
a) the phasing of oscillations in the fast and slow variables ($\gamma = 1.5$, $\beta = 2.7$);
b) the phasing of oscillations in the fast variable and the formation of a "cellular" distribution in the slow variable ($\gamma = 2$, $\beta = 2.7$);
c) the formation of a stationary distribution ($\gamma = 3$, $\beta = 2.7$). The null isoclines for corresponding parameters are shown at the top.

and a contrasting dissipative structure, which does not change with time, is formed (see Chapter 7).

The space-time evolution of cellular structures is shown in the plane r, t in Fig. 6.7. These figures were constructed as follows. Threshold values $x_{\text{th}} = x_{\text{cr}}$ and $y_{\text{th}} = y(y_{\text{cr}})$ were chosen. Regions with $y < y_{\text{th}}$ are cross-hatched. The excitation fronts for the fast variable are shown by a heavy line and $x_{\text{th}} = x_{\text{cr}}$ by a dashed line. Regions where the slow variable exceeds a threshold value are shown by dots. Light regions in the plane r, t correspond to the values of x and y below the threshold.

Thus far we have considered a system with a very fast travelling front (TF) ($v \sim \varepsilon^{-1}$, $\varepsilon \ll 1$). If the characteristic times of the variables x and y are comparable, then the spatial phase inhomogeneity of auto-oscillations is smoothed off. This agrees well with the conclusions derived from formula (6.8) for the synchronization band with the coupling in the fast variable. Note once again that the "complexity" of the desynchronization regimes considered here strongly depends on the form of the initial perturbation. It is important to take into account the relation between the observation time and the time of the transient process. If the observation time is less than the time of the transient process, then the latter can be thought of as a quasi-stochastic motion. It is interesting, however, to ascertain whether stochastic oscillations are really possible in such systems. In distributed relaxation-type systems, such a consid eration faces many mathematical difficulties. We shall therefore consider a simpler case.

2. *Stochastic regimes in a set of two relaxation-type auto-oscillatory systems with diffusion coupling*

In this chapter, we report the results of Sbitnev (1979), Pikovsky and Sbitnev (1981) on the stochastization of the oscillations of two coupled oscillators as their eigen-frequency difference increases. These authors used the following discrete model:

$$\frac{\mathrm{d}x_j}{\mathrm{d}t} = \varepsilon(-x_j + \tanh(y_{0j} + p_j \cdot y_j)), \tag{6.24}$$

$$\frac{\mathrm{d}y_j}{\mathrm{d}t} = -y_j + \tanh(x_{0j} + h_j y_j - b_j x_j) + d_j(y_k - y_j) \tag{6.25}$$

$$(j = 1, 2;\ k = 3 - j) \qquad (d_y \equiv 0).$$

These equations describe the activity of two interacting moduli composed of the populations of exciting and checking neurons. The threshold character of the dependence of pulse generation probability on membrane potentials is approximated by the hyperbolic tangent in Eqs. (6.24)—(6.25). The quantities x_{0j} and y_{0j} are the thresholds of response of exciting and checking neurons, and h_j, b_j, and p_j are the parameters characterizing the action of neurons on each other within one population.

At $d_y = 0$, the system decays into two noninteracting moduli, each being an auto-oscillator. To study the dynamics of the model, one should first describe the properties of a separate oscillator. Figure 6.8 shows the dependence of the frequency of complete revolutions of the representative point in the limit cycle f_p on the parameter h. The following parameter values were chosen: $p = 5$, $b = 0.5$, $y_0 = 4$, $x_0 = 0.8$ and $\varepsilon = 0.005$. At the bifurcation point, $h = 1.1$ and for $h > 1.1$, a limit cycle appears in the phase plane.

The frequency was calculated by the following method. The intersections of the isocline of the vertical tangents $x = \tanh(x_0 + hy - bx)$ with the phase trajectories were constructed for only those instants which satisfy the condition $\mathrm{d}y/\mathrm{d}t = 0$. On this secant, the function $y(t)$ attains alternatively its maximum and minimum values. The minimum values of the function $y(t)$ were traced ($\mathrm{d}^2y/\mathrm{d}t^2 > 0$ for $\mathrm{d}y/\mathrm{d}t = 0$) which occurred at successive times $\{t_1, t_2, t_3, \ldots, t_{m+1}\}$.

Using this sequence we find the following sequence of periods: $\{\tau_1, \tau_2, \ldots, \tau_m\} = \{t_2 - t_1, t_3 - t_2, \ldots, t_{m+1} - t_m\}$. We then introduce a quantity $f_m = m/(t_{m+1} - t_1)$, which is a harmonic mean of the following set of numbers $\{1/\tau_1, 1/\tau_2, \ldots, 1/\tau_m\}$. If the

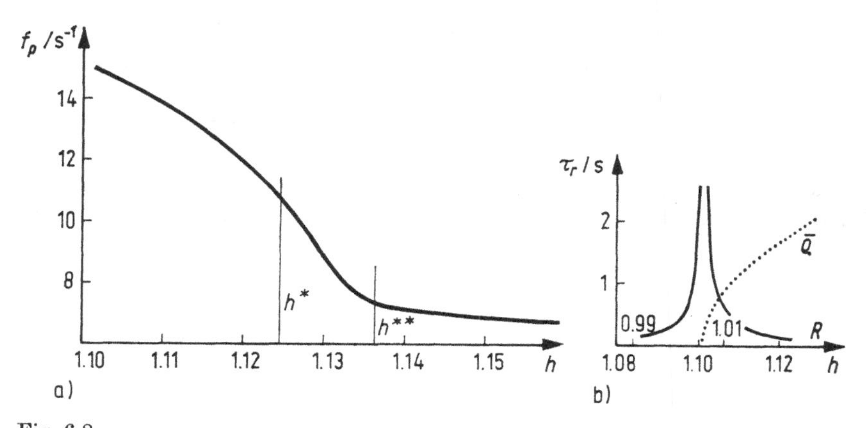

Fig. 6.8
Characteristics of a point generator of system (6.24)—(6.25):
a) the dependence of pulse repetition rate on h parameter;
b) the relaxation time τ_r (solid line) and the amplitude of auto-oscillations $\bar{\varrho}$ (dashed line)

periods are equal, then $f_m = f_p$. Three regions can be distinguished in the diagram $f_p = f_p(h)$ (Fig. 6.8). In the region $h < h^*$, the dependence is approximated by a curve that is obtained by applying a standard method of truncated equations (Butenin, Neumark, Fufaev 1976) to the system (6.24)—(6.25). The region $h > h^{**}$ can be described analytically on the basis of developed asymptotic methods (Mishchenko, Rozev 1975). It is difficult to obtain an analytical dependence in the region (h^*, h^{**}). As is shown by numerical experiments, the system exhibits complex behaviour in this region.

Such a behaviour can be described by introducing a parameter $\Delta(h_1, h_2) = f_m(h_1, h_2) - f_p(h_1)$. In a homogeneous system, where $h_1 = h_2 = h$, the two auto-oscillators show identical behaviour under homogeneous initial conditions: $f_m(h, h) = f_p(h)$ and $\Delta(h, h) = 0$ at any $h \in (1.10 \div 1.16)$. If the initial conditions are different, then different solutions should be expected for single and coupled auto-oscillators. However, it was found that the solutions of two coupled auto-oscillators are synchronous nearly in the whole region h. The system with homogeneous parameters "forgets" with time those inhomogenieties that occurred under the initial conditions. The calculations show that for such a system, there is a region $h \approx (1.126 \div 1.136)$ where the values $\langle\Delta(h, h)\rangle$, the mathematical expectation of the function $\Delta(h, h)$, are small but do not vanish. In this region, the oscillators can remember an inhomogeneous initial action for a long time. Such behaviour is much like desynchronization in the strong-relaxation distributed systems considered above (for $\delta > 1$, see (6.22)). The fact that the region where in-phase homogeneous oscillations are absent coincides with the region (h^*, h^{**}; see Fig. 6.8) corresponding to the transient region between quasilinear and relaxation oscillations, can be explained by the peculiarities of the model considered here.

Interactions between oscillators in an inhomogeneous system where the parameters h_1 and h_2 vary in the region ($h^* < h_i < h^{**}$) for a fixed coupling coefficient $d_x = 0.01$

were investigated. It was found that synchronization, beating, multimode operation and randomness are possible, depending on the relation between the parameters h_1 and h_2. Figure 6.9 shows the ranges of values of h_1 and h_2, which correspond to possible operating conditions (6.24)—(6.25). According to the paper by Sbitnev (1979), randomness is observed only in regions 1a and 1b (Fig. 6.9). There are two types of boundaries between stochastic regions: a boundary with a smooth variation of the state and a boundary with a jump-like transition. The latter boundary mainly separates stochastic regimes and synchronous oscillations.

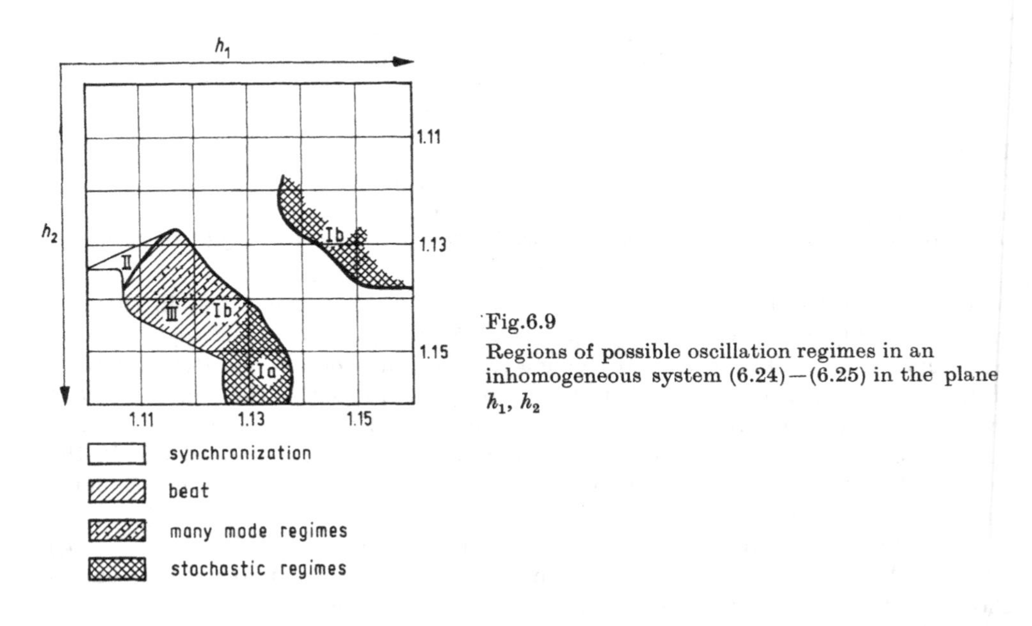

Fig.6.9
Regions of possible oscillation regimes in an inhomogeneous system (6.24)—(6.25) in the plane h_1, h_2

Since the transition from beating to randomness is smooth, the boundary between them is shown approximately. Simple beats evolve smoothly, via multifrequency regimes, to a continuous spectrum of oscillations, which is an indication of stochastic motion. Typical characteristics of stochastic motion are given in Fig. 6.10. This brief discussion of the data obtained by Sbitnev (1979) can be supplemented with analogous calculations, which show that complex motions, including stochastic ones, are possible in a system of coupled auto-oscillators. Thus, either regimes with a periodic rearrangement of the spatial structure of desynchronization or regimes associated with quasistochastic motions can occur in the inhomogeneous system.

3. *The order level*

It is useful to keep in mind that the synchronization and desynchronization leading to the formation of complex space-time structures are the manifestation of one and the same process. There is not a sharp boundary between these regimes and they can

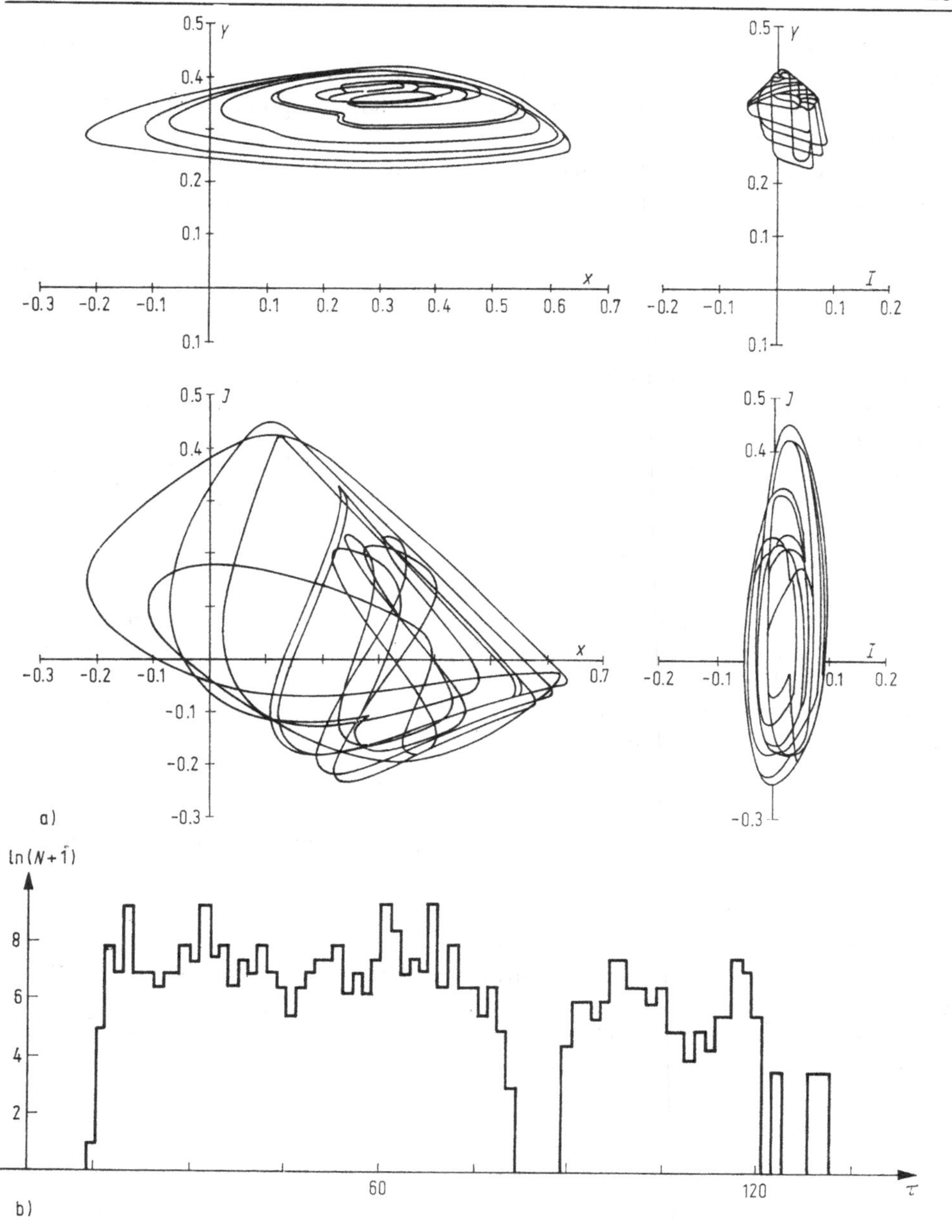

Fig. 6.10

Stochastic motions in system (6.24)–(6.25) at $h_1 = 1.13$ and $h_2 = 1.141$;

a) projections of the phase trajectories in the plane $X = \frac{x_1 + x_2}{2}$, $Y = \frac{y_1 + y_2}{2}$, $J = \frac{x_1 - x_2}{2}$, $I = \frac{y_2 - y_1}{2}$;

b) a histogram of pulse repetition intervals

smoothly convert into each other. Therefore it is reasonable to introduce a quantitative criterion for the degree of randomness of autowave processes. Such a criterion was found by Gaponov-Grekhov and Rabinovich (1983) who proposed the order level for dynamic models.

It is well known that each nonlinear structure can be described by a mathematical image in the form of an attractor in the phase space. The order level was taken in the form

$$P = \frac{m - D}{m - 1}, \qquad (6.26)$$

where m is the dimension of the phase space, and D is the fractal dimension of the attractor, which indicates to what extent the phase space is filled with phase trajectories. As an illustration of this, we consider simple attractors, which can be limiting trajectories: a point of equilibrium ($D = 0$), a limit cycle ($D = 1$), an ergodic winding on an M-dimensional torus ($D = M$). If D approximates m, i.e., $m \simeq D$, then the system has a most disordered structure and P tends to zero. If the dimension of the attractor, at rather large m, slightly exceeds two, i.e., if the attractor's trajectories are localized in a skin layer, then most variables in the system can be related accurately to each other by algebraic formulae. In this case, the degree of order is rather high ($P \simeq 1$). Thus, the parameter P seems to be most informative when chaos-to-chaos transitions are considered (see, for example, Afraimovich, Rabinovich and Ugodnikov 1983).

However, it is unclear, how the order level in investigating autowave structures can be determined in distributed systems.

6.4 A synchronous network of auto-oscillators in modern radio electronics

Characteristic of modern radio electronics is the microminiaturization of equipment. This means that designers can concentrate hundreds and thousands of simplest elements on very small "linear" areas (of fractions of mm^2). Specifically, semiconductor generators and Josephson junctions can serve as such elementary units. Josephson junctions generate millimetre and centimetre wave length oscillations at liquid helium temperatures. In any case, there are the following problems to be solved by designers:

1. Sufficient power can be obtained only by combining the oscillators into a synchronous network with common load. The power increases proportionally to the number N of oscillators.

2. Correspondingly, the network should have the output resistance matched to the load resistance.

3. Usually, microminiaturization is associated with a high scatter of the parameters of some elements, specifically, of the partial frequencies of generators.

The self-synchronization of generators is the clue to solving these three problems simultaneously. Increased stability of the frequency of a self-synchronized network of N auto-generators is another advantage. The frequency stability increases in proportion to $\sqrt{N}$ (see Malakhov and Maltsev 1971; Kanavets and Stabinis 1972). If a narrower generation band is needed, then a generator with extremely stable frequency is added. The theory of self-synchronization of ensembles of generators as used in transmitters is considered in ample detail by Dvornikov and Utkin (1980). The purpose of the present considerations is to show ways of increasing generated power.

A variety of investigations of controllable coupled generators in active structures are conducted at the Faculty of Physics, Moscow University (see Dudnik, Kuznetsov and Minakova 1984). Control of the oscillation spectrum of an ensemble of generators and the study of multifrequency processes and conditions for transition to single-frequency autonomous and nonautonomous regimes call for new research methods. Specifically, it is reasonable to study a volume element composed of four oscillators that are coupled by twelve monodirectional channels of "each with all" type. Such an element permits one to investigate simpler and varied cases of coupling. It is important that different types of oscillators can be used and studied by a computer experiment.

An interesting result is obtained when three of the oscillators have their parameters fixed and the fourth oscillator is retunable. In this case, the dependence of the regions of synchronous regime on coupling coefficients d_{i4} and d_{4i} is not symmetrical. For d_{i4} the bandwidth of single-frequency regime is equal to the bandwidth of external synchronization of three oscillators by the fourth one. As d_{i4} increases, the bandwidth of synchronous frequencies remains practically the same for resistive coupling. The synchronization band broadens and can be one order above the bandwidth of synchronous frequencies. Thus, oscillators that are more and more diverse in partial frequencies can be involved in synchronous operation only by varying the parameter d_{i4}. Analogous relations have been obtained for two-frequency regimes. Note that it is possible to simulate complex distributed auto-oscillatory systems with different types of leading centres on the basis of four-unit elements. Moreover, systems that ensure spatially in-phase coherently emitting antennae can be constructed of networks of oscillators. Some authors suggest that oscillators are combined into units, which, in turn, are integrated into networks.

Networks composed of hundreds of Josephson junctions seem to be the most promising (Jain, Likharev et al. 1984). These junctions represent links arising between two thin superconducting films of, for example, plumbum or nyobium. Tunnelings that can be formed between the films can also be used in future. Presently there exist in-phase devices that consist of chains of one hundred or more Josephson generators. The basic problem which arises in designing such systems is the choice of optimal coupling between elements. The junctions can be integrated into a two-dimensional network. The phase synchronization is then accomplished by the wave propagating in this network. Note that the truncated equations for the phase differences in networks of Josephson junctions are similar to those considered in Section 6.2. (See also Kuzmin, Likharev, Ovsyanikov 1981).

Chapter 7
Spatially inhomogeneous stationary states: dissipative structures

Spontaneous breaking of symmetry and the formation of stationary structures in spatially homogeneous systems with local positive feedback, Eqs. (1.1)—(1.3), were first indicated by A. Turing (1952) in his pioneering work on one of the crucial problems in biology of development — morphogenesis. The title of the work we cite was "Chemical foundations of morphogenesis". After a decade, in the Sixties, such structures have been investigated further by members of I. Prigogine's Brussels School (Glansdorff and Prigogine, 1971). In their studies the phenomenon was provided with a name, "dissipative structures" (DS), which is relevant to thermodynamical aspects of the problem: the structures exist owing to an influx of energy, substance etc. from the environment and the dissipation within the system. During the past decade, various aspects of the theory of DS and its applications have been investigated in many laboratories (Vasilev, Romanovskii and Yakhno 1979, Kerner and Osipov 1978, 1980, 1983, Belintsev 1983, Nicolis and Prigogine 1977, Romanovskii, Stepanova and Chernavskii 1984; Murray 1977).

The theory of DS has been applied to various domains of science. In physics, models of concrete objects have been constructed and discussed (see, e.g., work by Kerner and Osipov 1978, 1980, 1983). In biology, in spite of the longer history of the topic, the problem is still treated at a "conceptual" level, there are no models of concrete systems (Meinhardt and Gierer 1980, Vasilev, Romanovskii and Chernavskii 1982, Belintsev 1983). Although the levels at which the models are investigated in physics and biology are essentially identical, the concrete models and their applications to experimental data are far from having been elaborated to the same degree in these two fields. Anyway, qualitative results in the theory concerning DSs have been the most important until now, and this is the subject of the present Chapter.

Essential theoretical points involved here are as follows. First, the conditions under which equations of the type (1.1)—(1.3) have stationary inhomogeneous equations. Second, methods of construction of stationary inhomogeneous solutions making use of small parameters which are present in the systems in view. Quasiharmonical spatial distributions correspond to the case of a "weak" nonlinearity. In contrast DS arise in the case $D_y \gg D_x(\partial F_x/\partial x|_{x=\bar{x}} \geqq 0)$. Third, the role of mutual diffusion in the formation of DS, as well as some aspects of the problem of the reduction of multicomponent systems to basic models. Fourth, for the structures localized in bounded domains of infinite homogeneous space, we determine the types of systems which ensure their stable existence. Fifth, an analysis of the stability of the obtained solution, this is an important and difficult item in the study of the models. It is still impossible to analyse the dynamics of formation of DS exhaustively enough. Nevertheless, a number of interesting facts are known, and these facts will be discussed here. The problems relevant to the formation and existence of DS, with due account for fluctuations and external noise, require special methods, so they are treated in the next Chapter.

7.1 Conditions of existence of stationary inhomogeneous solutions

As we have already mentioned many times, analysis of the simplest discrete models provides important information on the character of AWPs, yet it is quite simple and spectacular. Therefore we prefer to consider the conditions under which DS exist having such models in mind (Vasilev 1976). The argumentation we give is rigorously extended to distributed models, but this extension exploits elements of functional analysis (Volpert and Ivanova 1981).

Let us rewrite the equations of the simplest discrete model (Sections 3.6) in terms of normal coordinates (3.33),

$$\dot{\Delta}_i = F_{\Delta_i}(\Delta, \mathbf{S}); \quad \dot{S}_i = F_{S_i}(\Delta, \mathbf{S}); \quad i = 1, 2, \ldots, n, \tag{7.1}$$

where the dot means the time derivative, $\Delta_i = ({}^1\tilde{x}_i - {}^2\tilde{x}_i)/2$; $S_i = ({}^1\tilde{x}_i + {}^2\tilde{x}_i)/2$. ${}^j\tilde{x}_i$ are variables in the j-th sector ($j = 1, 2$), referred to the m-th singular point of the local equations, F_{Δ_i} are transformed functions of local kinetics F_i, cf. Eq. (1.5).

One needs a natural assumption concerning the functions F_i: they must be "physically realizable", in other words, the local system with these functions $F_i(x)$ must not have infinitely increasing solutions. A more accurate requirement is also less restrictive. The functions $F_i(x)$ must lead to the existence of separable surfaces, which are boundaries of a volume, such that integral curves having initial points within this volume do not leave it at any t. As an m-th singular point must be in this volume, one can state that the models must be "physically realizable" in a domain of the phase space having its centre at the m-th singular point.

The main statement of this Section is as follows. The sufficient condition of existence of stationary DS ($\Delta_i \neq 0$) for the model (7.1) with physically realizable kinetics (F_i) is the DS-instability (cf. Section 3.3) of the trivial solution. In other words, the free term in the characteristic equation (3.6) must be negative. This condition is essentially similar to the requirements for trigger local systems: $q_0(0) < 0$ is the requirement that the singular point is a saddle. Below we give an outline of the proof that the conditions are indeed sufficient.

Let us consider Δ_i, S_i and S_l ($i \neq l$) as implicit functions of Δ_l, which are fixed by the equations $\dot{\Delta}_i = 0$, $\dot{S}_i = 0$, and $\dot{S}_l = 0$. The expression for the total derivative of $\dot{\Delta}_l$ by Δ_l is of the form

$$\frac{\mathrm{d}\dot{\Delta}_l}{\mathrm{d}\Delta_l} = \sum_{i=1}^{n} \left(\frac{\partial F_{\Delta_l}}{\partial \Delta_i} \frac{\mathrm{d}\dot{\Delta}_i}{\mathrm{d}\Delta_l} + \frac{\partial F_{S_l}}{\partial S_i} \frac{\mathrm{d}S_i}{\mathrm{d}\Delta_l} \right). \tag{7.2}$$

An appropriate substitution leads to

$$\frac{\mathrm{d}\dot{\Delta}_l}{\mathrm{d}\Delta_l} = \frac{D(F_{\Delta_1}, \ldots, F_{\Delta_n}; F_{S_1}, \ldots, F_{S_n})}{D(\Delta_1, \ldots, \Delta_n; S_1, \ldots, S_n)} \Bigg/ \frac{D(F_{\Delta_1}, \ldots, F_{\Delta l+1}, \ldots, F_{S_n})}{D(\Delta_1, \ldots, \Delta_{l-1}, \Delta_{l+1}, \ldots, S_n)}. \tag{7.3}$$

The Jacobians at the point $\Delta_i = S_i = 0$ which are present in expression (7.3) contain elements of the linearized system (3.4):

$$\left.\frac{\mathrm{d}\dot{\Delta}_l}{\mathrm{d}\Delta_l}\right|_{\Delta,S=0} = \frac{\det\{M_{ij} - 2d_{ij}\}}{\{M_{ij} - 2d_{ij}\}_{ll}}; \tag{7.4}$$

where the denominator is the minor of the ll-th element of the linearized system matrix.

If the i-th variable is chosen in such a way that the single positive feedback is broken as soon as this variable is "frozen", then $\mathrm{d}(\dot{\Delta}_l)/\mathrm{d}(\Delta_l)|_{\Delta,S=0} \geqq 0$ only if the free term in the characteristic equation is negative. In particular, for $n = 2$ the minor is $(M_{ii} - 2d_{ii})$ $(i \neq l)$, and it is negative for the system with a single autocatalysis. The existence of values $\Delta_l > 0$ (< 0) for which the condition $F_{\Delta_l}(\Delta_l) = \dot{\Delta}_l \leqq 0$ $(\geqq 0)$ is satisfied is guaranteed by the "physical realizability" condition for $F_l(\tilde{x})$. The function $\dot{\Delta}_l(\Delta_l)$ is continuous in Δ_l if all the partial derivatives $\partial F_i/\partial x_j$ $(x, j = 1, \ldots, n)$ are continuous. Moreover, the break, the disconnection in the feedback circuit guarantees that the function $\dot{\Delta}_l(\Delta_l)$ is single-valued. Thus we conclude: i) the function $\dot{\Delta}_l(\Delta_l)$ is continuous, ii) if $q_0(2) < 0$ then $\mathrm{d}(\dot{\Delta}_l)/\mathrm{d}(\Delta_l)|_{\Delta_1=0} > 0$, iii) for some $\Delta_l > 0$ (< 0) one has $\dot{\Delta}_l < 0$ (> 0). Consequently, in the region $\Delta_l > 0$ this function acquires at least one zero value[1]), i.e., the equation $\dot{\Delta}_l(\Delta_l) = 0$ has a nontrivial solution.

If more than one positive feedback is present in the local system (the same is true if the chosen l-th variable is neither autocatalytic $(n = 2)$, nor cross-catalytic $(n \geqq 3)$, then the behaviour of the function $\dot{\Delta}_l(\Delta_l)$ is different from the picture we have drawn. Let the j-th variable be autocatalytic, and $\{M_{jj} - 2d_{jj}\}_{ll} > 0$. Then for some Δ_l one has $\mathrm{d}\dot{\Delta}_l/\mathrm{d}\Delta_l = 0$. This fact can be proven calculating a derivate of the equality $F_{\Delta j} = 0$ and making use of the "physical realizability" condition. At this point one has $\mathrm{d}\dot{\Delta}_l/\mathrm{d}\Delta_l \to \infty$, and this is a turning point of the curve $\dot{\Delta}_l(\Delta_l)$. In particular, if $\{M_{ij} - 2d_{ij}\}_{ll} > 0$ and $\det\{M_{ij} - 2d_{ij}\} > 0$, then $\mathrm{d}\dot{\Delta}_l/\mathrm{d}\Delta_l|_{\Delta_l=0} = 0$. However, because of the turning point a branch of the function $\dot{\Delta}_l(\Delta_l)$ escapes to the second quadrant, and one cannot be sure that the equation $\dot{\Delta}_l(\Delta_l) = 0$ has nontrivial solutions. Inversely, at least a single nontrivial solution is guaranteed if there is a turning point and $q_0(2) < 0$. This completes the proof that the above existence conditions are sufficient.

Some simplest forms of dependence $\dot{\Delta}_l(\Delta_l)$ are shown in Fig. 7.1. Of course, more complicated forms are also possible; for instance, turning points may appear in the second and fourth octants of the graph, as in Fig. 7.1a.

We will not dwell on general reasoning and instead illustrate the most essential aspects with examples of some concrete models.

---

[1]) In general, the function $\dot{\Delta}_l(\Delta_l)$ may have several "independent" branches. However, only states relevant to one of them are meaningful from the physical point of view.

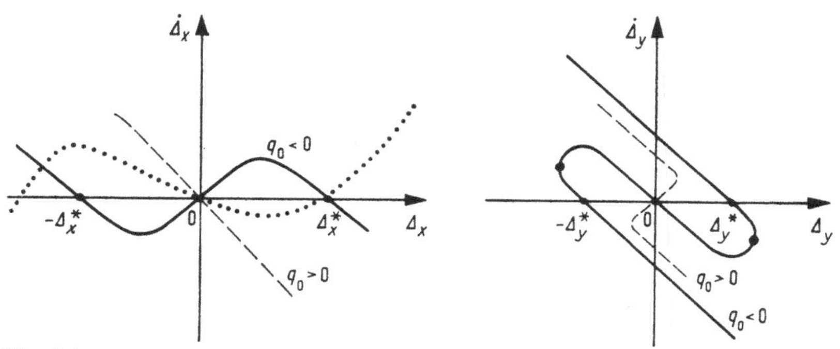

Fig. 7.1
Conditions of existence for stationary homogeneous solutions.
a) Dependence $\dot{\Delta}_x(\Delta_x)$, x represents an autocatalytic variable; the dotted line represents a case of a "physically non-realizable" system.
b) Dependence $\dot{\Delta}_y(\Delta_y)$, y represents a non-autocatalytic variable.

Example 1: Brusselliator. The existence conditions obtained above for DS are only sufficient. This fact is shown with the example of the Brussellator model (2.54). In the case of two "boxes" stationary inhomogeneous solutions can be found from the following equation (here for simplicity $d_{ij} = 0$ if $i \neq j$),

$$(\Delta_x^2)^2 - (\Delta_x^2)(2A^2 - 2d_{yy} - B/\varkappa_2) + A^2 q_0(2)/(2\varkappa d_{yy}) = 0,$$
$$\Delta_y = -x_2\Delta_x; \quad S_y = (\Delta_x^2)(2A^2\varkappa_2 - B/A)/(A^2 + \Delta_x^2); \quad S_x = 0, \tag{7.5}$$

where

$$\varkappa_2 = (1 + 2d_{xx})/(2d_{yy}); \quad q_0(2) = 2d_y(B - B_{\text{cr2}}); \quad B_{\text{cr2}} = 1 + A^2\varkappa_2 + 2d_{xx}.$$

If $A > 2\sqrt{d_y}$ and $q_0(2) \gtrless 0$, there are two solutions ($\Delta_{x,1}^* > 0$ and $\Delta_{x,2} > 0$, Fig. 7.2a, c). So stationary inhomogeneous states can exist even if the DS-instability conditions do not hold, and these conditions are not necessary. Evidently, for $q_0(2) > 0$ the homogeneous solution is stable, and DS arise in the "hard" regime because of large perturbations.

If $A < 2\sqrt{d_y}$ the solutions (7.5) exist only if the DS-instability conditions are satisfied: $q_0(2) < 0$, Fig. 7.2b, d. A comparison of the "state diagrams" (Figs. 7.2c and d) shows they have an important common feature: in both cases the homogeneous solution becomes unstable at the point B_{cr2} (the bifurcation point of the inhomogeneous solution). It is possible, however, that some separate branches, disconnected from the homogeneous state, may appear in the "state diagrams".

Example 2: The Van-der-Pol oscillator system with hard self-excitation,

$$\frac{\mathrm{d}}{\mathrm{d}t}x = -y - \delta_1 x + \delta_3 x^3 - \delta_5 x^5; \quad \frac{\mathrm{d}}{\mathrm{d}t}y = x. \tag{7.6}$$

The DS-instability conditions cannot hold for any positive values of the system parameters $\left(q_0(2) = 1 + 2\delta_1 d_{yy} + 4d_{xx}d_{yy} > 0\right)$. Meanwhile, stationary inhomogeneous solutions do exist, as is easily verified by inserting (7.6) into the simplest

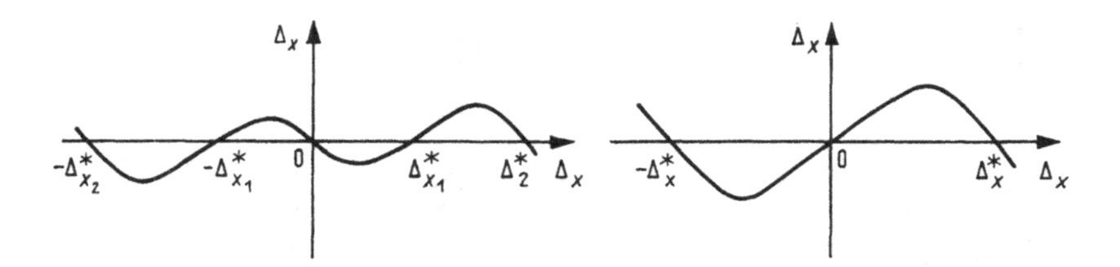

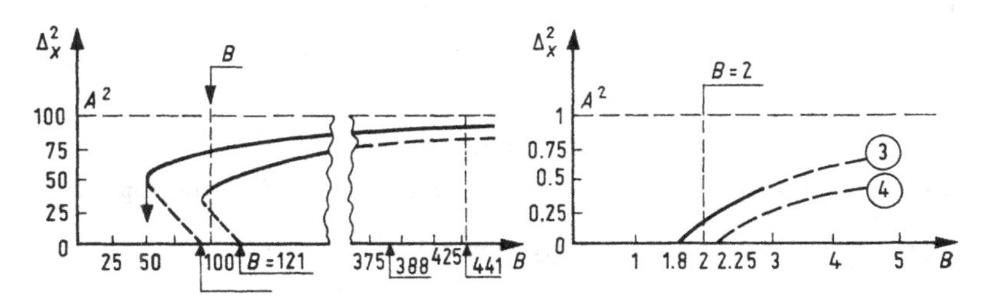

Fig. 7.2
Cases of hard (a) and c)) and soft (b) and d)) excitations of the stationary inhomogeneous states in the discrete Brussellator (2.54): c) $A = 10$, $d_x = 4$, $d_y = 5$; d) $A = 1$, $d_x = 0.1$, $d_y = 1$

discrete model (7.1). One should bear in mind, however, that the model (7.6) is essentially different from the "Brussellator" model.

Actually, the necessary existence condition for DS ($d_{ij} = 0$, $i \neq j$) is the existence of two domains in the phase space of the local system; the integral curves must have "opposite directions" in these domains: $F_i > 0$ for $x_i > \bar{x}_i$, and $F_i < 0$ for $x_i < \bar{x}_i$ (see Fig. 3.8). The local "forces" (F_i) act as a factor separating the representing points, while the "diffusion" forces induce convergence of these points. Stationary dissipative structures result from equilibrium of the forces of both types. In the case of the Brussellator-type models the domains have a common point, namely the singular point of the local system. In the model (7.6), the domains are always separated by a finite distance. Therefore the DS-instability conditions derived from the analysis of trajectories near the singular point never take place in the system, although DS do exist.

It is noteworthy that in the Brussellator-type models pre-critical DS ($q_0(2) > 0$ for $B > B_{cr2}$, see Fig. 7.2c) exist owing to both quadratic (!) and cubic terms in the reduced system (in general, both even and odd terms are necessary.[1]) For pre-critical values of the parameters the "opposite direction" domains also have a common point, but there is no DS-instability. As the parameters move farther into the pre-critical domain, the "diffusion" forces diminish and become too small to retain the representing points; the separation between the points increases and they leave the "opposite direction" domains. After that the points converge.

---

[1]) This kind of nonlinearity is compulsory for genuine chemical systems where $F_i|_{x_i=0} \geqq 0$ providing non-negativity of the variables, and $F_i|_{x_i \to \infty} < 0$, so the kinetics is physically realizable.

The analysis based upon the concept of "opposite direction" domains enables one to verify quite simply the necessary conditions for the existence of stationary inhomogeneous solutions for concrete models, as well as to draw some general conclusions. For instance, if a two-component system has a local limit cycle (hard or soft) one can always find some values of d_{11} and d_{22} for which DS exist. Recall that this statement is not true, in general, for multicomponent systems ($n \geqq 3$), cf. the example (3.34) in Section 3.

Example 3: A system with physically non-realizable kinetics. It is important to emphasize the role of the physical realizability in the formulation of the existence conditions for DS. Let us consider the following system:

$$\frac{\mathrm{d}}{\mathrm{d}t}\, x = y + A_1 x + A_2 x^2; \quad \frac{\mathrm{d}}{\mathrm{d}t}\, y = B_1 x + B_2 y. \tag{7.7}$$

Such kinetics cannot be realized for any values of the model parameters. We omit simple algebra and present only the main results. Stationary inhomogeneous solutions exist in the "two-box" system even for such a choice of the parameter signs where the system is not active, and DS-instability is always impossible. On the other hand, for values of the parameters ($A_1, \ldots, B_2$) which permit DS-instability the latter does not guarantee the existence of DS-type solutions. Of course, the model given in (7.7) is rather exotic, but it makes a clear illustration of the fact that not every formally stationary inhomogeneous solution can be considered as an image of DS, as such a solution may have nothing to do with the active properties of the medium. A solution is "formal" in this sense if the representing points go into domains of the local system phase space which contain physically non-realizable trajectories. Dealing with such solutions one may obtain results which are absurd from the point of view of DS theory.

Example 4: The existence condition for the first-kind boundary conditions (impenetrable ends) (1.4). For the first-kind boundary problem one should consider no less than three "boxes"; in the extreme boxes the variables must be fixed. For instance, for spatially homogeneous problems one should set: ${}^1x_i = {}^3x_i$ ($i = 1, \ldots, n$). In this case we can also show that the sufficient DS existence conditions formulated above are valid. For the Brussellator for instance an inhomogeneous solution can be found from the following equation:

$$(x - \bar{x})^2 - (B/A\varkappa_2 - 2A)\,(x - \bar{x}) + q_0(2)/\varkappa_2 d_{yy} = 0. \tag{7.8}$$

The state diagram is like that shown in Fig. 7.2c, and pre-critical DS are absent only for $B = 2A^2x_2$. Note that DS always exist for $q_0(2) < 0$.

The existence of pre-critical DS is not a necessary attribute of the first-kind boundary problem. This is seen if one looks at the "four-box" model (${}^1x_i = {}^4x_i = \bar{x}_i$). In this case, the diagram of states symmetrical with respect to the $\bar{x}_i$ mode may be without pre-critical branches. The other mode has properties similar to those of (7.8). The equation of the states is rather complicated, yet one can show that a symmetrical mode solution always exists for $q_0(3) < 0$, and an asymmetrical mode solution is present for $q_0(2) < 0$. It is also notable that an inspection of representing points in the phase plane makes it possible to verify that the symmetrical mode corresponds to even modes of the distributed system, while the asymmetrical mode corresponds to odd modes.

Concluding the outline of properties of the simplest discrete models, we would like to stress some points. In studying systems of N elements coupled by diffusion, one can also introduce normal coordinates (as for $N = 2$). The proof of the existence of a nontrivial solution if at least one quantity $q_0(k) < 0$ ($k = 2, 3, \ldots$) is negative, is still valid in all parts. This fact enables one to extend rigorously the sufficient conditions stated above to discrete systems containing any finite number of elements. Of course, this is not a rigorous proof for distributed systems where the number of elements is infinite, but the approach may be of heuristic interest (cf. Sections 7.2 and 7.7). We will not develop this subject further but instead look at the application of the discrete models from another point of view.

The projection of a stationary solution from (x_i, r) to the phase space (x_i) is a curve. The curve is a straight line for harmonic distributions. The simplest discrete model has a projection which is a pair of points. As was mentioned above, the possibility of obtaining nontrivial stationary solutions is due to the presence of "opposite direction" domains in the phase space. Evidently, if no stationary position exists for two points, it is all the more impossible for a continuous line. Bearing in mind the fact that we have only sufficient existence conditions for DS, in dealing with complicated multicomponent systems we can start from an analysis of their discrete models. Such an analysis would be useful as a way to understanding special features of nonlinearity in the model which are relevant to DS.

In the following Sections we shall consider quasi-harmonical DS in distributed systems. It is remarkable that the equations for the mode amplitudes bear a qualitative resemblance to those for solutions of the discrete models (cf. also Section 5.3). Therefore the qualitative results of this Section and the next are in a sense complementary and overlap in part.

7.2 Bifurcation of solutions and quasi-harmonical structures

The problem of constructing stationary solutions of the basic model (1.1) to (1.3) which are periodical in the space coordinate and satisfy given boundary conditions, is similar to conventional problems relevant to oscillations in nonlinear systems. In fact, in nonlinear systems the period of oscillations depends on the amplitude. In the boundary problems on the segment $[0, L]$ considered the period must be exactly $2L/k$ (k is an integer). Therefore one has to find amplitude and shape of periodical solutions which provide this requirement. Hence the following scheme arises for constructing the desired solutions.

(i) Take the stationary version of the model and extract from it a generating system which has solutions (modes) with periods $2L/k$,

ii) put the modes obtained into the complete system, make the period variable and obtain a dependence of the mode amplitude on its period,

(iii) determine the mode amplitude for which the period equals $2L/k$.

The method proposed employs an iterative procedure and may be effective if small parameters are present in the system. In the following we consider the case of weak nonlinearity; systems with a small parameter in the diffusion term of the equation have some specific features and are considered separately in Section 7.4.

The simplest way to get periodical solutions is to construct harmonic balance equations (Polyakova 1974). Though this method is not mathematically rigorous, it leads to a qualitatively correct understanding of the behaviour of the solutions near the bifurcation points. Most rigorous are the methods based upon the theory of bifurcations for solutions of nonlinear equations (Vaynberg and Trenogin 1969). This approach to the construction and analysis of solutions for the Brussellator model with periodical boundary conditions has been developed in a series of studies (Babskii and Markman 1977, Markman 1980, Babskii, Markman and Urintsev 1982, Dombrovskii and Markman 1983). Such boundary conditions are responsible for certain peculiar properties of the characteristic structures and we do not treat them here. A detailed description of applications of the bifurcation theory to the first- and second-kind boundary problems for the Brussellator model is given in the book by Nicolis and Prigogine (1977). Below we consider a simplified version of the method used for the construction of periodical solutions (Vasilev 1976) based upon the ideas of Bogolyubov and Mitropolskii (1957). This method is intermediate between the harmonic balance and the bifurcation theory approach as to its rigorousness and, on the other hand, its clarity.

Let us write the basic model (1.3) as follows ($\partial/\partial t = 0$):

$$L_k(B_{\mathrm{cr}})\,\tilde{\boldsymbol{x}} \equiv \left(\hat{\mathrm{D}}\,\frac{\partial^2}{\partial r^2} + \hat{\boldsymbol{M}}'(B_{\mathrm{cr}})\right)\tilde{\boldsymbol{x}} = \hat{\boldsymbol{M}}''(B_{\mathrm{cr}} - B)\,\tilde{\boldsymbol{x}} + F(\tilde{\boldsymbol{x}}), \tag{7.9}$$

where $\tilde{\boldsymbol{x}} = \{\tilde{x}_1, \ldots, \tilde{x}_n\}$ is the vector of the reduced variables, $\hat{\boldsymbol{M}}' - \hat{\boldsymbol{M}}'' = \hat{\boldsymbol{M}}$, $\hat{\boldsymbol{M}}$ is the matrix of the reduced local system (1.5). The matrix M depends on a parameter B. It is assumed that for $B = B_{\mathrm{cr}}$ the linear operator $\hat{L}_k(B_k)$ has a zero eigenvalue and the corresponding eigenfunction $\boldsymbol{x}_k$ has a period $2L/k$. For instance, in the case of the second-kind boundary conditions (1.4) one has $\boldsymbol{x}_k = \boldsymbol{C}_k \cos(\pi k r/L)$ and for the first-kind boundary conditions $\boldsymbol{x}_k = \boldsymbol{C}_k \sin(\pi k r/L)$, where $\boldsymbol{C}_k$ is a constant vector which is determined up to an arbitrary factor, namely the structure amplitude (A_k). For definiteness we shall assume that $\boldsymbol{C}_k = A_k \boldsymbol{C}_k'$, where $\boldsymbol{C}_k'$ has the first element equal to one ($C'_{1k} = 1$).

Near the bifurcation point $\left((B_{\mathrm{cr}} - B) \ll B_{\mathrm{cr}}\right)$ the structure amplitude is relatively small. It is apparently natural to search for a solution written as a series,

$$x = A_k x_k + A_k{}^2 x_{\mathrm{I}} + A_k{}^3 x_{\mathrm{II}} + \cdots. \tag{7.10}$$

According to the bifurcation theory, the following expansion is used as a measure for the distance of the solution to the bifurcation point (cf. Eq. (3.30)):

$$B - B_{\mathrm{cr}} = A_k \gamma_{\mathrm{I}} + A_k{}^2 \gamma_{\mathrm{II}} + A_k{}^3 \gamma_{\mathrm{III}} + \cdots. \tag{7.11}$$

Using the distance ($B - B_{\mathrm{cr}}$) directly as a small parameter one cannot represent the variety of forms with sufficient completeness: only powers with integer degrees are possible while even fractional degrees are not represented. Nevertheless one can simplify (7.11) with due account for the special properties of the problem in view.

Let us insert (7.10) into (7.9) and, collecting the terms with equal powers of A_k obtain systems of differential equations for, etc. These equations have the following structure

$$\hat{L}_k x_{\mathrm{I}} = F_{\mathrm{II}}(A_k \boldsymbol{x}_k,\, A_k \gamma_{\mathrm{I}} + \cdots), \quad \hat{L}_k \boldsymbol{x}_{\mathrm{II}} = F_{\mathrm{III}}(A_k \boldsymbol{x}_k + A_k^2 \boldsymbol{x}_1,\, A_k \gamma_{\mathrm{I}} + \cdots). \tag{7.12}$$

Secular terms are present in the solutions $\boldsymbol{x}_{\mathrm{I}}$, $\boldsymbol{x}_{\mathrm{II}}$ of the system (7.12), they have periods $2L/k$ and satisfy the boundary conditions only if the right-hand sides are orthogonal to eigenfunctions $\boldsymbol{x}_k^+$ of the operator $\hat{L}_k^+$ which is conjugate to $\hat{L}_k$.

$$\int_0^L \boldsymbol{F}_{\mathrm{I}}(A_k \boldsymbol{x}_k,\, A_k \gamma_{\mathrm{I}} + \cdots)\, \boldsymbol{x}_k^+ \, \mathrm{d}r = 0. \tag{7.13}$$

Integrating one gets algebraic equations for the determination of $\gamma_{\mathrm{I}}, \gamma_{\mathrm{II}}, \ldots$ Inserting the solutions into (7.11) one can find P_k as functions of $(B_{\mathrm{cr}} - B)$ and then evaluate these quantities from Eq. (7.12). Although the procedure described solves the problem concerned, its scrupulous application may lead to rather awkward calculations which cannot be performed in all cases (we mean, for instance, summing up infinite series with constant terms). The use of simple physical arguments concerning the behaviour of the solution may enable one to make the analysis simpler without introducing essential errors into the results. Let us consider this point for the examples of quasi-harmonical DS and the Brussellator model. With the reduced variables ($\tilde{x} = x - A$, $\tilde{y} = y - B/A$) the model equations are

$$\begin{aligned} D_x \frac{\mathrm{d}^2}{\mathrm{d}r^2} \tilde{\boldsymbol{x}} + (B_k - 1)\, \tilde{x} + A^2 \tilde{y} &= (B - B_k)\, \tilde{x} - F(\tilde{x}, \tilde{y}), \\ D_y \frac{\mathrm{d}}{\mathrm{d}r^2} \tilde{y} - B_k \tilde{x} - A^2 \tilde{y} &= -(B - B_k)\, \tilde{\boldsymbol{x}} + F(\tilde{x}, \tilde{y}), \end{aligned} \tag{7.14}$$

where

$$F(\tilde{\boldsymbol{x}}, \tilde{y}) = (B/A)\, \tilde{x}^2 + 2A\tilde{x}\tilde{y} + \tilde{x}^2\tilde{y}; \quad B_k = 1 + \varkappa_k A^2 + \mathcal{K}^2 D_x;$$

$$\varkappa_k = (1 + \mathcal{K}^2 D_x)/(\mathcal{K}^2 D_y), \quad \mathcal{K}^2 = \pi^2 k^2/L.$$

Note that with such kinetics one has for (1.3) $q_0(\mathcal{K}^2) = (B_k - B)\, \mathcal{K}^2 D_y$ and for $B - B_k$ the left-hand sides of (7.14) vanish if $2L/k$-periodical functions are substituted, say,

$$x_k = -y_k/\varkappa_k = A_k \cos(\pi k r/L) \quad \text{or} \quad A_k \sin(\pi k r/L). \tag{7.15}$$

A typical k dependence of B_k is shown in Fig. 7.3; the minimum is attained at $k = \bar{k}$. If $B > B_{\bar{k}} = \left(1 + A(D/D_y)^{1/2}\right)^2$, the DS-instability conditions are fulfilled at least for a single mode.

Next we will consider the second-kind boundary problem (1.4). Any solution can be written as a series,

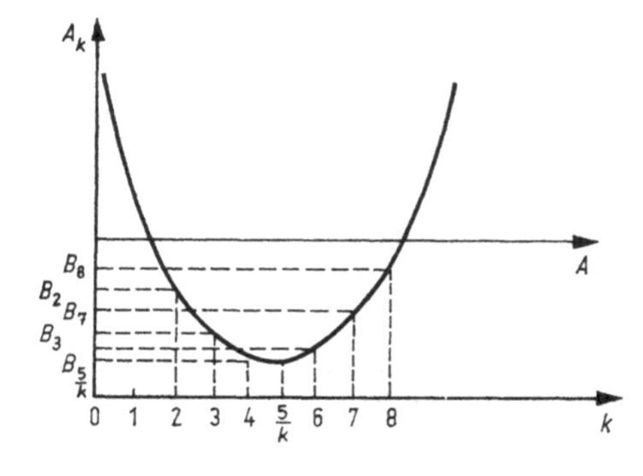

Fig. 7.3
Variations of critical values of the parameter B with the wave-length (I/k) for a Brussellator (2.54)

$$x = \sum_{i=0}^{\infty} a_i \cos(\pi ir/L); \quad y = \sum b_i \cos(\pi ir/L). \tag{7.16}$$

The same expansions hold for the functions $x_{\rm I}$, $x_{\rm II}$, ... in (7.10). The substitution of (7.16) into (7.14) gives a relation between a_i and b_i: $a_i = \varkappa_i b_i$ where $\varkappa_i = (1 + i^2\mathcal{K}^2 D_x)/(i^2\mathcal{K}^2 D_y)$. Next we use Eq. (7.12) and obtain $\boldsymbol{x}_{\rm I}, \boldsymbol{x}_{\rm II}$ etc. explicitly. Quadratic terms of the nonlinear function $F(x, y)$ in (7.14) which are proportional to $A_k^2 \cos(kr) = 1/2\, A_k^2(1 - \cos(2kr))$, are present in the equations for $x_{\rm I}$; they include the zero-th and the second harmonics of the structure with the period $(\pi L/k)$. The cubic terms in Eq. (7.14) contribute to the first and third harmonics, and $\boldsymbol{x}_{\rm II} \sim \cos(3\pi kr/L)$. Thus the expansion of the solution (7.10) is written as follows,

$$\begin{aligned} \tilde{x}(r) &= A_k \cos(kr) + A_k^2\alpha_2 \cos(2kr) + A_k^3\alpha_3 \cos(3kr) + \cdots, \\ \tilde{y}(r) &= -A_k\varkappa_k \cos(kr) + A_k^2(\alpha_0 - \alpha_2\varkappa_2 \cos(2kr)) - A_k^3\varkappa_3\alpha_3 \cos(3kr) + \cdots, \end{aligned} \tag{7.17}$$

where the coefficients of the harmonic amplitudes are obtained from the linear equations (7.12) by means of the harmonic balance method,

$$\begin{aligned} \alpha_0 &= (2\varkappa_k A^2 - B)/(2A^3); \quad \alpha_2 = (2\varkappa_k A^2 - B)/\big(2A(B_k - B_{2k})\big), \\ \alpha_3 &= \big[\alpha_2\big(B/A + A(\varkappa_k + \varkappa_{2k})\big) - \varkappa_k/4\big]/(B_k - B_{3k}). \end{aligned} \tag{7.18}$$

Equations for the amplitudes A_k stem from the orthogonality, Eq. (7.13), of the right-hand sides of Eq. (7.14) where $\tilde{x}$ and $\tilde{y}$ are given by the expressions in (7.17), to the vector x^+,

$$A_k[q_0(k)/(2k^2D_y) + a_1A_k^2 + a_2{}^3A_k^4 + \cdots] = 0, \tag{7.19}$$

where

$$\begin{aligned} q_0(k) &= (B_k - B); \quad a_1 = 3\varkappa_k/8 - \alpha_0 A + [A(\varkappa_k + \varkappa_{2k}) - B/A]\,\alpha_2, \\ a_2 &= A\alpha_2\alpha_3(\varkappa_{2k} + \varkappa_{3k})/2 + \alpha_2^2(\varkappa_k + \varkappa_{2k})/4 + \alpha_3(\varkappa_k + \varkappa_{3k})/8 - \alpha_0\alpha_2/8. \end{aligned} \tag{7.20}$$

The coefficient a_2 is always positive, while for $B = B_k$ the coefficient a_1 can be either positive or negative. Respectively, two types of bifurcations are possible at $B_k\big(q_0 \sim (B_k - B)\big)$, cf. Fig. 7.4.

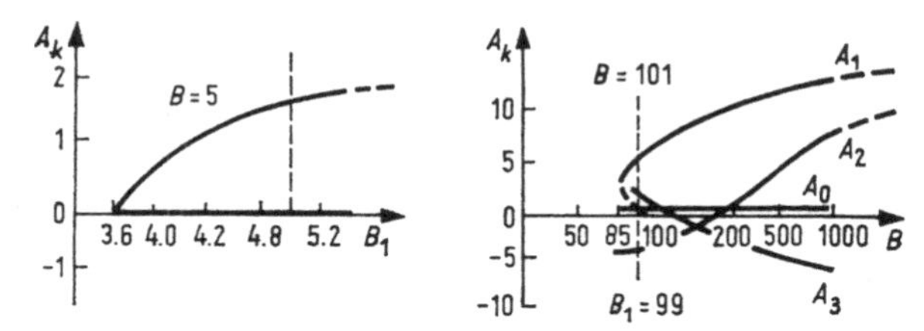

Fig. 7.4
Two types of the variations of the amplitudes of the stationary harmonical components (A_k) with the values of the parameter B for a Brusselator close to a bifurcation point. The dashed curves show unstable states,
a) $A = 2$, $D_x = 0.08$, $D_y = 0.4$, $L = 1$;
b) $A = 10$, $D_x = 0.75$, $D_y = 1$, $L = 1$.

Equation (7.19) is just the expansion of (7.11) for the Brusselator model: $q_0 \sim (B_k - B)$; $\gamma_{\rm I} = a_1$; $\gamma_{\rm III} = 0$; $\gamma_{\rm IV} = a_2$; $\gamma_{\rm V} = 0$ ... Equations of this type are specific for problems with physically realizable kinetics and the condition of absence of fluxes at the boundaries. Actually, as the problem is symmetrical with respect to the substitution $A_k \to -A_k$, all coefficients of odd powers of A_k are zero. The coefficient $a_1 = \gamma_{\rm II}$ is non-zero only if quadratic terms are present in F_i, so that the possibility of pre-critical structures for $B < B_k$ is determined by just these terms. Meanwhile, the existence of an upper branch in the dependence A_k(B) (Fig. 7.4b), i.e. a finite-amplitude solution at $B = B_k$, is due to cubic terms in F_i. Recall that the cubic terms provide physical realizability of the model, and quadratic terms its chemical meaning, i.e. the non-negativity of the variables (Section 7.1).

For $(B_k - B) \to 0$, equation (7.11) has a solution $A_1{}^2 \sim q_0/a_1 \to 0$, so that the point $q_0 = 0$ $(B = B_k)$ is indeed a bifurcation point where a solution with the period $2L/k$ separates. The bifurcation type is determined completely by the sign of a_1, so that it is established in the first approximation (in general, in the first approximation with nonzero a_1). For $a_1 < 0$ the existence of overcritical structures (Fig. 7.4b; $B > B_k$, $q_0 < 0$) is ensured only in physically realizable models. For solutions having relatively high amplitudes A_k the expansion in (7.11) is invalid. Nevertheless, one can prove the non-negativity of all coefficients for the highest powers of A_k; that is to say, solutions of equation (7.11) with large A_k exist for $B > B_k$ $(q_0 < 0)$. It is reasonable to expand in powers of (A_k/A), and $A_k < A$ for any DS. We shall not dwell on this aspect, as its meaning is quite clear from the simple arguments presented in Section 7.1.

The present approach makes it possible to determine subsequently coefficients in the harmonic series (7.16). The problem can be solved directly by inserting these series into (7.14). Collecting the terms in the same harmonics, one gets an infinite system of the harmonic balance equations which is used to determine a_i and b_i. Expansions (7.10) and (7.11) provide a regular method to truncate or reduce the system (the terminology adopted in applications of the Bogolyubov-Mitropolskii method). However, the shortest series can be used for qualitative estimates of the model solutions. We mean binomial expressions containing only the main and the zeroth harmonics. In the case of the Brussellator one has the following equation for the amplitude,

$$A_k[(3/4A^2)\,A_k^4 - (3/\varkappa_k + \mathcal{K}^2D_y)/(2A^2))\,A_k^2 + q_0/2\varkappa_k\mathcal{K}^2D_y] = 0. \qquad (7.21)$$

Even with this equation one can see the main properties of the phase graphs for DS in the model considered. Note that exchanging $k^2 \rightleftarrows 2$ one gets from Eq. (7.21) the equation for Δ_x^2 in the simplest discrete model (7.5). Seemingly, the use of such "minimal" approaches in investigations of potential possibilities of systems with respect to DS is quite reasonable.

Let us discuss in brief the first-kind boundary problem for Eq. (7.14). In this case "cos" in the harmonical series (7.16) must be replaced by "sin". Since in this case there is no zeroth harmonic, the expansion of $\sin^2(\pi kr/L)$ is an infinite series. The equations for A_k are different for even and odd k. In fact, for even k the solution is symmetrical under the substitution $A_k = -A_k$, as the regions $x < \bar{x} = A$ and $x > \bar{x} = A$ contain equal numbers of half-waves. Therefore $\gamma_{\rm I}$, $\gamma_{\rm III}$, $\gamma_{\rm V}$ etc. in (7.11) are zero. For odd k such a symmetry cannot be maintained in systems with quadratic terms in the nonlinear functions. A quadratic equation similar to Eq. (7.8) for the discrete model[1]) ($k^2 \rightleftarrows 2$) is obtained for the evaluation of A_k in the first approximation (7.10) to (7.13). The equation indicates that two structures with different amplitudes exist for every $B > B_k$ (and certain $B < B_k$). Detailed calculations following a scheme close to that presented here are to be found in published work (Nicolis and Prigogine 1977, Vasilev 1976). An important point should be emphasized. In the analysis of linearized models (Chapter 3) the boundary conditions were not essential, whereas they are quite important for properties of models which contain even weak nonlinearities. It is the boundary conditions that are responsible for the appearance of a new type of state diagram (for odd k).

Usually, the existence conditions hold for several different solutions. Other types of AWPs can take place in the same systems. Therefore, in order to predict which DS can exist one has to investigate the stability of the solutions found.

7.3 Multitude of structures and their stability

Analysis of the stability of structures with respect to small perturbations is essentially a search for a spectrum of perturbations increasing with time for the trivial solution of the linear inhomogeneous set of equations: $\partial \boldsymbol{x}/\partial t = \hat{L}\boldsymbol{x} + \boldsymbol{\Phi}(r)\,\boldsymbol{x}$, where $\hat{L}$ is a linear differential operator, and $\boldsymbol{\Phi}(r) = \partial\boldsymbol{\Phi}/\partial\boldsymbol{x}|_{x=\bar{x}(r)}$. The perturbations must satisfy the boundary conditions for the problem considered, but may have an arbitrary form within the interval. Any such perturbation can be expanded as an infinite series over the harmonical modes; e.g. they are the functions $a_i(t)\cos(\pi ir/L)$, cf. Eq. (7.16), for the second-kind boundary problem (see also Section 5.4). The equa-

[1]) Of course, values of coefficients in these equations are somewhat different, but the qualitative deductions concerning the role of the model parameters in the behaviour of the solutions are valid. Meanwhile, for these equations the harmonical balance method is not at all effective.

tions for $a_i(t)$ are coupled here, unlike the situation for the stability analysis in the case of a homogeneous state. No universal procedure has been developed to find out what are positive eigenvalues in the set of equations for $a_i(t)$. There are only some methods applicable to special structures, but they are rather complicated.

For certain fixed types of perturbations, stability is tested in a relatively simple way, yet it is possible to indicate some regions in the space of parameters where the structures considered are instable. Notably, the situation here is the same as in the theory of nonlinear oscillations where the stability of limiting cycles is analysed on the basis of reduced equations: one gets conditions which are necessary, but not sufficient, for stability of such a solution in the complete system. Nevertheless, with an appropriate choice of dangerous perturbations one is able to gain a fairly complete understanding of the character of instable structures and branching solutions. In this Section, we deal with problems relevant to "interactions" between solution pairs and consider an example where an accurate analysis of the structure stability is possible.

Ordinary differential equations for the mode amplitudes $(a_{jk}(t))$ can be obtained using the methods mentioned in Section 7.2; e.g. one can apply the harmonical balance method. The derivative operator $\partial/\partial t$ is incorporated in $\boldsymbol{F}(\boldsymbol{x})$. If one takes the zeroth harmonic and some other mode, one gets equations coinciding qualitatively, as was often mentioned above, with the equations for the simplest discrete ("two-box") model written in terms of the Δ, s coordinates (Sections 5.4, 7.1). To make the discussion simpler, we refer mainly to the discrete problem.

As in the case of homogeneous states, it is appropriate to distinguish between two main classes of bifurcation points (cf. Section 3.3). First, there are points at which an odd number of roots with positive real parts appears in the characteristic equation. Such instabilities shall be called exponential (or trigger), and second, the points where an even number of such roots appears. Instabilities of this type will be called oscillatory. Both homogeneous and inhomogeneous synchronous auto-oscillations can appear in the system.

The exponential instability is specific for the case where different stationary states are competing. The simplest example of such instability is easily understood using the graph of $\dot{\Delta}_x(\Delta_x)$ dependence (Fig. 7.2a). For the solution with smaller amplitude $\Delta_x{}^*$ one has $\partial\dot{\Delta}_x/\partial\Delta_x > 0$, and the equations for $\dot{\Delta}$, $\dot{s}$ (cf. Section 5.4) allow small independent motions of Δ_x (independent of Δ_y, s_x and s_y). As a consequence, the solution with smaller amplitude is always instable when the bifurcation of pre-critical $(B < B_{cr})$ solutions takes place (see Figs. 7.2a and 7.4b). In the pre-critical region, structures can be excited in a hard regime; in other words, such systems are of the trigger type.

It is instructive also to consider systems with "physically non-realizable" kinetics (Section 7.1, Example 3). The existence of stationary structures in such systems is guaranteed only for pre-critical magnitudes of the parameters (where $q_0 > 0$). Solutions ramifying at the critical points have exponential instability, as can be easily verified analysing the dependence $\dot{\Delta}_l(\Delta_l)$ (Fig. 7.1a, dashed—dotted line). Note that, as a rule, all inhomogeneous stationary solutions are unstable in such models.

The following statement is true for oscillatory instabilities in stationary DS: the instabilities take place only for those values of the parameters which correspond to auto-oscillating local kinetics. We omit the proof, as it is rather cumbersome though not too difficult, and turn to a discussion of concrete models.

Example 1. A system of Van-der-Pol oscillators with a cubic nonlinearity

$$\frac{\mathrm{d}}{\mathrm{d}t}x = -y + \delta_1 x - \delta_3 x^3; \quad \frac{\mathrm{d}}{\mathrm{d}t}y = x. \tag{7.22}$$

A limiting cycle for this system exists for any positive values of δ_1 and δ_3. For such local kinetics, there is a single type of the dependence $\dot{\Delta}_x(\Delta_x)$ as shown in Fig. 7.1a (solid line); $\Delta_x^* = \sqrt{-q_0(2)/6\delta_3 d_y}$; $\Delta_y^* = \Delta_x^*/2d_y$; $s_x = s_y = 0$, where $q_0(2) = -2d_y(\delta_1 - 2d_x)$. The system linearized with respect to the solution $\Delta_x^* \neq 0$ has a block-diagonal matrix, and the characteristic equation is reduced to a pair of equations,

$$\begin{aligned} &\lambda^2 - (\delta_1 - 3\delta_3\Delta_x^{*2})\,\lambda + q_0(0) = 0; \\ &\lambda^2 - (\delta_1 - 3\delta_3\Delta_x^{*2} - 2d_x - 2d_y) + q_0(2) = 0. \end{aligned} \tag{7.23}$$

Equations (7.23) show that if $\Delta_x^{*2} < \delta_1/(3\delta_3)$, the stationary solution Δ_x^* is instable. This condition can be represented in quite a general form, if one takes into account that $x_u = \pm\sqrt{\delta_1/(3\delta_3)}$ are boundaries of the increment region for the oscillator, i.e. the region in the phase space where $(\partial F_x/\partial x + \partial F/\partial y) > 0$, and the oscillator gains energy when the representing point is moving in this region, this is just the effect responsible for the existence of the self-sustained oscillations (Strelkov 1972).

The next statement is true for any system with diffusion couplings. If all the representing points of a stationary solution are contained within an increment region of the system phase space, the solution is instable. We do not present a proof here, but note that the condition is sufficient for instability of DS. The Brussellator model is an example evidencing that this condition is not necessary: for a Brussellator, as well as for some other models, stationary solutions having oscillational instability are projected into the increment regions, at least in part.

The model with a cubic nonlinearity enables one to see a substantial difference between the losses of stability in homogeneous and inhomogeneous states. It is found that for any combination of roots with positive real parts in equations (7.23) the only attractor is the homogeneous limiting cycle $\big(\Delta(t) \equiv 0\big)$. Only the regime of the excitation of such auto-oscillations depends on the type of instability for the solution Δ_x^*. This is also true, in general, for other systems. However, another type of auto-oscillations is also possible in models having quadratic (even) nonlinearities; namely, quasi-synchronous self-sustained oscillations around an inhomogeneous state. No analysis of stability against small perturbations makes it possible to predict what type of motion will exist in the nonlinear system.

Example 2. Auto-oscillations in the Brussellator model, Eqs. (2.54) and (7.4). In this model, for $B > B_{\mathrm{osc}} = 1 + A^2$, the local kinetics exhibits auto-oscillations.

In this spatially homogeneous case synphase auto-oscillations are always stable. Moreover, for sufficiently large values of B (see Figs. 7.2, 7.4) it is the only stable regime.

If $B_{cr} < B_{osc}$ and $A^2 < 4dy$ (i.e. in the absence of pre-critical structures, see Fig. 7.2b and compare Fig. 7.2a) two new limiting cycles appear at the bifurcation taking place as stability of the stationary inhomogeneous state is lost: one cycle is stable, the other unstable. The bifurcation parameters can be evaluated for $d_y \gg d_x$ as follows: $B_{bif} \simeq (d_x + A^2) + \sqrt{(d_x + A^2) + (1 + 2d_x)} > B_{osc} > B_{cr}$. An example of stable inhomogeneous auto-oscillations taking place for $B \gtrless B_{bif}$, which was calculated by means of numerical integration of the complete system, is shown in

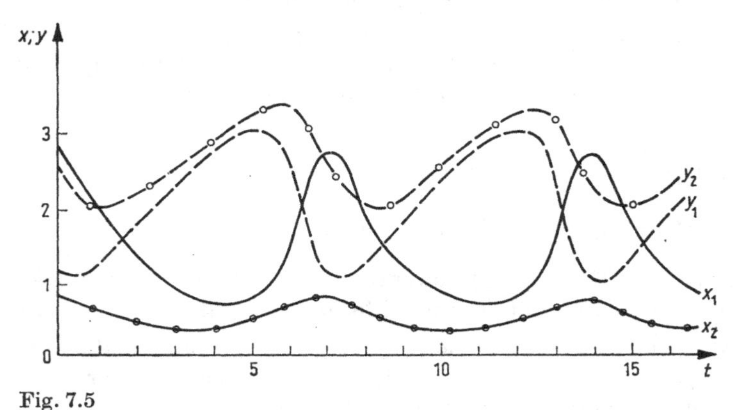

Fig. 7.5
Quasi-synchronous auto-oscillation around the inhomogeneous state of the discrete Brusselator (2.54): $A = 1$, $B = 2.5$, $d_x = 0.1$, $d_y = 1$ (see also Fig. 7.2)

Fig. 7.5. It is seen that the variables x_1 and x_2, y_1 and y_2 have almost synphase variations with respect to the states $x_{1,2}^* = \bar{x} \pm \Delta_x^*$. In order to get an analytical construction of such a limit cycle one can use the Bogolyubov-Mitropolsky (1974) averaging method (see Section 5.4). The small parameter is $\varepsilon = (B - B_{bif})^{1/2}$, and at least two terms must be taken into account in the expansion of the solution in powers of ε. Omitting the details, we present the final solution:

$$\Delta_x(t) = \Delta_x^* + 0(\varepsilon^3); \quad s_x(t) = s_x^* + \varepsilon a \cos \omega t + \varepsilon^2(a_0' + a_2 \cos 2\omega t) + \cdots,$$

where $a \simeq a' \pm \sqrt{(a'^2 + b) \operatorname{sign} (B - B_{bif})}$ and $B > 0$ for $\varepsilon^2 = B - B_{bif}$.

However, with an increase in $(B - B_{bif})$ the quantity b becomes negative, and these limiting cycles disappear finally. Thus this is a bifurcation of the fusion-of-cycles type.

According to the above arguments, which are confirmed by numerical experiments, homogeneous auto-oscillations are the only stable regime in the system. In general, such limiting cycles appearing at $B = B_{cr}$ do not lose their stability, as a rule, for all values of the system parameters. Usually, they have the largest domain of attraction in the system phase space.

No stable inhomogeneous oscillations are possible in the precritical parameter region ($B \ll B_{cr}$). There may be either stationary states (Fig. 7.2a), or a homogeneous limiting cycle. In the case of this type of bifurcation, the inhomogeneous stationary states have, as a rule, considerably larger regions of stability against small perturbations, as compared with situations like that in Fig. 7.2b. However, in this case also, the domain of attraction rapidly contracts as B increases with respect to B_{osc}. For instance, if the system parameters acquire the values indicated in Fig. 7.2a deviations of the dynamical variables from the stationary magnitudes cannot exceed 0.5% already for $B = 2B_{osc} \simeq 1/2B_{max}$ (DS are absolutely unstable if $B \geqq B_{max}$), while the deviations are less than 10^{-2}% for $B \simeq 0.9B_{max}$.

As we have seen, the multitude of bifurcations for the Brussellator model is considerably larger than the model with purely cubic nonlinearity (Example 1). Seemingly, the properties of the former model are typical for similar "chemical" systems. A detailed accurate analysis of bifurcations in an ecological model with $d_y \to \infty$ performed previously (Bazykin et al. 1980, Bazykin and Khibnik 1982) confirms this statement. It is not difficult to extend these results to the cases of distributed systems with quasiharmonical DS, and similar results can be obtained independently for systems of other types[1]). One should bear in mind that we mean only some aspects of relations between stationary DS and synchronous auto-oscillations.

In distributed systems, the DS-instability conditions are fulfilled, as a rule, simultaneously for several wave lengths (Fig. 7.3) corresponding to ramification of stationary solutions having different shapes (periods in space). Can all of them be indeed realized? One would be able to answer this question only after an investigation of their stability, in particular with respect to perturbations having the forms of the alternative solutions. Remaining within such a statement of the problem, we will consider forms of dependence of the solution amplitudes (A_k) on the system length (L). This problem is in itself of interest in view of some applications to biology (Vasilev, Romanovskii and Chernavskii 1982). It reveals also the multitude of realizable DS.

Example 3. Hysteresis transitions between DS induced by changes in the system length. Interactions between waves for which the ratio of the periods is 1/3 are most effective in systems with cubic nonlinearities. Therefore the most important set of dynamical equations for the Van-der-Pol model (7.22) is that involving the amplitudes of the first and third modes (A_1 and A_3). Taking the solution in the form $x(t, r) = A_1 \cos(\pi r/L) + A_3 \cos(3\pi r/L)$, Eq. (5.16), and averaging over the periods $L/2$ and then $3L/2$, one gets

$$\begin{aligned} \frac{\mathrm{d}}{\mathrm{d}t} A_1 &= \frac{A_1}{4} (\alpha_1 - 3A_1^2 - 3A_1A_3 - 6A_3^2), \\ \frac{\mathrm{d}}{\mathrm{d}t} A_3 &= \frac{A_3}{4} (\alpha_3 - 3A_3^2 - 3A_3A_1 - 6A_1^2), \end{aligned} \tag{7.24}$$

---

[1]) For instance, pulsations of contrast ($D_y \to \infty$, Section 7.4) DS in the Brussellator model have been considered by Kerner and Osipov 1982, 1983.

where $\alpha_1 = -4q_0(\pi^2/L^2) \cdot L^2/\pi^2 D_y$, $\alpha_3 = -4q_0(9\pi^2/L^2)\, L^2/9\pi^2 D_y$, and q_0 is the free term in the characteristic equation (3.6). For simplicity we put $\delta_1 = \delta_3 = 1$.

A graphical method is appropriate for the qualitative analysis of solutions of (7.24). The zero-isoclines are shown in Fig. 7.6 for the following cases:

(i) $-q_0(\pi^2) \sim \alpha_1 > 0$ and $-q_0(\pi^2 9) \sim \alpha_3 < 0$ (Fig. 7.6a),
(ii) $\alpha_1 > 0$ and $\alpha_3 > 0$,
(iii) $\alpha_1 < 0$ and $\alpha_3 > 0$.

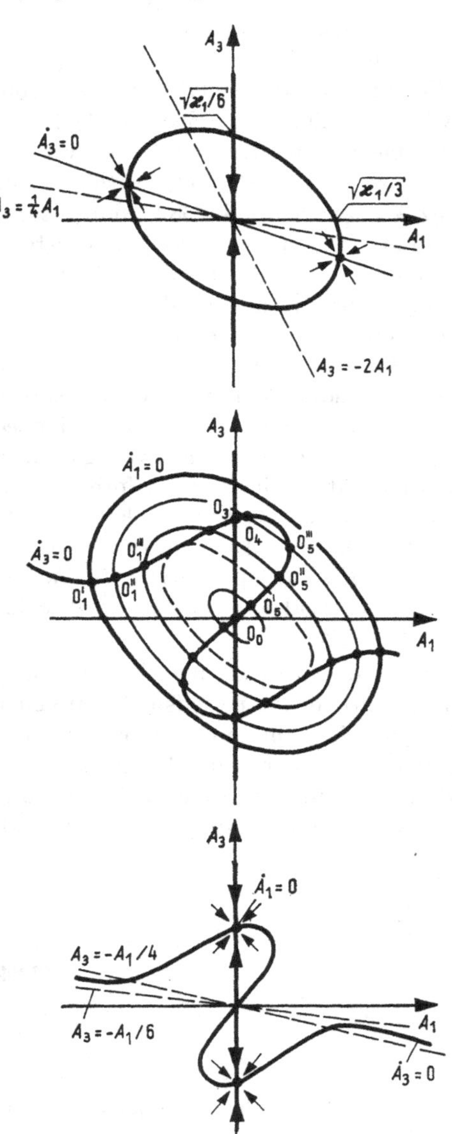

Fig. 7.6
Phase plane of the two-mode model (7.24) of the distribution Van-der-Pol systeme (7.22). The hysteresis transitions between first and third mode

The family of enclosed ellipses (Fig. 7.6 b) corresponds to decreasing values of $\varkappa_1$. The singular points O_1 and O_5 are stable nodes, O_0, O_2 and O_4 are saddle points, and O_3 is a saddle point if it is inside a certain ellipse, otherwise it is a stable node. Using graphs of the L dependence of $\varkappa_1$ and $\varkappa_3$ (Fig. 7.7 c) and taking into account the changes in positions, number and character of the singular points which are due to the variations of L (Fig. 7.6) one can easily construct the desired L dependence for A_1 and A_3 (Fig. 7.7 b). The $1 \rightarrow 3$ transition occurs at a value of L for which the ellipse (dashed line in Fig. 7.6 b) makes contact with the A_3-zero-isocline within the second and fourth quadrants. The inverse $3 \rightarrow 1$ transition occurs when this contact takes place within the first or third quadrants.

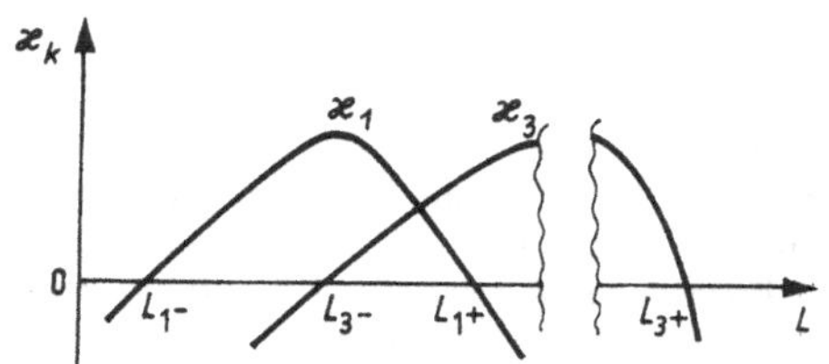

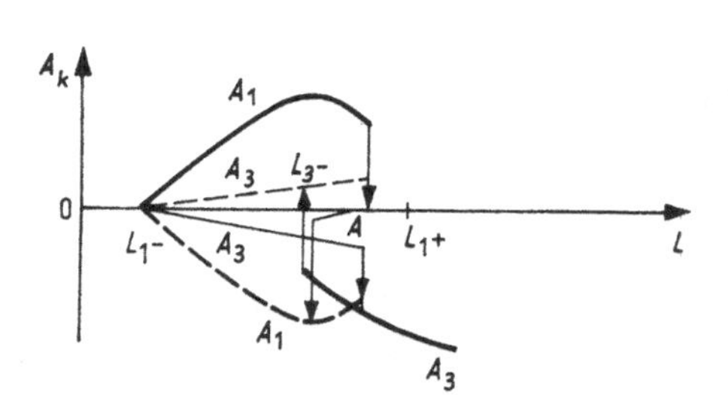

Fig. 7.7
Variations
a) of the linear increments of the modes ($\varkappa_k$) and
b) of the harmonic components (A_k) of DS with the length (L) of the Van-der-Pol system (7.22)

Thus a hysteresis loop appears as L changes.[1] Note that the "polarization" of DS is inversed ($A_1 \rightarrow -A_1$) when the loop is encircled completely ($1 \rightarrow 3 \rightarrow 1$). The multitude of DS is a general property of kinetic systems. Actually, the hysteresis is due to the fact that the transformation of the first harmonic to the third one, arising from the nonlinearity, is not equivalent to the inverse transformation. Naturally, the orders of modes between which transitions are possible, are determined completely by the type of the nonlinearity.

In the Brussellator model, the transitions concerned proceed according to another scheme: $1 \rightarrow 2 \rightarrow 4 \rightarrow 8 \rightarrow \cdots$ or $3 \rightarrow 6 \rightarrow 12 \rightarrow \cdots$ etc. A closed chain of transitions is possible only if the self-excitation regions of the modes do not overlap at all (Fig. 7.8 c). A series of numerical experiments has been performed to elucidate this point (Vasilev and Romanovskii 1977); the results are shown in Fig. 7.8, they are in complete agreement with predictions of the above qualitative analysis. The conclusion holds not only for quasiharmonical DS, but also in the general case.

---

[1]) One should bear in mind that the approach described is relevant to the particular model. In fact, we disregard the existence of a powerful attractor, the homogeneous limiting cycle. The final state in the chain of transitions may be not a stationary dissipative structure, but the homogeneous limiting cycle.

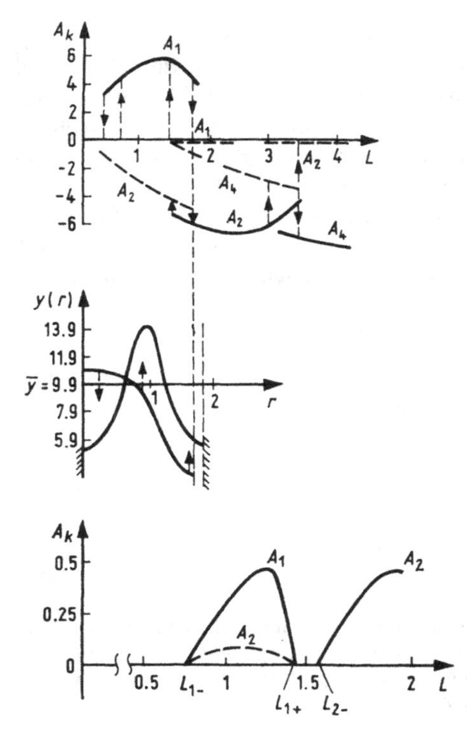

Fig. 7.8

a) Hysteresis transitions between DS with different shapes upon a change in the length of Brussellator (2.54). A_k are the amplitudes of the harmonic components ($A = 10$, $B = 99$, $D_x = 0.75$, $D_y = 1$);

b) Initial and final states in the transition from $k = 1$ to $k = 2$;

c) possibility of soft transitions between DS with different shapes in Brussellator ($A = 5$, $B = 18$, $D_x = 0.04$, $D_y = 1$)

Example 4. The role of the boundary conditions. DS-type solutions are also possible in one-dimensional systems homogeneous in the coordinate (Chafee 1974). Note that the system must be homogeneous including the boundary conditions. For the third-kind boundary conditions, $D_x\, \partial x/\partial r|_{r=0,r=L} = -c(b - x)$, the DS-type solutions are unstable (Lifshits et al. 1981). These authors have also analysed bifurcations due to variations of the length; it was found that only the first mode is stable. So the possibility of such systems to generate structures is rather poor. The variable is distributed smoothly in the pre-critical region of the parameters, and a single "step" is formed in the over-critical region; the further into this region the steeper is the step.

The last example of the simplest models indicates quite clearly that the boundary conditions are essential for the solution of the stability problem for DS. This fact must be taken into account in extensions of the qualitative analysis of DS models to other systems. Other conclusions drawn from the discussion of the examples given above are valid for a wide class of system and for various types of DS.

7.4 Contrast dissipative structures

Systems in which the DS instability conditions are fulfilled for a large number of modes ($k_{\max}/k_{\min} \gg 1$) deserve a special treatment in the theory of DS, since sequences of multiple modes interacting effectively due to the system nonlinearity may appear among the instable modes. The procedures proposed for constructing solutions based on the evaluation of a linearized generating system (cf. Section 7.2) are rather ineffective in this case; they can serve only to localize the bifurcation points. In the theory of auto-oscillations, such a situation is specific for relaxational systems; in analogy to the latter one can treat systems with small parameters in the spatial derivatives (D_{ii}). To be more precise, the necessary condition is $D_y(\partial F_x/\partial x/D_x(\partial F_y/\partial y) \gg 1$ corresponding to $k_{max}/k_{min} \gg 1$ (see Section 3.3). We shall assume for simplicity that the specific time scales for the variables x and y have close orders of magnitude, so that we have $D_x/D_y \ll 1$ for the systems concerned.

DS appearing in such systems are called "contrast" DS, as the space covered by DS of this type is divided into domains separated by sharp boundaries. Two types of contrast DS are known; they differ in the distributions of the variables. The corresponding models are solved by means of schemes which are different but have some common features.

Let us begin with some general aspects of constructing solutions for systems of stationary equations,

$$D_x \frac{d^2}{dr^2} x + F_x(x, y) = 0; \quad D_y \frac{d^2}{dr^2} y + F_y(x, y) = 0 \tag{7.25}$$

provided with boundary conditions at the end-points of an interval $[0, L]$. Suppose we are dealing with second-kind boundary conditions, Eq. (1.4), then the relations

$$\int_0^L F_x(x, y)\, dr = \int_\Omega F_x(x, y(x))\, dx = \int_0^L F_y(x(y), y)\, dy = 0 \tag{7.26}$$

hold for any solution of (7.25) (here Ω is the projection of the solution onto the (x, y) phase plane).

It is appropriate to start from an elementary cell of DS. The size of the cell must be about equal to the diffusion length in the "passive" variable y, and the second-kind boundary conditions (1.4) are assumed. This is a "half-wave" cell, it is an analogue of the $k = 1$ mode in the case of quasi-harmonical DS. Evidently, such cells having a length of L_p can be used to build DS in the whole interval, the length of which is L. The cells in view can be investigated in order to evaluate the number of stable solutions depending on the shapes of F_s and F_y, to find the minimal and maximal values of L_p for which DS exist.

Case 1: Rectangular DS. DS of this type are typical for systems having an N-shaped zero-isocline for the auto-catalytic variable x (where $D_x \ll D_y$), see Fig. 7.9. A modified "Oregonator" model (5.4), obtained from that in (2.57), may serve as an example for such a system. (Note that here we have interchanged x and y in (5.4) in order to make the notation more suitable for comparison with other models, so that x is the auto-catalytic variable.)

Suppose $D_x = 0$, so that we get the zeroth approximation in the parameter D_x/D_y. Evidently, for any stationary solution of (5.4) its projection must be on the x-zero-isocline; the projection is shown by a heavy line in Fig. 7.9. The presence of a discontinuity (dashed line), i.e. the existence of two segments lying above and below

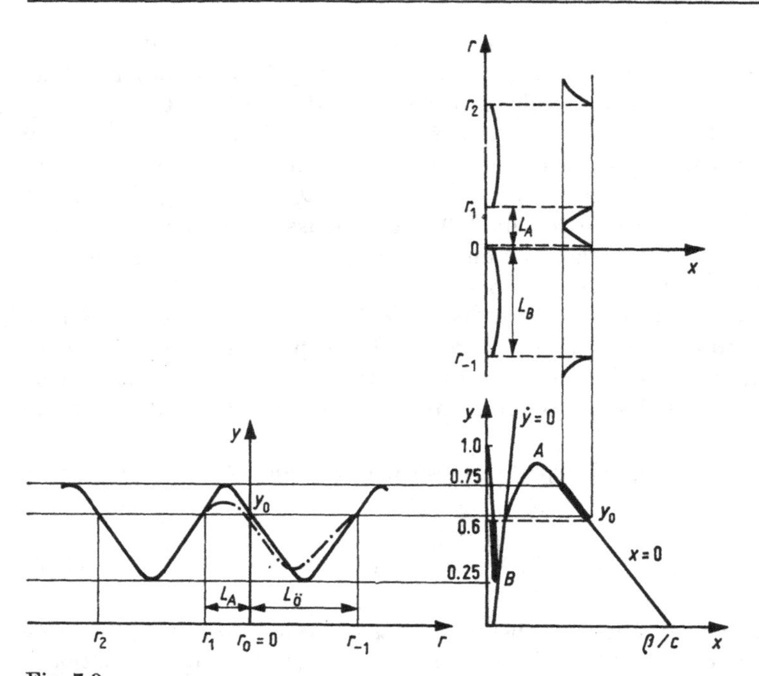

Fig. 7.9
Contrast DS in the Oregonator model (5.4). The scheme of solution construction for the case $D_x \equiv 0$

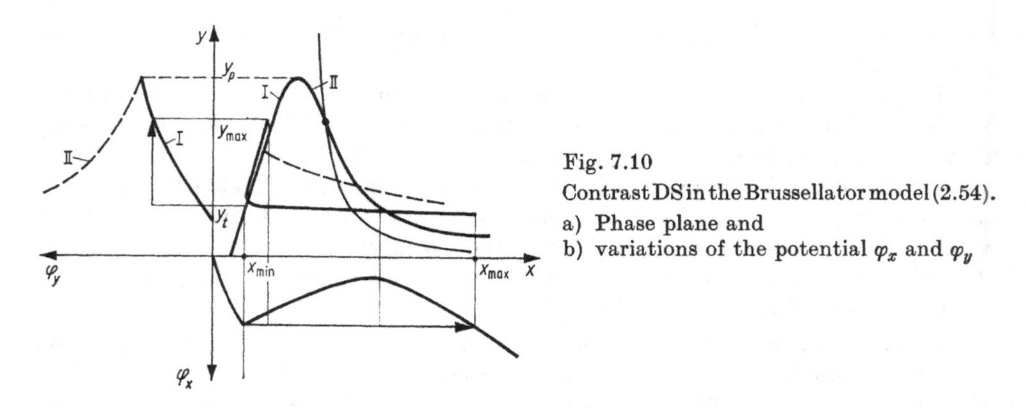

Fig. 7.10
Contrast DS in the Brussellator model (2.54).
a) Phase plane and
b) variations of the potential φ_x and φ_y

the y-zero-isocline, is a consequence of condition (7.26): $\int F\big(x(y), y\big)\,\mathrm{d}y = 0$, and the continuous solutions corresponding to the segment AB of the y-zero-isocline are unstable.[1]) In the regions where solutions are continuous, there is a dependence $x(y)$ determined by the algebraic equation $F_x(x, y) = 0$. The function $x(y)$ has three

[1]) This statement seems evident if one takes into account the fact that the solutions, in fact, satisfy a homogeneous system with a single variable.

branches $x_A(y)$, $x_B(y)$, and $x_{AB}(y)$ corresponding to the segments A, B and AB of the y-zero-isocline (Fig. 7.10).

The function $y(r)$ must be continuous and satisfies equations (7.25) at the end-points of the interval (L_p), $\mathrm{d}^2y/\mathrm{d}r^2 + F_y(x(y), y) = 0$, as well as the second-kind boundary conditions (1.4). The boundary conditions for x are also fulfilled in this case, as

$$\left.\frac{\mathrm{d}x}{\mathrm{d}r}\right|_{r=0;L_p} = \frac{\mathrm{d}y_{A;B}(x)}{\mathrm{d}x}\left(\frac{\mathrm{d}y}{\mathrm{d}r}\right)_{r=0;L_p} = 0. \tag{7.27}$$

The continuity in y means that the fluxes between the branches $y_A(r)$ and $y_B(r)$ are equal

$$\left.\frac{\mathrm{d}y_A}{\mathrm{d}r}\right|_{y=y_p} = \left.\frac{\mathrm{d}y_B}{\mathrm{d}r}\right|_{y=y_p} = 0. \tag{7.28}$$

For $D_x = 0$, the place where y_p has a discontinuity may be chosen arbitrarily in the phase plane, within the segments A and B. Therefore a continual set of solutions can be constructed in the interval L_p. For instance, one can use the following procedure for this purpose.

(i) First, one has to solve the Cauchy problem with the initial conditions $\mathrm{d}y/\mathrm{d}r_{r=0} = 0$ and $y(0) = y_{\max}$ for the segment A on the interval of a length L_A (in the following $y_{\max}$ and L_A will be considered as parameters for the family of solutions).

(ii) One solves the Cauchy problem with the initial conditions $y_B(0) = y_A(L_A)$ and $\mathrm{d}y_B/\mathrm{d}r|_{r=0} = \mathrm{d}y_A/\mathrm{d}r|_{r=L_A'}$ for the segment B also, and omits the only solution having zero derivative, say, at the point $r = L_B$. Such a point always exists for solutions of the second-order equations $\big(Fy(x(y), y)\big)$.

Remark. The solutions $y_A(r)$ and $y_B(r)$ may exhibit oscillations. One should retain only monotonic segments, as oscillating solutions are absolutely unstable (see example 4 in Section 7.3). Evidently, one should discard solutions, projections of which go outside the segments A and B. Nevertheless, there still remains a large stock of solutions.

(iii) Having performed the first and the second step, one gets a family of solutions, among which those with $L_A + L_B = L_p$ are now to be selected. Thus the problem for the interval of length L_p is solved for $D_x = 0$ and $D_y \neq 0$.

The procedure proposed here to obtain solutions shows that one has a continuum of solutions for the interval L_p. An evident consequence of this fact is the existence of a multitude of contrast DS forms for an interval $L \gg L_p$. Restrictions on the selection of solutions, which were formulated in the above remark, suggest that there is a maximal length of the elementary cell (the period of DS) which is determined, first, from the requirement of the absence of oscillations in y on the smooth segments A and B (hence $L_p \approx (D_y)^{1/2}$), and second, from the maximal amplitude $y(r)$ ($y_{\min} < y(r) < y_{\max}$) determined by the local kinetics of the system. So it is impossible to estimate L_p without analysis of the complete nonlinear model, as was the case for quasiharmonical DS.

One of the parameters specifying the solutions for $D_x = 0$ may be the coordinate of the discontinuity. Clearly, solutions belonging to families with different values of L_p can be matched by continuity at $y = y_{max}$ and $x = x_A(y_{max})$ (or y_{min} and $x_B(y_{min})$). Therefore for systems with $L \gg L_p$ one can get contrast DS with randomly distributed distances between the peaks, $L = L_{p,1} + L_{p,2} + L_{p,3}$, while shapes of the peaks are also different. It should be emphasized that this result is based on the existence of discontinuities in the distribution of the autocatalytic variable x for $D_x = 0$.

In the next approximation, taking into account the first-order terms in D_x/D_y, we find that the distribution $x(r)$ must be continuous; the region where this variable has a sharp variation is about l in length, where $l = L_p(D_x/D_y)^{1/2} \ll 1$. In this region, the variable $y(r)$ is a constant y_p, up to corrections of order of magnitude $\epsilon = D_x/D_y$. One can verify this statement using the perturbation techniques (Section 3.4). In the regions where the distributions are smooth, the solution can be expanded in powers of $\epsilon: x(r) = x_A(r) + \epsilon x_1 + \epsilon^2 x_2 + \cdots$, where $F_x(x_A, y_A) = 0$. Inserting this expansion into Eq. (7.25) one gets $x_1 \simeq (\mathrm{d}^2 x_A/\mathrm{d}r^2)/(\partial F/\partial x)|_{x_A, y_A}$. As x_A satisfies the second-order differential equation, x_1 is finite throughout the segment A. Using these results we can evaluate the integrals present in the conditions necessary for existence of the solutions, Eqs. (7.26),

$$\int_0^L F_x(x, y)\,\mathrm{d}r = \int_0^{L_A - l/2} F_x\,\mathrm{d}r + \int_{L_A - l/2}^{L_p} F_x(x, y)\,\mathrm{d}r \simeq \epsilon + \sqrt{\epsilon} \int_{x_{Min}}^{x_{Max}} F_x(x, y)\,\mathrm{d}x = 0\,.$$

Thus for $\epsilon \to 0$, (7.26) is reduced to the following equation for y_p:

$$\int_{x_{Min}}^{x_{Max}} F_x(x, y)\,\mathrm{d}x = 0\,. \tag{7.29}$$

It is noteworthy that the solution of (7.29) coincides with the zero of the function $v(y_p)$, i.e. of the dependence of the TP front velocity on the slow variable (Section 3.7). For systems with monotonic dependence $v(y_p)$, there is a single position of contrast DS fronts, so that only cells with identical L_p can be matched in the interval. If a system has a non-monotonic $v(y_p)$ (Section 4.7) there are two stable positions of the fronts. It is only for discrete models (Sections 3.6 and 3.9) that a continuum of y_p values exists, for which $v = 0$ (Fig. 3.16). Because of this fact, in numerical experiments employing finite-difference schemes solutions which are specific for the case $D_x = 0$ may appear for the basic models (1.3) even with low numbers of elements, N. Genuine random distribution can take place in media with non-monotonic $v(y_p)$; for a model with two values of v_p one can encounter a random mixture of two DS elements, which is determined by the initial conditions.

To complete the construction of the first approximation, one has to specify the matching condition in y, (7.28), in the region of sharp variation. To this end one can use the continuity condition (1.1); the result is

$$I_A - I_B = D_y(\mathrm{d}y_A/\mathrm{d}r - \mathrm{d}y_B/\mathrm{d}r)|_{y_A = y_B = y_p} = \int_{x_{min}}^{x_{max}} F_y(x, y_p)\,\mathrm{d}x\,, \tag{7.30}$$

where the integral (the source in the sense of Eq. (1.1)) provides a contribution of the order of $\sqrt{\epsilon}$. The substitution of Eq. (7.30) for (7.28), as well as higher-order approximations in ϵ, do not change the character of the solutions qualitatively. They essentially modify the spatial dependence of the transition between the regions of smooth and sharp distributions.

Now the scheme appropriate for construction of contrast DS is as follows. The zeroth approximation scheme is applied to obtain a family of solutions for L_p, then a single solution is extracted from this family by means of Eq. (7.29), the shape of which is made more precise using (7.30).

Case 2. Spiking contrast DS. In models of the Brussellator type, Eqs. (2.54), the x-zero-isocline has a single extremum (Fig. 7.10), so the contrast DS considered above can not be constructed[1]). The sufficient conditions for the existence of DS are fulfilled in a very wide region of the model parameters, $B > B_{cr} = 1 + \pi^2 k^2 D_x/L^2$ (cf. (7.14), Fig. 7.3, for the case $D_x/D_y \approx 1$). The question arises of what shape DS have in such systems for $\epsilon = (D_x/D_y) \to 0$.

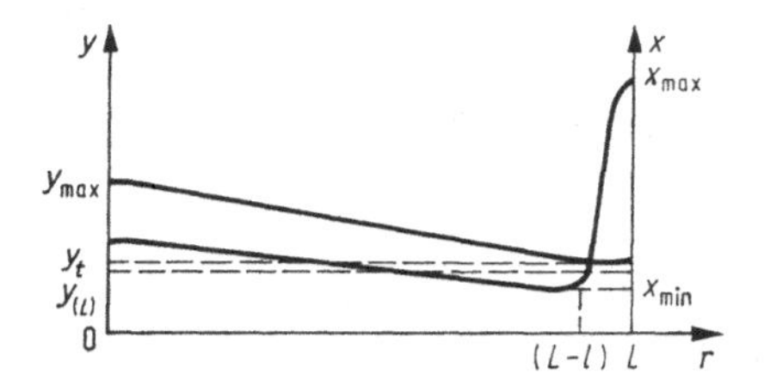

Fig. 7.11
Shape of contrast DS in the Brusselator model (2.54)

The main arguments relevant to case 1 are valid here as well. The differences between the two cases are due to the behaviour of solutions in the regions where the distributions vary sharply. Putting $y(r) = y_t = \text{const}$, one gets an equation

$$\frac{d^2}{dr^2}\,x + A(B+1)\,x + x^2 y_t = \frac{d^2}{dr^2}\,x + \frac{d}{dx}\,\varphi_x(x) = 0\,, \tag{7.31}$$

where $\varphi_x = \int F_x(x, y)\,dx$ is a potential, the potential φ_y is introduced similarly. The introduction of the potential functions φ_x and φ_y makes it possible to exploit mechanical analogies and to obtain a qualitative understanding of the solutions in the smooth and sharp segments (Kerner and Osipov 1978, 1980).

The function $\varphi_x(x, y_t)$ indicates that there is a potential well (Fig. 7.10). Only motions starting from the saddle point of $\varphi(x, y)$ are interesting, and conditions (1.4) must hold at the boundary point x_{max}, $dx/dr = 0$. The function $\varphi_y(y)$ has two branches corresponding to two solutions $y(x)$ of the quadratic equation $F_x(x, y) = 0$. It is easily seen that in the potential φ_y the motions corresponding to the second branch are unstable, so we disregard them in the following.

---

[1]) Such a situation is specific for the so-called systems with triangular limiting cycles (Romanovskii et al. 1975). Note that substituting $x \to x'$, $x + y \to y'$ in the local model one gets an N-shaped x-zero-isocline, but the diffusion matrix is not diagonal in the distributed system, so the shapes of contrast DS are changed qualitatively.

The use of the potentials enables one to integrate the equations of motion in separate segments and to obtain the families of solutions satisfying the second-kind boundary conditions at the end-points of the interval L_p. The families are parametrized by y_{max} and y_t. To complete the construction of the solution, one has to match the variables and their derivatives.

With an accuracy up to corrections of order ϵ, one has $y_t = y(L_p)$, where $y(r)$ is the solution on the smooth segment, and $y(0) = y_{max}$, $dy/dr|_{r=0} = 0$ (Fig. 7.11). The continuity conditions, (1.1) or (7.26), lead to the following condition

$$\int_{L_{pl}}^{L_p} F_y(x, y)\, dr = D_y \frac{d}{dr} y \bigg|_{r=L}, \tag{7.32}$$

where $l \sim \sqrt{\epsilon}\, L_p$; l must be evaluated precisely from the obtained solution in the sharp region. Both sides in Eq. (7.32) have the order of $\sqrt{\epsilon}$, so it can be used only in the first and higher-order approximations. Condition (7.26) for $F_x(x, y)$ is accurate up to terms in ϵ in the zeroth approximation, to ϵ^2 in the first approximation, etc. Thus in the first approximation one can get a solution in the interval L_p, and the length of the latter determines the parameters x_{max}, y_t, y_{max} unambiguously.

Basic properties of contrast DS

Contrast DS arise under conditions imposed on the local kinetics which are much weaker than those in the case of quasi-harmonical DS. For example, in the Brusselator model such DS appear already for $B \simeq 1$, while active properties of the system do not manifest themselves in other processes. The contrast DS already have large amplitudes near the first bifurcation point. The maximum possible period is due to the front (spike) width; it can be estimated from the magnitude of the diffusion coefficient in the auto-catalytic variable, while the exact value depends only on the rate of the auto-catalysis in the system. The maximum possible period depends on D_y, as well as on features of the local kinetics, e.g. on positions of extrema of zero-isoclines in the phase plane. For long systems ($L \gg L_p$) there are many stationary solutions with periods L/k ($k = k_{min}, k_{min} + 1, \ldots$). The number of possible solutions goes to infinity as $L \to \infty$, but the set of solutions remains separable.

The stability of contrast DS against small perturbations has been investigated exhaustively by Kerner and Osipov (1978, 1980, 1982a, 1982b). They have developed a method which can be used to estimate eigenvalues for the corresponding linear inhomogeneous problems. Analogies to familiar quantum-mechanical problems were found to be fruitful (see also Romanovskii, Stepanova and Chernavskii 1984). We are not in a position to describe here in detail the results of these studies; only some points will be presented now. Among one-spike DS, only those are stable, which have the spike at the boundary of the system. The one-spike DS may have oscillational instability in short specimens ($l \ll L \ll L_p$). Supposedly, quasi-homogeneous auto-oscillating DS arise in this case (cf. Fig. 7.5). The stability of DS having various shapes is shown both for spiking and rectangular DS. This means that any contrast DS belonging to a set can be formed depending on the initial conditions.

In our opinion, one should in this respect bear in mind the following. In real systems kinetic processes are always associated with fluctuations. They play an especially important role near the bifurcation points where the system states are separated by relatively low thresholds. This is the case for most of the DS which are possible in principle. Effects of fluctuations in formation processes have been investigated previously for some objects which are simpler than contrast DS (see Chapter 8); a few numerical experiments have been performed for contrast DS (Balkarey, Evtikhov and Elinson 1983). Nevertheless, we feel it is not out of place to present some qualitative arguments concerning the behaviour of contrast DS in fluctuation fields.

Suppose certain "complicated" initial conditions lead to a structure. Fluctuations initiate transition to another, and then other transitions follow until the system arrives at a dissipative structure which has the largest attracting region. Actually, in long systems ($L \gg L_p$) DS having close shapes are separated by low thresholds. The terms involved in the matching conditions (7.30) or (7.32) have the order of magnitude of $\sqrt{\epsilon} \ll 1$, so if the conditions do not hold, times specific for transitions to "regular" states are as large as $1/\epsilon$, while formation times for spikes (pulses) are ≈ 1 by order of magnitude. One can suppose that there is a more dense set of quasi-stable DS in the system. Evolution of the system from an "accidental" initial distribution to a final structure proceeding via these states can be almost continuous, and its duration time may be relatively short.

Formation of DS in fluctuation fields deserves an accurate theoretical investigation, and this is a subject of forthcoming work. On the other hand, deterministic models describe the basic process of formation of DS which also leads to a principal state.

As a rule, formation of DS proceeds as "self-building": after the appearance of the first spike the second one arises at a distance of L, followed by the third spike at a distance of L from the second one, and so on. This phenomenon had already been observed in early experimental work dealing with DS (Zhabotinsky and Zaikin 1973, Zhabotinsky 1974). A qualitative explanation can be proposed for the effect.

Let us suppose that a spike has been initiated in the distribution of a variable x for a long system which was originally in a stationary homogeneous state. (See Fig. 7.12, where projections of the distributions on the phase plane for the Brusselator model are presented for consecutive time moments.) The spike projections move as indicated by the arrows ($F_x > 0$, $F_y < 0$). As D_x is small, projections of the spike elements move away from the x-zero-isocline. Since D_y is large, in the L_p scale the projections of adjacent regions move downwards from the singular point, and entering the domain with $F_x < 0$ progress to the left branch of the x-zero-isocline. Only motions to the extremum point of the x-zero-isocline are possible along this branch. In the regions whose projections have attained the extremum, the next spike forms, then the process is repeated, until the structure occupies the whole system.

The quantity y_t can be estimated for systems with $\tau_x/\tau_y \ll 1$, where it acquires a value corresponding to the maximum possible L_p for $y_{\max} = y_{\text{extr}}$.

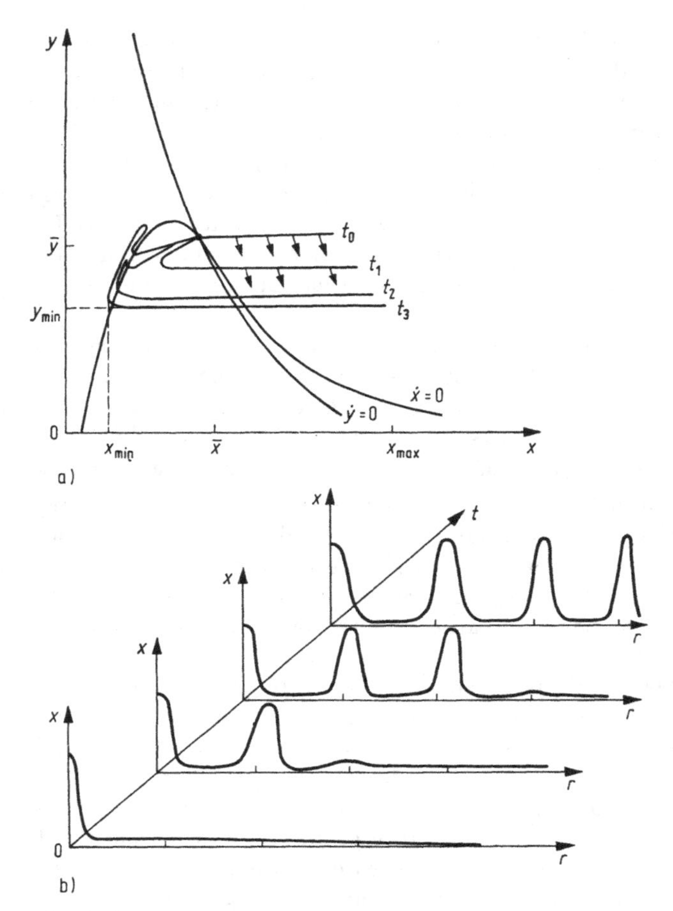

Fig. 7.12
Establishment of DS in a Brussellator by means of the process of self-building up

Only the formation times for the first and second spikes depend on the initial conditions, in the subsequent evolution the system forgets the initial conditions completely, and the region occupied by the structure expands with a constant velocity. Thus the formation of DS is itself an AWP, like the propagation of TFs or waves of phase transitions.

If DS expand in a system of finite length L, the last period is, in general, different from the auto-model estimate ($L/L_p = k$ — an integer). Then a new process begins, the equalization of periods throughout the structure. Times specific for this stage can be much larger than the self-building times. Note that the second stage may not be observable in numerical experiments, if the periods formed after self-building up of the DS have only negligible differences. In this case one should test the continuity conditions for the medium, Eq. (1.1), written down as (7.26). If it was found that they are not fulfilled with the computer accuracy, one should continue the computations with an increasing number of elements in the discrete model. If "random" forces are switched on in the model for some time period, the process of formation of periodical DS is completed more rapidly. Naturally, the number N must be a multiple of k.

Self-building up of DS also occurs in the case of rectangular structures. The x-zero-isocline has two extremal points, so processes of two types are possible. The first type behaves as that described above, but in the second type self-building is as follows. The pulse front moves towards the free medium, so that the pulse plateau expands. Then a dip appears in the plateau, and the newly born structure element leaves the place of its origin, and its plateau expands. Thus, the formation of rectangular DS is also an AWP.

The DS self-building up mechanisms are effective in a wide region of the system parameters. For instance, self-building has been observed in numerical experiments performed for contrast DS (Balkarey, Evtikhov and Elinson 1983) in systems having monostable, auto-oscillational, as well as trigger local kinetics. It occurs also in the process of formation of quasiharmonical DS, i.e. in systems where $D_x \lesssim D_y$ (Vasilev, Romanovskii and Yakhno 1979). There are two stages in the formation of such DS: (i) a sequence of spikes appears relatively quickly, (ii) distributions in the system variables become smooth and acquire a quasi-harmonical form. This process takes place not only for pre-critical values of the system parameters ($B < B_{cr}$, cf. Fig. 7.4b), it is also probable in situations where the DS self-excitation conditions are fulfilled ($B > B_{cr}$, Fig. 7.4). Actually, for small values of the mode amplitudes the rate of their increase is exponentially small, and the self-building waves have enough time to reach the system boundary, if the system is not too long, before the instability of the medium breaks down the homogeneous state of the system. Of course, the latter remark is quite relevant to all types of DS. (See also Elenin et al. 1983).

Comparing contrast and quasiharmonical DS one can note that the differences between them are quantitative rather than qualitative. A study of both types of DS, which are just certain limiting cases, enables one to extend the results obtained here to arbitrary DS, which cannot be simulated with simple models.

7.5 Dissipative structures in systems with mutual diffusion

As shown in Chapters 3 and 5, for models with non-diagonal diffusion matrix the requirements on the local kinetics are essentially weaker, yet the system is able to generate AWPs. This statement can be demonstrated in a most spectacular way for DS. Let us consider the model of Eq. (2.55):

$$\mathrm{d}x/\mathrm{d}t = Ay - Bx - x^2, \quad \mathrm{d}y/\mathrm{d}t = Bx - py .$$

This is one of the simplest examples of a thermodynamically open system. The parameters have a simple physical meaning: A specifies the influx of energy (substance) into the system, and p the dissipation of energy to the environment. The ratio A/p determines the rate of positive feedback in the system.

The phase plane for the local system, (2.55), is shown in Fig. 7.13 for the case $A > p$. The singular point $\bar{x} = 0$, $\bar{y} = 0$ is a saddle-point, and the point $\bar{x} = (A/p - 1) B$, $\bar{y} = B\bar{x}/p$ is a stable node. If $A < p$, the singular point is a stable node, and there are no singular points in the first quadrant. Clearly, the model has only trivial solutions for any values of the parameters. Moreover, such a picture of the phase plane is, seemingly, an indication that the system cannot generate AWPs.

Let us analyse the stability of the homogeneous state, assuming that $A > p$ (Section 3.3). Since $M_{11} = -B(2A/p - 1) < 0$, and $M_{22} = -p < 0$, the stationary state is stable for any D_{xx} and D_{yy}, if $D_{xy} = D_{yx} = 0$, (Eq. (3.8)). In the general case, the free term of the characteristic equation reads ($k^2 = \pi^2 K^2/L^2$)

$$q_0(k^2) = k^2(D_{xx}D_{yy} - D_{yx}D_{xy}) + k^2[D_{xx}p + D_{yy}(2A/p - 1) + D_{xy}B + D_{yx}A] + (A - p) \cdot B. \tag{7.32a}$$

For stable diffusion processes one has $\det (D_{ij}) > 0$, and since $A > p$, a DS-instability is possible only if D_{yx} and D_{xy} are negative $\left(q_0(k^2) < 0\right)$. Besides, if D_{yx}, $D_{xy} = 0$, there is a possibility of finding an A_{cr} such that the system state, which is homogeneous in space, is unstable for $A > A_{cr}$.

For example, take $B = 0.1$, $p = 1$, $D_{xx} = 2.4$, $D_{yy} = 1.2$, $D_{yx}(\bar{y}) = -0.8$, $D_{xy}(\bar{x}) = -0.4$. The DS-instability conditions hold for $k^2 \in (k^2_{\min}, k^2_{\max})$ and $A > A_{cr} = 6$.

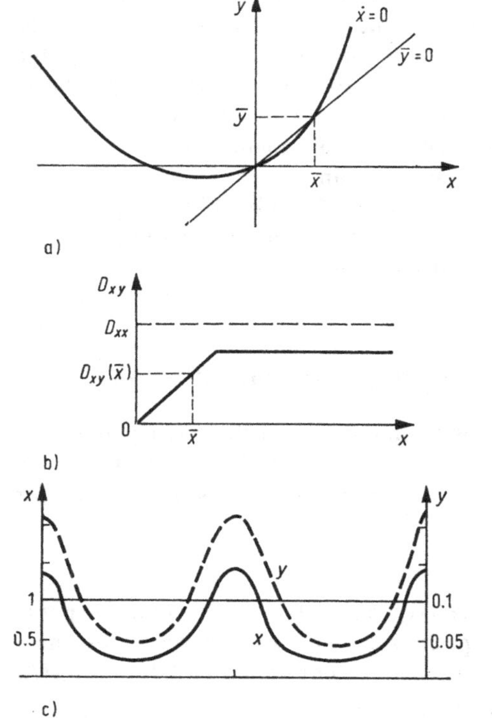

Fig. 7.13
DS in the model with mutual diffusion (2.55).
a) Phase plane;
b) typical dependence of D_{xy} upon "its" variable x;
c) final spatial distribution

The length of the interval of allowed wave numbers increases rapidly with A, i.e. as the influx rises.

In order to analyse stationary solutions, the system (2.55) has been integrated numerically for $A > A_{cr}$. The dependence of the mutual diffusion coefficients on the system variables has been taken as a piecewise-linear function (Fig. 7.13b). Such functions $D_{xy}(x)$ are reasonable from the point of view of the diffusion theory (see Section 2.2), and they lead to the existence of stationary solutions with finite amplitudes (Fig. 7.13c).

The forms of dependence of D_{xy} and D_{yx} are essential for the existence of bounded solutions. If D_{xy} does not vanish at $x = 0$, the latter can be negative, and the solution would increase infinitely. Linear forms of dependence (and others of this type) also lead to infinite solutions, and at some level one gets det $(D_{ij}) < 0$. Only forms of dependence that are reasonable from the physical point of view $\left(D_{xy}(0) = 0, |D_{xy}(x)| < D_{xx}\right)$ provide stable stationary solutions, DS.

Systems of type (2.55) cannot be reduced to models with diagonal diffusion matrix even for constant D_{yx} and D_{xy}: introducing a basis of eigenvectors of the corresponding linearized problem (as in Section 5.4) one can obtain representations of the problem having similar forms, but in the case of system (2.55) one gets a model with physically non-realizable local kinetics.

It seems probable that in the model (2.55) DS have properties different from those considered above. Actually, even near the first bifurcation, solutions of (2.55) are rather complicated, so that quasi-harmonical DS are not observable at all. Contrast DS do not exist either in this model. As to the stationary distributions in the variables, the DS considered here are intermediate between contrast and quasi-harmonical DS. No special features have been detected in the behaviour and characteristics of these DS. The existence of DS in systems with such a simple local kinetics is of interest in itself.

7.6 Localized dissipative structures

The homogeneous systems considered in preceding sections are completely occupied by DS. The DS self-building AWPs lead to complete occupation even in the cases where the homogeneous state is stable against small perturbations, but there are periodical DS solutions. The problem considered in this Section is to determine the conditions under which systems of the type (1.1) to (1.3) have localized solutions. The minimal conditions are as follows. Localized DS must be independent of the system size, be stable, and interact "weakly" with other DS of this type, if the separation between them is above a threshold. The problem has not been solved exhaustively, we shall discuss only some of its aspects.

Localization of the region occupied by DS may be due to a spatial inhomogeneity of the system itself. A detailed investigation of the Brussellator model, Eqs. (2.54), is given in the monograph by Nicolis and Prigogine (1977). To generate an inhomogeneity, the diffusion equation was introduced for a system parameter (B or A);

the equation was provided with non-zero first-kind boundary conditions and was independent of the model variables. The most important result was that DS occupy the domain (with a half-wave accuracy) where magnitudes of the system parameters satisfy the DS existence conditions for the homogeneous system, so that DS are localized because of spatial inhomogeneity. As in homogeneous systems, there is a multitude of states; which one of them will occur depends on the initial conditions. Systems of this type cannot be considered as autonomous. One may say that in this case DS just reveal the distribution of sources in the medium.

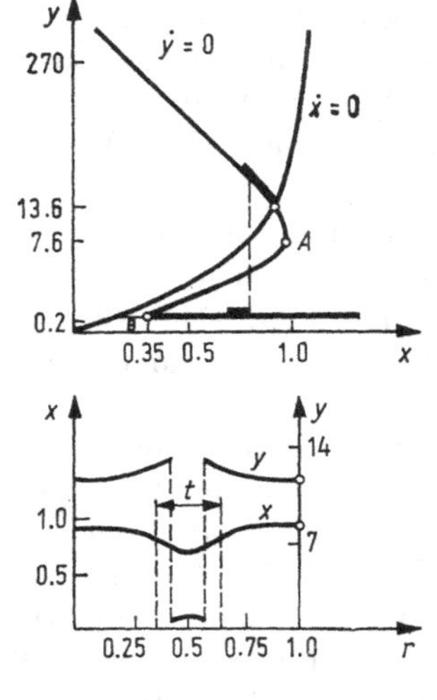

Fig. 7.14
Typical localized DS in the Orengonator system (5.4). ($D_x \equiv 0$ (!), $\nu = 1$, $\gamma_0 = 0.5$, $p = \beta = 500$, $\varkappa = 1$, $q = 8$, $\tau = 0.5$, $c = 1$, $D_y = 0.05$); $z \equiv 0$.

Quasi-harmonical DS are, evidently, impossible in homogeneous autonomous media, so one should consider contrast DS. Let us turn back to the discussion of the Oregonator model, Eq. (5.4), which was given in Section 7.4. Now we suppose that the singular point is in the segment A of the zero-isocline for the auto-catalytic variable x. Such local kinetics is a sufficient condition for stability of the homogeneous state of any D_x and D_y.

Next we put $D_x = 0$ and suppose that as a result of an intensive localized perturbation a small part of the system goes into the state corresponding to the segment B. This region cannot return to the initial state. Actually, the local kinetics tends to decrease the magnitude of y in this region, but the flux from other parts of the system which are still in the state A compensates this trend. Since $D_x = 0$, no $B \to A$ transitions are possible for $y(r) > y_B$. Thus one has a distribution like that shown in Fig. 7.14.

It is not difficult to perform an analytical construction for the solution considered. The "case-1" method of Section 7.4 is quite applicable here; and one can draw the same conclusions: for any magnitudes of the model parameters it is possible to find an interval of the system lengths where arbitrarily shorter perturbations generate inhomogeneities, and any combination of the inhomogeneities along the system length are possible, provided that spacings between them are above a threshold. Thus for

$D_x = 0$ the allowed infinite sets of solutions satisfy the requirements imposed on localized DS.

If D_x is not exactly zero, yet $\epsilon = D_x/D_y \ll 1$, one cannot take $y(r) = y_p$ in the transition region (where the distributions have sharp variations), since conditions (7.29) must be fulfilled. Using once more the reasoning of Section 7.4, one can show that a single inhomogeneous stationary solution exists in the system of length L.

A unique solution can also be found for the Brussellator-type models at $D_x \neq 0$, see Fig. 7.15. This is a one-spike solution. One can prove the stability of the solution in the region where the model parameters do not vanish, but this question is irrelevant at present.

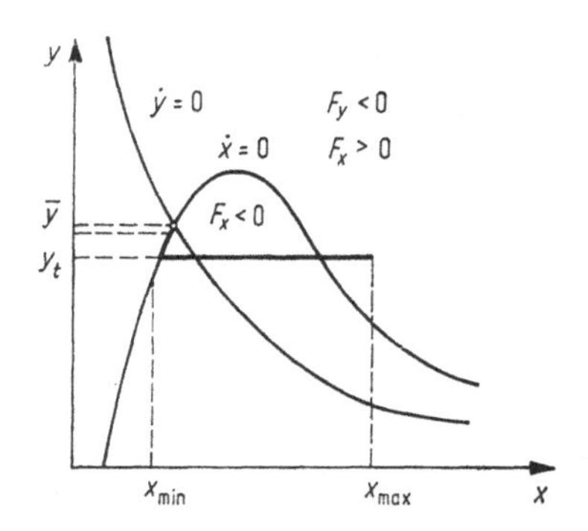

Fig. 7.15
The phase plane of typical localized contrast DS in the Brussellator for the case of infinite length of the system

Solutions of this type cannot be considered as localized, even though the area under the spikes is very small; in fact, the structure occupies the system completely, as no additional spike can be put into the system. Nevertheless, such models may be of some interest in view of localized DS. Actually, if conditions (7.29) or (7.32) do not hold, the discrepancy in the solutions does not exceed $\sqrt{\epsilon} = (D_x/D_y)^{1/2} \ll 1$ by order of magnitude, whereas the times of transition processes are large, of order $\approx 1/\epsilon$. Therefore such states can be treated as metastable localized DS.

Turning back to the case $D_x = 0$, one should note that it does not occur in continuous systems. However, it represents adequately enough the situation which is relevant to cellular systems in biology (see Section 2.5). An x-dependence of D_x makes continuous models more similar to the systems with $D_x = 0$. We do not know of any results on systems having complicated local kinetics and $D_x(x)$, the more interesting is therefore a study of the simplest type of such systems, those with a single component. Probably, a combination of these systems with those considered above will enable one to considerably extend the known variety of AWPs.

7.7 Self-organization in combustion processes

"This order of the world, which is common to all, was not created by one of the gods or a man, rather it always was, is now and always will be the eternal living fire, forever leaping up and dying down" (Heraclitus).

How could one not be reminded of this striking dictum of the ancient philosopher when writing on structures which form in a nonlinear burning medium? The prediction of these structures and their study using the simplest mathematical models are extremely interesting even though they are nonstationary and do not persist for any length of time.

The following considerations involve results obtained by A. A. Samarskii, S. P. Kurdyonov and their followers G. G. Elenin, N. V. Zmitrenko, A. P. Mikhailov, V. A. Galaktionov and others. These authors derived and comprehensively investigated the rules governing nonlinear burning as a structure (see for example Samarskii et al. 1976, 1977; Elenin, Kurdyumov 1977; Galaktionov et al. 1981; Kurdyumov 1982; also literature cited therein).

The simplest mathematical model is based on the one-dimensional quasilinear heat conduction equation[1])

$$\frac{\partial T}{\partial t} = \frac{\partial}{\partial r}\left[\varkappa(T)\,\frac{\partial T}{\partial r}\right] + Q(T), \tag{7.33}$$

where

$$\varkappa(T) = \varkappa_0 T^\sigma, \quad Q(T) = q_0 T^\beta,$$

$$\varkappa_0 > 0, \quad q_0 > 0, \quad \sigma > 0, \quad \beta > 1.$$

The constants $\varkappa_0$, q_0 can be set equal to unity without loss of generality. This means passing to other units of time, length and temperature. Furthermore the nonlinear dependence of the heat source and of the coefficient of heat conductivity on temperature is essential. Certain σ and β correspond to interesting and unusual properties of the process considered. An initial temperature profile $T_0(r)$ symmetrical to the origin of co-ordinates causes the inflammation. The further spatial temperature distribution is obtained by solving the following initial and boundary value problem of equation (7.33):

$$T(r, 0) = T_0(r), \quad 0 \leqq r < +\infty, \quad 0 \leqq T_0(r) < \infty, \quad \lim_{r\to\infty} T_0(r) = 0,$$

$$T_r'(0, t) = 0, \quad \lim_{r\to\infty} T(r, t) = 0, \quad \lim k(T)\, T_r' = 0. \tag{7.34}$$

The solution of (7.33—7.34) need not be smooth where $T(r, 0) = 0$. The temperature gradient can be infinite on the heat wave front. Two cases are distinguished in this connection:

A) The initial distribution is finite

$$T(r, 0) = \begin{cases} T_0(r) > 0 & \text{for} \quad 0 \leqq r < a, \\ 0 & \text{for} \quad r \geqq a \end{cases} \tag{7.35}$$

and the temperature is defined on the variable length interval $l = r_F(t)$. The equation of motion of the heat front is found from continuity conditions for the temperature $T(r_F(t), t) = 0$ and the heat flux $k\big[T\big(r_F(t), t\big) \cdot T_r'\big(r_F(t), t\big)\big] = 0$.

[1]) The case $q_0 = 0$ was investigated earlier (Zeldovich, Kompaneets 1950), the case $q_0 < 0$ may be found in the paper of Martinson and Pavlov (1975).

B) The initial distribution does not vanish in the whole coordinate space $T(r, 0) = T_0(r) = 0$.

The distinction between the problems A and B is not of principal importance but is used here for convenience in investigating and formulating the results.

As noted above, stationary dissipative structures (DS) have the property of spatial symmetry. The travelling pulses behave like invariant or approximately invariant solutions of the travelling wave type. We look for a stable invariant or approximatively invariant solution for the regimes of the "front deepening" (or for metastable DS). The nonstationary burning process is in a regime of front deepening if the temperature becomes infinite for at least one space point in a finite time interval. In particular the regimes of front deepening are described by solutions of the automodel type:

$$T(r, t) = g(t, \tau)\, \theta(\xi, \tau) \not\equiv 0, \tag{7.36}$$

where $\xi = r\varphi^{-1}(t, \tau)$, if $g(t, \tau) = (1 - t\tau^{-1})^{-\gamma}$, $\varphi(t, \tau) = (1 - t\tau^{-1})^{\alpha}$, $\gamma = (\beta - 1)^{-1}$, $\alpha = 0.5\gamma(\beta - \sigma - 1)$. The function $\theta(\xi, \tau) \neq 0$ is the nontrivial solution of the boundary value problem for the automodel profile:

$$(\theta^{\sigma}\theta_{\xi}')_{\xi}' - \frac{\alpha}{\tau}\,\theta_{\xi}' - \frac{\gamma}{\tau}\,\theta + \theta^{\beta} = 0. \tag{7.37}$$

For the case A we get the boundary conditions:

$$\theta_{\xi}'(0) = 0, \quad \theta(a) = 0, \quad \theta^{\sigma}(a) \cdot \theta_{\xi}'(a) = 0, \tag{7.38}$$

and for the case B

$$\theta_{\xi}'(0) = 0, \quad \lim_{\xi \to \infty} \theta(\xi) = 0. \tag{7.39}$$

The problems (7.37—7.38) and (7.37—7.39) are equations for the eigenvalues τ and define the desired eigenfunctions $\theta(\xi, \tau)$ which describe the temperature distribution in the regimes of front deepening. The sign of the eigenvalue τ determines the time behaviour of the regime. Even $\tau < 0$ and its eigenfunctions define automodel solutions with front deepening. Negative $\tau < 0$ together with its eigenfunctions define automodel solutions which exist for arbitrary time $t > 0$.

The eigenfunction analysis of the nonlinear problem yields the following conclusions:

1. If $1 < \beta < \sigma + 1$, then the burning process is always in the regime of front deepening.
2. If $1 < \beta < \sigma + 1$, then the automodel problem B has no nontrivial solutions. The problem A has a unique positive eigenvalue for an arbitrary fixed value of a. This eigenvalue corresponds to a unique finite eigenfunction. This automodel solution describes a heat wave of the so-called HS regime. The front of the heat wave is distributed with increasing velocity. The temperature becomes infinite in the whole space in a finite time, which is equal to the eigenvalue τ (time of deepening, see Fig. 7.16).

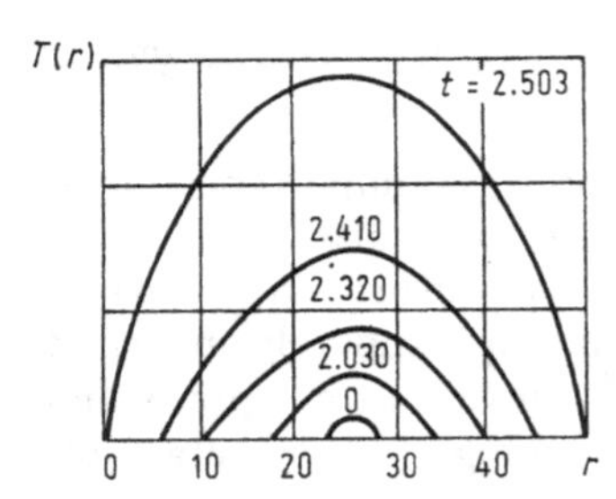

Fig. 7.16
Propagation of the heat wave in the HS regime ($1 < \beta < \sigma + 1$) (Samarskii, Elenin et al. 1977)

3. If $\beta = \sigma + 1$, then problem B has no nontrivial solutions, but problem A has a solution in the regime of front deepening only if $a > L_S/2$, where L_S is the fundamental length of the S regime of deepening. This solution is obtained by direct integration of (7.37) with the boundary conditions (7.38):

$$\theta(\xi, \tau) = \begin{cases} \theta_0 \cos^{2/\sigma} (\pi r / L_T{}^S), & 0 \leqq r \leqq L_T{}^S/2, \\ 0, & r \geqq L_S/2, \end{cases} \tag{7.40}$$

where

$$L_T{}^S = 2\pi\sigma^{-1}(\sigma + 1), \quad \theta_0 = [2(\sigma + 1)\, \sigma^{-1}\tau^{-1}(\sigma + 2)^{-1}].$$

It follows from (7.40) that the burning process is localized on the fundamental length $L_T{}^S$. The lifetime of this solution is defined by the amplitude θ_0 of the initial automodel temperature distribution.

4. If $\sigma + 1 < \beta < \sigma + 3$, then problem A has no nontrivial solutions. Problem B has a solution in this range for arbitrary $\tau > 0$. Here every value of τ does not correspond to a unique eigenfunction, but to a finite number of eigenfunctions with qualitatively different structure (Elenin, Kurdyonov 1977). In this LS regime the temperature increases to infinity at the origin of co-ordinates. The fundamental length of the LS regime $L_T{}^{LS}$ does not only depend on the medium parameters σ and β, but also on the initial excitation energy $W_0 = aT_{0m}$, where $T_{0m} = \max T_0(r)$ (Kurdyomov 1982)

$$L_T{}^{LS} = (\alpha^2 \omega_0{}^{b-1-\beta})^{1/b+3-\beta}, \quad \alpha = \pi \sqrt{\frac{2(\beta + \sigma + 1)}{\sigma(\beta - 1)}} \tag{7.41}$$

$$(\beta \neq \sigma + 3).$$

It is necessary for inflammation in the regime with front deepening that the magnitude of the excitation region is not less than a certain resonance length ($a \geqq L_T{}^*$). (See Fig. 7.17.) For resonance inflammation the time the temperature needs to reach infinite values can be estimated by the following expression:

$$t_t = \frac{\beta + \sigma + 1}{(\sigma + 1)(\beta - 1)\, T_{0m}^{\beta - 1}}. \tag{7.42}$$

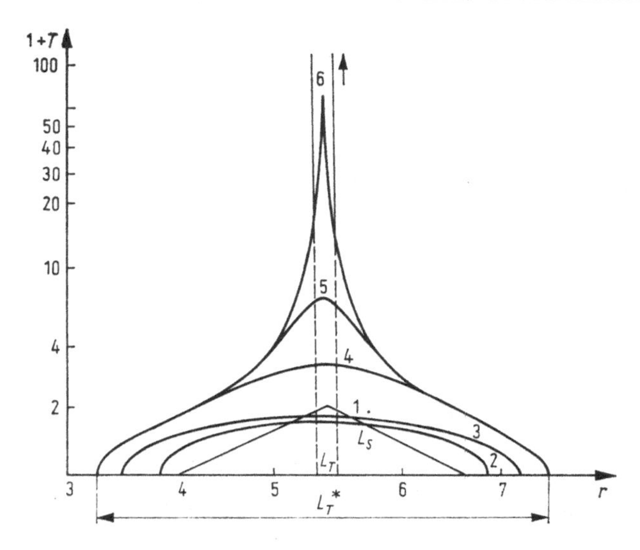

Fig. 7.17
Avalanche-increase of the accentuation regime in the boundary of the resonance region (Kurdyumov 1982)

5. If $\beta > \sigma + 3$, then two burning regimes of the medium are possible. The first is the regime with front deepening and localization of heat realized by initial resonance temperature. The second is the regime with non-resonance excitation, where a temperature wave with damped amplitude is observed.

The statements 1 to 5 are based on the investigations of automodel solutions. There is a question: Can the automodel solutions arise in a finite time?

It was shown by analytical investigations as well as by computer experiments that the automodel solutions of the HS and S regimes and also the first eigenfunction of the LS regime are asymptotically structurally stable against perturbations of the initial distribution.

The simultaneous independent burning of a few structures in the medium is possible, if their localization regions do not overlap. This is especially valid for resonance inflammation structures with the same deepening time t_f and the same inflammation time t_0. However, if the structures have different deepening times then the fastest outshines the others very rapidly. There are two important examples. The initial temperature distribution is symmetrical with respect to the origin of coordinates as shown in Fig. 7.18a. The distance between the maxima of the excitation is larger than their resonance length L_T*. In this case the evolution of burning in the medium yields the formation of two structures by localization of the temperature in the vicinity of each maximum. The temperature profiles $T(r, t)$ for different times evolving from the initial distribution (cf. Fig. 7.18a) are shown in Fig. 7.18b. It is evident that the burning of the "background" and of the medium in the region of the structures which evolves from a smaller initial excitation is a very slow process compared with the burning of the medium inside the structure caused by a large temperature excitation.

If the resonance lengths of two initial excitations overlap then an interaction occurs between the structures and they change into a convergent temperature wave.

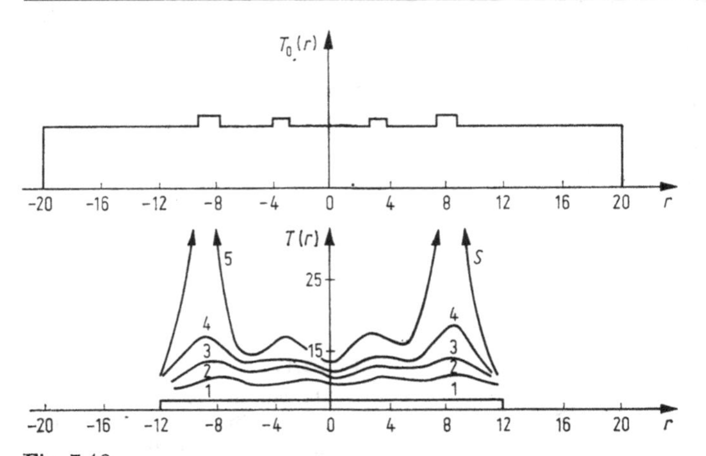

Fig. 7.18

a) Initial temperature distribution for the case $\beta = 3.18$, $\sigma = 2$, $\varkappa_0 = q_0 = 1$;
b) development of the structures from this initial distribution (Kurdyumov 1982)

These situations are illustrated by results of computer experiments given in Fig. 7.19a, b. However, if the initial excitations are at a distance larger than the resonance length $L_T{}^*$ then independent metastable structures are formed in the points of these excitations (Fig. 7.20a, b). The dotted lines in Fig. 7.19b and Fig. 7.20b mark the movement of the temperature wave maximum. This was calculated using automodel solutions. The results of large computer experiments for two- and three-dimensional structures or metastable DS with front deepening are given. There are even critical distances for two-dimensional interacting structures which guarantee their inde-

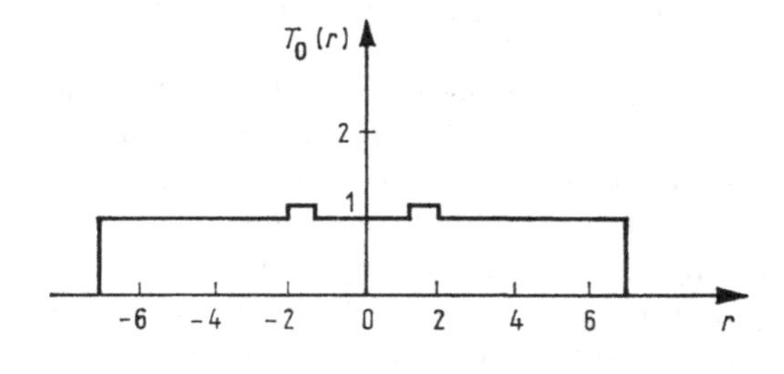

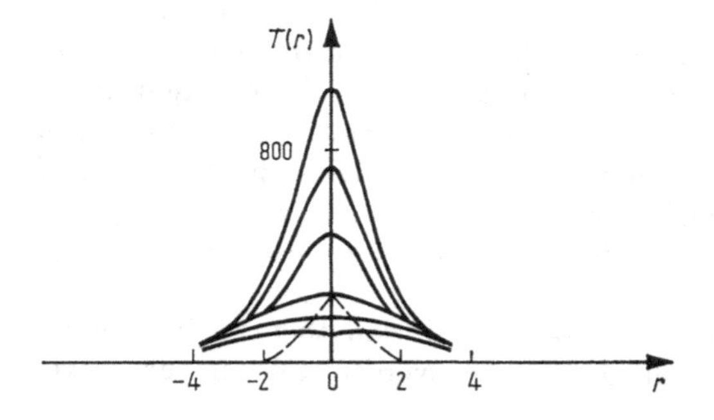

Fig. 7.19

a) Initial temperature distribution (the perturbations are symmetrical with respect to the centre of the system, distances between them are smaller than the resonance length $L_T{}^*$) and
b) propagation of the burning wave after such initial conditions (Kurdyumov 1982)

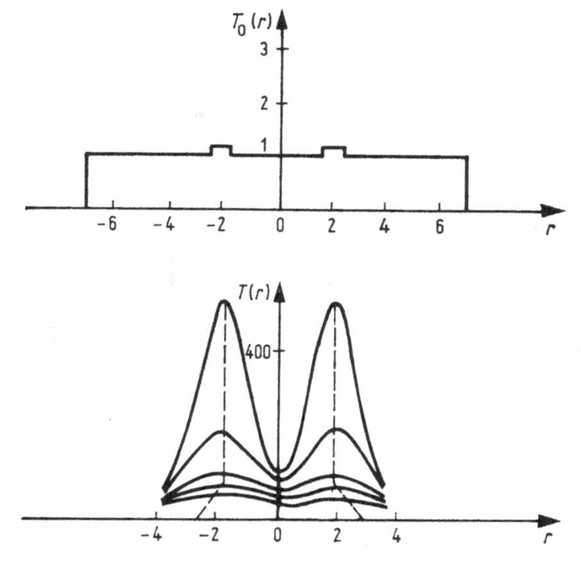

Fig. 7.20

a) Distance between the initial perturbations exceeds the resonance length (L_T*), and

b) "Torches" or regions with the accentuation arising after such perturbations (Kurdyumov 1982)

pendent existence. The most stable structures are the symmetrically inflamed ones. An example for such structures is shown in Fig. 7.21. Non-symmetric initial configurations degenerate much faster.

Thus, one can speak about "long-living" complex structures only if the initial excitations are identical and consequently, if they have the same deepening times. Because real systems always have stochastic background perturbations even small variations of the temperature necessarily lead to the destruction of the complex structures. However one has to take into account that the structure exists at the onset of structure formation because of variations with different amplitudes. One can neglect all other structures if only an arbitrary one (the fastest-increasing) reaches the final deepening regime. Note that even DS are unstable in one-component systems (see Section 7.2). However, the influence of the second component leads to the stabilization of the DS. An open question is how systems with front deepening will behave if there is not only temperature conduction but also a diffusion of the burning material. However, this question is of great interest. One can state already that the structures will burn holes into the actual burning medium, i.e. destroy its homogeneity. In other words, the burning process with deepening and localization inevitably covers its track in the evolution of the medium in which the fire originates. In conclusion we note the following. It is most probable that the prediction of the described phenòmena is important for the investigation of the processes of spatial self-organization and evolution in cosmology, in laser plasmas and even in thermonuclear reactors. Presumably nonlinear diffusion plays an important role in the DS models of living tissues and describes active transport through biological membranes. If there are autocatalytic terms in these models one can quite expect that they behave like systems with front deepening or describe metastable DS.

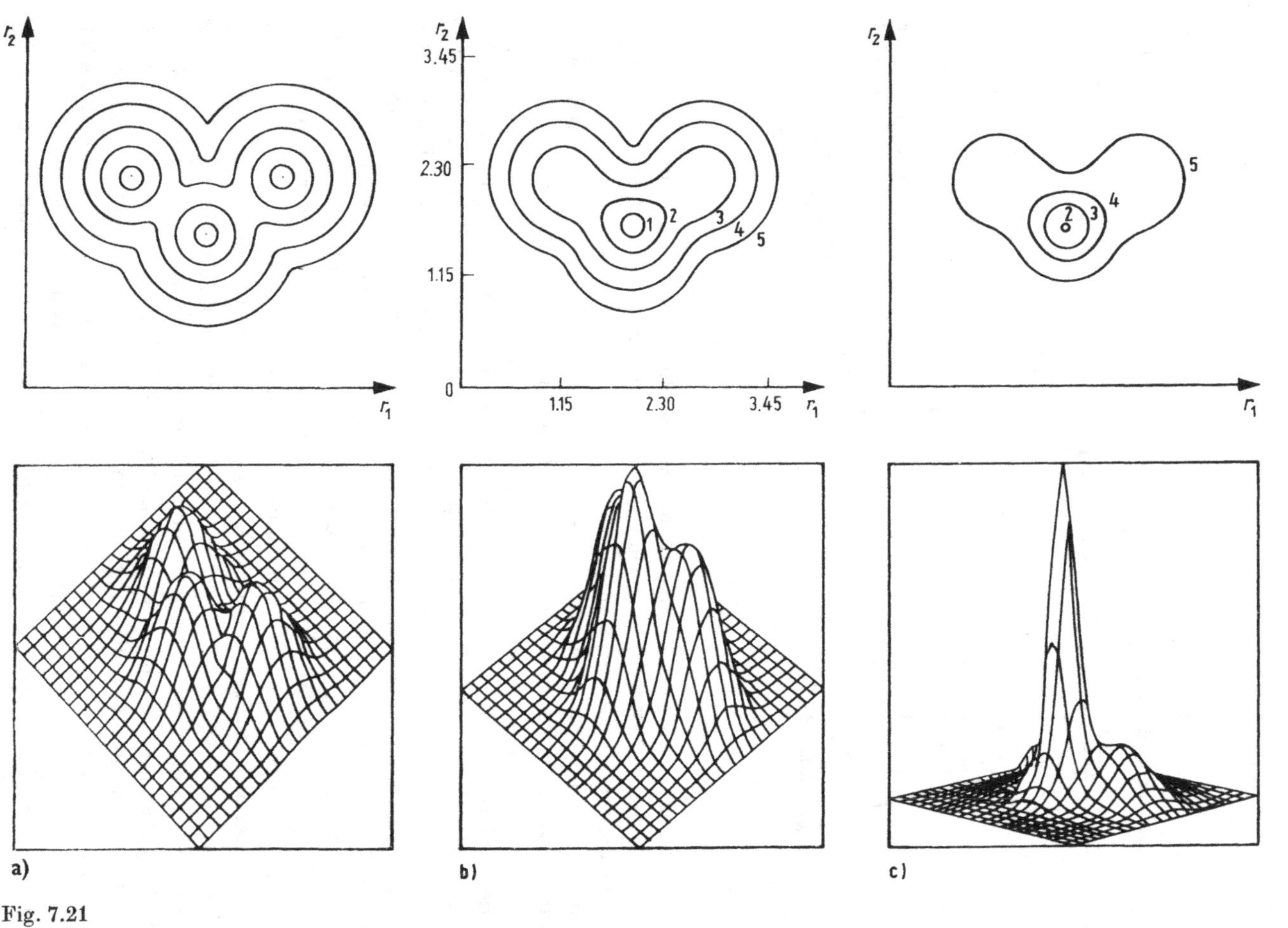

Fig. 7.21
Interaction of two-dimensional heat structures. Above, the level lines and below, the corresponding heat surface. The figures correspond to three moments of dimensionless time: $t_1 = 0, t_2 = 5.94 \times 10^{-2}, t_3 = 6.04 \times 10^{-2}$ (Kurdyumov, Malinetskii et al. 1980)

Chapter 8
Noise and autowave processes

8.1 Sources of noise in active kinetic systems and fundamental stochastic processes

The fundamental kinetic equations (1.1) for AW processes are obtained by statistical averaging. Indeed, the "concentrations" or "velocities" of the kinetic variables are certain averaged values. In principle an arbitrary deterministic model corresponds to a stochastic model whose kinetic interactions are described by the probabilities of elementary events that occur in small time intervals in small elementary volumes. Noise arising in discrete interactions will become small after averaging if the number of interacting objects (molecules, living organisms) is large. We refer to such types of noise as "natural noise" in the following. However they can also be important in systems with a small number of interacting objects, some examples are ecological communities, systems of genes and organelles in a living cell. Even small amounts of noise can have a substantial effect in systems with concentrations near bifurcation points. Natural noise also has a pronounced effect on multistationary systems. It is interesting to note that in relaxation-type auto-oscillatory chemical reactions there may be time intervals during which the concentration of reactants decreases by many orders of magnitude. The part played by natural fluctuations increases sharply (Zhabotinskii 1974; Romanovskii et al. 1975).

Besides natural noise rather intense external noise can occur in kinetic interactions. External noise is produced by fluctuations of external parameters such as temperature, illumination etc.

Finally, various auto-stochastic processes can be sources of noise in active distributed systems (Gaponov, Rabinovich 1978; Kuznetsov 1982). These consist of the onset of strange attractor regimes, the appearance of chemical and hydrodynamic turbulence etc. The latter can be a consequence of chemical transformations or biological activity of microorganisms (Levandovskii et al. 1975). Auto-stochastic sources are most diversified in nature but they may all be of significant intensity. Moreover, complicated transient processes can have a similar appearance to autostochastic processes (cf. Section 6.3). One should bear in mind that the "lifetimes" of open systems are often comparable to the duration of transient processes.

A rigorous theory of fluctuations in active continuous media is still far from being complete. However it is possible to estimate the contribution of noise to the qualitative behaviour of models (Bharuche-Reid 1969; Nicolis, Prigogine 1979; Haken 1980; Romanovskii et al. 1975; Klimontovich 1980, 1982). We assume we can take internal and external noise into account in such a way that the basic equations may be extended by additive and parametric or multiplicative terms. The problems arising here are as follows:

1. Determining the statistical characteristics of small random deviations of AW processes.
2. Finding the mean lifetimes of a system in one of its possible states.
3. Determining the characteristics of fluctuations of kinetic variables in the vicinity of the bifurcation points, or specifically, investigating fluctuations arising during the formation of dissipative structures.
4. Determining the part played by noise at the points of instability of stationary states or AW regimes, for example of a stationary one.
5. Determining the conditions for points of instability of secondary and higher order in kinetic systems or the conditions for the rise of autostochastic regimes.

8.2 Parametric and multiplicative fluctuations in local kinetic systems

1. First we consider the possible sources of natural or internal fluctuations if complete internal mixing occurs in the system and the diffusion terms in (1.1) can be neglected. As an example we consider the well-known Volterra system which describes the coexistence of predator and prey populations.

Let the prey population consist of x individuals (e.g. hares) and the predators (e.g. foxes) number y at a certain instant of time. Over a time interval $\mathrm{d}t$ the following events can occur in the community:

a) birth of one hare with the probability $\alpha x \,\mathrm{d}t$,

b) birth of one fox with the probability $\beta xy \,\mathrm{d}t$ (occurs only if food, x hares, is available),

c) killing of a hare with the probability $\gamma xy \,\mathrm{d}t$,

d) natural death of a fox with the probability $\delta y \,\mathrm{d}t$.

Let us denote the probability that there are x hares and y foxes at an instant t by $w(x, y, t)$. Over a time $\mathrm{d}t$ the probability will evolve according to

$$\begin{aligned}\frac{\partial w(x, y, t)}{\partial t} &= \alpha[(x-1)\,w(x-1, y, t) - xw(x, y, t)] \\ &+ \beta x[(y-1)\,w(x, y-1, t) - yw(x, y, t)] \\ &+ \gamma y[(x+1)\,w(x+1, y, t) - xw(x, y, t)] \\ &+ \delta[(y+1)\,w(x, y+1, t) - yw(x, y, t)]. \qquad (8.1)\end{aligned}$$

This is a so-called master equation. Note that the first term in the right-hand side is the product of the probability $w(x-1, y, t)$ and the transition probability for $\{x-1, y, t\} \to \{x, y, t\}$. The other terms are obtained in similar fashion. Eq. (8.1) is valid for all x and y and $w(x, y, t) = 0$ is assumed for $x < 0$ and $y < 0$. Eq. (8.1) can be solved in a straightforward manner using a computer, for instance by the Monte-Carlo method (cf. Section 8.4). However there are other methods of studying stochastic models, one of which consists of deriving a set of equations from Eq. (8.1) for the mean and maximum values of the prey and predator numbers. We propose a method by which one can proceed from Eq. (8.1) to a model with fluctuation terms. To do this we compare Eq. (8.1) with the Kolmogorov-Fokker-Planck equation assuming that the number of individuals in the system is substantially larger than unity ($x \gg 1$, $y \gg 1$) (Romanovskii et al. 1975). Then terms of the form

$$(x \pm 1)\, w(x \pm 1, y, t) - xw(x, y, t)$$

can be approximated by a Taylor series expansion

$$\pm \frac{\partial}{\partial x} [xw(x, y, t)] \pm \frac{1}{2} \frac{\partial^2}{\partial x^2} [xw(x, y, t)] .$$

The master equation (8.1) then reads

$$\begin{aligned} \frac{\partial w(x, y, t)}{\partial t} = & -\frac{\partial}{\partial x} [(\alpha x - \gamma xy)\, w(x, y, t)] \\ & -\frac{\partial}{\partial y} [(-\delta y + \beta xy)\, w(x, y, t)] \\ & + \frac{1}{2} \frac{\partial^2}{\partial x^2} [(\alpha x + \gamma xy)\, w(x, y, t)] \\ & + \frac{1}{2} \frac{\partial^2}{\partial y^2} [(\delta y + \beta xy)\, w(x, y, t)] . \end{aligned} \tag{8.2}$$

This is a Kolmogorov-Fokker-Planck equation which in turn can be compared with a set of stochastic Langevin equations (Alekseev, Kostin 1974; Romanovskii et al. 1975).

$$\begin{aligned} \frac{\mathrm{d}x}{\mathrm{d}t} &= \alpha x - \gamma xy + \sqrt{\alpha x + \gamma xy}\, \xi_1(t) - \frac{\alpha + \gamma y}{4} , \\ \frac{\mathrm{d}y}{\mathrm{d}t} &= -\delta y + \beta xy + \sqrt{\delta y + \beta xy}\, \xi_2(t) - \frac{\delta + \beta x}{4} , \end{aligned} \tag{8.3}$$

where ξ_1 and ξ_2 are δ-correlated random functions of time and

$$\langle \xi_i(t)\, \xi_j(t + \tau) \rangle = \delta_{ij} \delta(\tau) ; \qquad (i, j = 1, 2) ,$$
$$\langle \xi_i \rangle = 0 .$$

Thus we have derived the well-known Volterra system with fluctuation terms taken into account. It is important to note that, firstly, the fluctuation "amplitude" depends on the numbers x and y, and secondly the last terms in the right-hand side of Eq. (8.3) also have a fluctuating nature.

From the Langevin equations (8.3) we obtain by averaging common Volterra equations for the mean densities of x and y (or mean predator and prey numbers at a given instant of time). Natural noise as well as other equations describing chemical reactions or interactions in biosystems can be introduced in much the same way.

We emphasize that the fluctuation terms vanish on averaging (for the averaging procedure see Stratonovich 1961 and Mitropolskii 1971). Thus

$$\left\langle \sqrt{\alpha x + \gamma x y}\, \xi_1(t) \right\rangle = \langle (\alpha + \gamma y)/4 \rangle, \qquad \left\langle \sqrt{\delta y + \beta x y}\, \xi_2(t) \right\rangle = \langle (\delta + \beta x)/4 \rangle. \tag{8.4}$$

Therefore internal or natural noise in kinetic systems does not lead to parametric excitation despite the multiplicative nature of these types of noise.

2. Conversely, external multiplicative noise can lead to loss of stability. This can be illustrated by the Lotka system. As is well known the chemical Lotka model may be represented as follows (Lotka 1925)

$$A \xrightarrow{k_0} y \xrightarrow{k_1} x \xrightarrow{k_2} P.$$

The initial substance A is in abundance. Following Lotka we write equations corresponding to his chemical model

$$\frac{dy}{dt} = k_0 A - k_1 x y, \qquad \frac{dx}{dt} = k_1 x y - k_2 x. \tag{8.5}$$

We assume that $x(t)$ and $y(t)$ are small deviations from the stationary densities

$$x_0 = \frac{k_0 A}{k_2}, \qquad y_0 = \frac{k_2}{k_1}. \tag{8.6}$$

Using the usual notation for oscillatory systems we write the linearized system (8.5) in the form

$$\frac{dy}{dt} = -2\delta y + \omega^2 x, \qquad \frac{dx_1}{dt} = y, \tag{8.7}$$

where

$$x_1 = x k_2 / k_1 k_0, \quad \delta = k_1 k_0 / 2k_2, \quad \omega^2 + k_1 k_0.$$

If the reaction velocities are such that $\omega_0 \gg \delta$, then the state (x_0, y_0) will be a stable focus in the phase plane (x, y). We then assume that external noise acts on the Lotka system in such a way that the frequency undergoes δ-correlated fluctuations. System (8.7) then reduces to

$$\frac{d^2 y}{dt^2} + 2\delta \frac{dy}{dt} + \omega^2 [1 + \xi(t)]\, y = 0. \tag{8.8}$$

This equation as well as the nonlinear equations with fluctuating frequency were studied some time ago in the context of radio electronics in a number of papers (Stratonovich, Romanovskii 1958; Stratonovich 1961). There has recently been a renewal of interest in multiplicative noise, this time as applied to kinetic systems

(Kabashima, Kawakubo 1980; Mikhailov, Uporov 1980). In this connection we briefly present a method with which the loss of stability in Eq. (8.8) was first observed (Stratonovich, Romanovskii 1958) for the case that $\xi(t)$ is not necessarily δ-correlated but rather exhibits a small correlation time τ_{cor}.

We define the amplitude and phase of the solution $y(t)$ by the equations

$$y(t) = A(t) \cos \Phi ,$$
$$\frac{\mathrm{d}y}{\mathrm{d}t} = -A(t) \cdot \omega \sin \Phi , \tag{8.9}$$
$$(\Phi = \omega t + \varphi(t)) .$$

The following exact equations can then be obtained from (8.8):

$$\frac{\mathrm{d}A}{\mathrm{d}t} \frac{1}{A} = \omega\xi \sin \Phi \cos \Phi - 2\delta \sin^2 \Phi ,$$
$$\frac{\mathrm{d}\varphi}{\mathrm{d}t} = \omega\xi \cos^2 \Phi - 2\delta \sin \Phi \cos \Phi . \tag{8.10}$$

We assume that the fluctuation intensity $\xi(t)$ is not too large, i.e. it does not lead to significant deviations in amplitude over the oscillation period $T = \omega/2\pi$. We then have

$$\langle \xi^2 \rangle^{1/2} \ll 1 \quad \text{if} \quad \tau_{\text{cor}} \gtrapprox \mathrm{T} ,$$
$$\omega^{1/2}[\varkappa(0) + \varkappa(2\omega)]^{1/2} \ll 1 \quad \text{if} \quad \tau_{\text{cor}} \lessapprox \mathrm{T} . \tag{8.11}$$

Here $\varkappa(2\omega)$ is half of the spectral density of $\xi(t)$ at a frequency 2ω:

$$\varkappa(2\omega) = \int_{-\infty}^{+\infty} \langle \xi(t)\, \xi(t + \tau) \rangle \cos 2\omega\tau \, \mathrm{d}\tau .$$

Averaging Eq. (8.10) over one period gives the following set of truncated equations

$$\frac{\mathrm{d}u}{\mathrm{d}t} = m_1 - \delta + \xi_1(t) ,$$
$$\frac{\mathrm{d}\varphi}{\mathrm{d}t} = m_2 + \xi_2(t) , \tag{8.12}$$

where

$$u = \ln A ,$$
$$m_1 = \omega\langle \xi \sin \Phi \cos \Phi \rangle = \frac{1}{8} \omega^2 \varkappa(2\omega) ,$$
$$m_2 = \langle \xi \cos^2 \Phi \rangle = 0 ,$$
$$\xi_1 = \omega\xi \sin \Phi \cos \Phi - m_1 ,$$
$$\xi_2 = \omega\xi \cos^2 \Phi ,$$
$$\int_{-\infty}^{+\infty} \langle \xi_1(t)\, \xi_1(t + \tau) \rangle \, \mathrm{d}\tau = \frac{2}{k_1} = m_1 ,$$
$$\int_{-\infty}^{+\infty} \langle \xi_2(t)\, \xi_2(t + \tau) \rangle \, \mathrm{d}\tau = \frac{2}{k_2} = m_2 = \frac{\omega}{4} \left[\varkappa(0) + \frac{1}{2} \varkappa(2\omega) \right] .$$

The symbols k_1 and k_2 are introduced for reasons of simplicity. If the evolution of the set of equations (8.12) is traced for a time greatly exceeding τ_{cor} then $A(t)$ and $\varphi(t)$ can be treated as Markov processes. This implies that the correlation functions for ξ_1 and ξ_2 can be substituted by δ-correlated functions, i.e. by $2k_1^{-1}\delta(\tau)$ and $2k_2^{-1}\delta(\tau)$ respectively and that the Kolmogorov-Fokker-Planck equation may be usefully employed. Thus the first equation of (8.12) corresponds to the equation for the probability density $w(u, t)$:

$$\frac{\partial w}{\partial t} = \frac{1}{k_1} \frac{\partial^2 \omega}{\partial u^2} - (m_1 - \delta) \frac{\partial \omega}{\partial u}. \tag{8.13}$$

The solution corresponding to the initial condition $w(u, 0) = \delta(u)$ reads

$$w(u, t) = \sqrt{\frac{k_1}{\pi t}} \exp \{-k_1[u - (m_1 - \delta) t]^2/t\}. \tag{8.14}$$

It is seen that the mean value $\langle u \rangle = \langle \ln A \rangle$ tends to $+\infty$ if the following condition is satisfied:

$$m_1 = \omega^2 \varkappa(2\omega)/8 > \delta. \tag{8.15}$$

On the other hand $\langle \ln A \rangle$ tends to $-\infty$ or $A \to 0$ if the inverse of (8.15) holds. It is easy to show that the probability limit is given by

$$\lim_{t\to\infty} P(A > b) \to 1, \tag{8.16}$$

where b is an arbitrary positive number. We stress once again that parametric noise excitation is conditioned by the spectral density of the fluctuations at a duplicate eigenfrequency, i.e. on average a fundamental parametric resonance occurs. Higher approximations should be introduced in calculating an excitation at other frequencies. In papers of Stratonovich and Romanovskii (1958) and Stratonovich (1961) the case of parametric excitation of a nonlinear oscillatory system (which may become auto-oscillatory) was also considered. Such a system is given by

$$\frac{\mathrm{d}^2 x}{\mathrm{d}t^2} + 2[\delta + \alpha x^2] \frac{\mathrm{d}x}{\mathrm{d}t} + \omega^2[1 + \xi(t)]\, x = 0. \tag{8.17}$$

The amplitude $A(t)$ can be found from the truncated equation of the form

$$\frac{\mathrm{d}A}{\mathrm{d}t} = (m_1 - \delta) A - \alpha A^3 + \xi_1(t) \cdot A. \tag{8.18}$$

Solving the Kolmogorov-Fokker-Planck equation corresponding to (8.18) we find the stationary amplitude distribution

$$w(A) = \frac{2}{\Gamma(1 - \delta/k_1)} \left(\frac{\alpha}{k_1}\right)^{1-\delta/k_1} A^{1-2\delta/k_1} \mathrm{e}^{-\alpha A^2/k_1}, \tag{8.19}$$

where Γ is the Gamma function.

If the excitation condition (8.15) is not satisfied then $w(A)$ has a nonintegrable singularity at zero, i.e., the system is not excited and fluctuates weakly around the stable equilibrium state. If $k_1 > \delta$ then the distribution $w(A)$ bears a similarity to the Rayleigh distribution and the mean amplitude has a finite value. It will be shown in Section 8.7 that transition from the non-excited to the excited state under the action of parametric noise can be treated as a phase transition.

We can thus infer that parametric instability does not occur in systems subject to natural noise. Such an instability arises in systems subject to external multiplicative noise. The excitation threshold depends on the noise intensity and the damping rate. We will dwell on this problem again when dealing with distributed systems (Section 8.7). We will now consider another qualitative effect which can be caused by both internal and external noise.

8.3 The mean lifetime of the simplest ecological prey-predator system

If fluctuations are neglected in a classical Volterra model then the time evolution of the prey and predator populations can be represented by closed trajectories depending on the initial conditions in the phase plane (Fig. 8.1). Thus, under certain

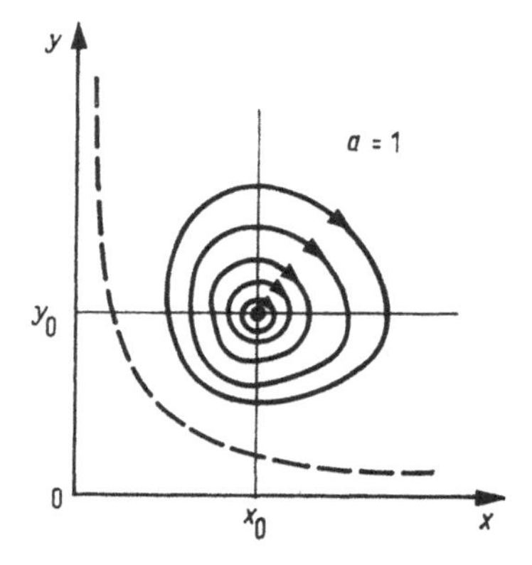

Fig. 8.1
Phase plane of the Volterra system. The boundary of the degeration is marked conditionally by a dotted line

conditions the integral curves may approach the coordinate axes. It is obvious that the conventional Volterra equations are meaningless if x and y are of the order 10^2 to 10^3. If internal fluctuations are taken into account there is always a possibility that prey or predator numbers will fall below a certain critical value beyond which a sufficient number of descendants will no longer be produced. We will consider the mean lifetime (T_e) of a Volterra system to be the time it takes on the average to reach that value. The critical boundary is shown by a dashed line in Fig. 8.1. For definiteness we assume that the initial position of the representative point coincides with the singularity of the system (which in this particular case is the "centre"). If fluctuations are neglected the Volterra system has an integral of motion given by

$$R = \alpha[\gamma y/\alpha - 1 - \ln(\gamma y/\alpha)] + \delta(\beta x/\delta - 1 - \ln(\beta x/\delta)]. \tag{8.20}$$

The equilibrium point (centre) corresponds to the lowest possible value $R = 0$ with $x_0 = \delta/\beta$ and $y_0 = \alpha/\gamma$. The occurrence of fluctuation terms in (8.3) leads to a variation of $R(t)$ in time. If the fluctuation density is small then $R(t)$ is a slowly varying function of time. A truncated equation for $R(t)$ can be obtained by averaging over one period. In this case T_e can be defined by the mean time it takes $R(t)$ to attain a boundary value R_b. We can also proceed from the two-dimensional Kolmogorov-Fokker-Planck equation (8.2). Using a special transformation of the variables and again averaging over a period we pass from the equation for the two-dimensional probability density $w(x, y, t)$ to the equation for a one-dimensional density $w(\varrho, t)$. Such a method was first proposed by I. K. Kostin (see also Alekseev, Kostin 1974). First, assuming

$$\tau = t\sqrt{\alpha\delta}, \quad x_0 = \delta/\beta, \quad y_0 = \alpha/\gamma,$$

$$x_1 = x\beta/\delta, \quad y_1 = y\gamma/\alpha, \quad a = \sqrt{\alpha/\delta}$$

we obtain Eq. (8.2) in the form

$$\frac{\partial w(x_1, y_1, \tau)}{\partial \tau} = \frac{\partial}{\partial x_1}[A(x_1, y_1)\, w] + \frac{\partial}{\partial y_1}[A_2(x_1 y_1)\, w] + \frac{1}{2}\frac{\partial^2}{\partial x_1^2}[B_1(x_1, y_1)\, w] + \frac{1}{2}\frac{\partial^2}{\partial y_1^2}[B_2(x_1, y_1)\, w], \tag{8.21}$$

where

$$A_1(x_1, y_1) = -ax_1(1 - y_1), \quad A_2(x_1, y_1) = y_1(1 - x_1)\, a^{-1},$$

$$B_1(x_1, y_1) = ax_1(1 + y_1)\, x_0^{-1}, \quad B_2(x_1, y_1) = y_1(1 + x_1)\, a^{-1} y_0^{-1}.$$

Let us introduce the variables

$$\varrho = \varrho(x_1, y_1), \quad \varphi = \varphi(x_1, y_1)$$

with

$$w(x_1, y_1) = w(\varrho, \varphi)\frac{\partial(\varrho, \varphi)}{\partial(x_1, y_1)}. \tag{8.22}$$

The integral of motion (8.20) has been chosen as the function $\varrho(x_1, x_2)$. For the variables x, y and a this integral has the form

$$\varrho(x_1, y_1) = a[y_1 - 1 - \ln y_1] + a^{-1}[x_1 - 1 - \ln x_1]. \tag{8.23}$$

The other function φ has been chosen in the form

$$\varphi(x_1, y_1) = -x_0[x_1 + \ln(x_1 - 1)^2] + y_0[y_1 + \ln(y_1 - 1)^2]. \tag{8.24}$$

On averaging (8.21) over one period T along the trajectories of the averaged Volterra system we obtain a Kolmogorov-Fokker-Planck equation which depends solely on ϱ

$$\frac{\partial w(\varrho, \tau)}{\partial \tau} = \frac{\partial}{\partial \varrho} [F(\varrho) w] + \frac{1}{2} \frac{\partial^2}{\partial \varrho^2} [G(\varrho) w], \tag{8.25}$$

where

$$F(\varrho) = -\frac{1}{2\pi} \int_0^T \left(\frac{1+x}{x x_0} + \frac{1+x}{y y_0} \right) dt,$$
$$G(\varrho) = \frac{1}{T} \int_0^T \left[\frac{(1+y)(x-1)^2}{a x x_0} - \frac{a(1+x)(y-1)^2}{y y_0} \right] dt. \tag{8.26}$$

The integrals in (8.26) have been taken along the trajectories corresponding to the given value of the parameter $\varrho = \varrho(x, y)$. The mean lifetime of the Volterra system can be roughly estimated from the mean time the Markov process ϱ described by Eq. (8.25) takes to reach the absorbing boundary. The boundary $\varrho = 0$ should be seen as a reflecting boundary. The value ϱ_b indicates that the populations of x and y can be equal to unity on a corresponding trajectory.

The value T_e for a one-dimensional Markov process is well-known (Stratonovich 1961) and is given by

$$T_e = 2 \int_0^{\varrho_b} \int_x^{\varrho_b} \frac{\exp\left\{ \int_x^y \frac{2F(\varrho)}{G(\varrho)} d\varrho \right\}}{G(x)} dx\, dy. \tag{8.27}$$

Representative values of T_e and corresponding values of the parameter $a = \sqrt{\alpha/\delta}$ and the populations x_0 and y_0 which define the position of the singularity $\varrho = 0$ are given in Table 8.1 (computing was provided by I. K. Kostin).

Table 8.1

Mean survival time (T_e) of a Volterra systems (expressed in periods)

$\bar{x}, \bar{y}$	a				
	0.125	0.50	1.00	2.00	8.00
10.10	0.13	0.45	0.76	0.45	0.13
10.100	0.64	2.2	1.4	0.65	0.17
10.1000	1.1	4.0	1.5	0.7	0.19
100.10	0.17	0.65	1.4	2.2	0.64
100.100	2.7	8.8	15	8.8	2.7
100.1000	9.7	33	29	12	2.7
1000.10	0.19	0.70	1.5	4.0	1.1
1000.100	2.7	12	29	33	9.7
1000.1000	35	130	220	130	35

The parameter a defines the extension of the elliptic trajectories near the equilibrium point (cf. Fig. 8.1). The value of $\sqrt{\alpha \cdot \delta}$ defines the orbital frequency (or period T) in the neighbourhood of (x_0, y_0). Lifetimes in Table 8.1 are expressed as fractions of the period T.

It follows from Table 8.1 that in societies with population numbers up to 1.000 internal fluctuations lead to a degeneration over a time of the order of 10 to 100 periods. The stability of the system is highest if $a \approx 1$, i.e. if the phase trajectories are nearly circular. Although the estimates we propose are rather crude, they do indicate that it is necessary to take internal noises into account in systems with a small number of interacting elements. This is illustrated below by distributed models.

8.4 Internal noise in distributed systems and spatial self-organization

What part can fluctuations, so diverse in nature, play in self-organization processes? In principle all that has already been said for local systems is also applicable to distributed systems. However spatial fluctuations have some special features. In this section we discuss the formation of dissipative structures arising from internal fluctuations. In other words we explore the behaviour of an open distributed system with a small number of interacting elements. Conceivably, the temporal and spatial fluctuations manifest their coherent properties near the self-organization threshold. This can be illustrated by a classical Brussellator, Eq. (2.54). Let a reactor in which a sequence of kinetic interactions occurs be divided into N equal volume elements. In order to describe the discrete system thus formed we pass from the densities to volume elements (or "boxes"): x_i, y_i, $A_i = A$; $B_i = B$ $(i = 1, 2, \ldots, n)$. Let us consider the $2N$-dimensional distribution function in a manner similar to that of the local model in (8.1):

$$W(\boldsymbol{x}, \boldsymbol{y}, t) = w(x_1, \ldots, x_n, y_1, \ldots, y_n, t).$$

Here W is the probability that at an instant t there are x_1 molecules of substance x, y_1 molecules of substance y in the first box, etc. The evolution of W over a small period of time $\mathrm{d}t$ is governed by the elementary reactions in each box and by diffusion transitions between adjacent boxes. The evolution due to elementary reactions is written as

$$\begin{aligned} &w(x_i, y_i \to x_i + 1, y_i) = Ax_i \, \mathrm{d}t, \\ &w(x_i y_i \to x_i - 1, y_i - 1) = Bx_i \, \mathrm{d}t, \\ &w(x_i, y_i) \to x_i + 1, y_i - 1) = x_i(x_i - 1) \, y_i \, \mathrm{d}t, \\ &w(x_i, y_i \to x_i - 1, y_i) = x_i \, \mathrm{d}t. \end{aligned} \tag{8.28}$$

and that due to diffusion as

$$w(x_i, x_{i+1} \to x_i - 1, x_{i+1} + 1) = w(x_i, x_{i+1} \to x_i - 1, x_{i-1} + 1)$$
$$= d_x x_i \, \mathrm{d}t,$$

$$w(x_1, x_2 \to x_1 - 1, x_2 + 1) = d_x x_1 \, \mathrm{d}t$$

$$w(x_{n-1}, x_n \to x_n + 1, x_{n-1} - 1) = d_x x_n \, \mathrm{d}t.$$

Transition probabilities for y can be written in much the same way. Here $d_{x,y} = D_{x,y} \times n^2/L^2$ and the length L of the reactor is assumed to be equal to unity.

The derivative $\mathrm{d}W(\boldsymbol{x}, \boldsymbol{y}, t)/\mathrm{d}t$ is defined in a manner similar to that of (8.1) (specifically by means of the balance of transition velocities). Consequently we obtain the following master equation:

$$\frac{\mathrm{d}w(\boldsymbol{x}, \boldsymbol{y}, t)}{\mathrm{d}t} = A\left[\sum_{i=1}^{n} w(x_{i-1} - 1, \hat{\boldsymbol{x}}, \boldsymbol{y}) - nw(\boldsymbol{x}, \boldsymbol{y}, t)\right]$$
$$+ \sum_{i=1}^{n} \{(x_i - 1)(x_i - 2)(y_i + 1) \cdot w(x_i - 1, y_i + 1, \hat{\boldsymbol{x}}, \hat{\boldsymbol{y}}, t)$$
$$- x_i(x_i - 1) y_i w(\boldsymbol{x}, \boldsymbol{y}, t) + B[(x_i + 1) w(x_i + 1, y_i - 1, \hat{\boldsymbol{x}}, \hat{\boldsymbol{y}}, t)$$
$$- x_i w(\boldsymbol{x}, \boldsymbol{y}, t)] + (x_i + 1) w(x_i + 1, \hat{\boldsymbol{x}}, \boldsymbol{y}, t)$$
$$- x_i w(\boldsymbol{x}, \boldsymbol{y}, t)] + d_x\{(x_1 + 1) w(x_1 + 1, x_2 - 1, \hat{\boldsymbol{x}}, \boldsymbol{y}, t)$$
$$+ (x_n + 1)[w(x_{n+1} - 1, x_n + 1, \hat{\boldsymbol{x}}, \boldsymbol{y}, t] - (x_1 - x_n) w(\boldsymbol{x}, \boldsymbol{y}, t)$$
$$+ \sum_{i=2}^{n-1} (x_i + 1)[w(x_{i-1} - 1, x_i + 1, \hat{\boldsymbol{x}}, \boldsymbol{y}, t)$$
$$+ w(x_i + 1, x_{i+1} - 1, \hat{\boldsymbol{x}}, \boldsymbol{y}, t)] - 2\sum_{i=2}^{n-2} x_i w(\boldsymbol{x}, \boldsymbol{y}, t) \tag{8.29}$$

+ terms corresponding to the diffusion of y.

Unlike $\boldsymbol{x}$ and $\boldsymbol{y}$ the vectors $\hat{\boldsymbol{x}}$ and $\hat{\boldsymbol{y}}$ in Eq. (8.29) do not contain those components which are the arguments of W. The boundary conditions (1.4) are included in the transition probabilities in the first and N-th boxes. Analytic or numerical solutions of the master equation (8.29) are highly problematic. We therefore have to perform computer experiments, or, to be more precise, Monte-Carlo simulations of the transition probabilities following the scheme of (8.28). We present below results of numerical simulations obtained by E. V. Astashkina (Astashkina, Romanovskii 1978, 1980). The scheme of (8.28) can be modified in such a way that it is able to describe a deterministic process. To do this the transition probabilities in the left-hand sides should be replaced by the real transitions, thus instead of $W(x_i \to x_i + 1) = Ax_i \, \mathrm{d}t$ we write $\mathrm{d}x_i = Ax_i \, \mathrm{d}t$ and obtain the distributed Brussellator equations in a discrete form. If the parameters of the deterministic and probabilistic schemes are identical then it is possible to compare both types of process. This renders the computer experiment much easier. As the parameters A, B, d_x, d_y were chosen for the numerical simulation the following measures were adopted:

1. The parameter values correspond to the saddle-type instability of the homogeneous state of the system in which dissipative structures or near-Turing bifurcations can exist (cf. Chapter 3).
2. One chooses $B < A^2 + 1$, $D_y > D_x$ to avoid time-periodic solutions.
3. The number N of boxes must be chosen such that the half wave length of the DS mode being investigated corresponds to 5 to 10 boxes.
4. The number of "molecules" in each box must be of the order of 10 to 100.

We now present the results of the computer experiments.

a) The velocities of two processes were compared in the simulation: the deterministic evolution from a nearly homogeneous initial distribution according to the distributed Brussellator equations and the evolution which is continuously accompanied by internal fluctuations according to (8.28). It became apparent that if the conditions for excitation of some DS modes are satisfied, then the fluctuations appreciably accelerate the initial stage of the process, since the fastest-growing modes are then involved. The relative influence of the fluctuations decreases since the system enters one or the other configuration.

b) The fluctuation level was studied as a function of the proximity to the Turing bifurcation point. If the system is far from the critical state $B \ll B_{cr}$ then the order of magnitude of relative fluctuations near the stable homogenous state is $\sqrt{\langle x\rangle}/\langle x\rangle$, $\sqrt{\langle y\rangle}/\langle y\rangle$, i.e. a Poissonian process occurs. As the system approaches the critical point ($B \lessapprox B_{cr}$) macroscopic fluctuations arise and the process becomes non-Poissonian. For certain periods the system is in a state near a homogeneous state and one close to a dissipative structure. The time for which the latter case occurs increases with B (Fig. 8.2). Actually a phase transition induced by internal noise takes place, or to be more precise the system attempts such a transition. The dissipative structure is only formed with probability equal to unity if the critical point is attained ($B = B_{cr}$). It is natural to expect that a complete phase transition will occur earlier, i.e. for $B \leqq B_{cr}$, if not only the internal but also the external noise is predetermined (see Section 8.5). If the excitation conditions of any DS mode are fulfilled and initial conditions are given near to this mode then the mode can be observed on the background of Poissonian fluctuations in the probabilistic model at relatively large molecule numbers in each box. An example of such a (third) mode and its corresponding spatial spectrum are given in Fig. 8.3.

c) "Metastable" states were investigated in the case when relations between the parameters permit the existence of some DS modes ($B > B_{cr}$). As we have already seen (Chapter 7), transitions exhibiting hysteresis between such modes are possible in the deterministic case. The stochastic model spends some time in one configuration and then some time in another as shown in Fig. 8.4. In this figure the permitted first and third modes coexist. We note that even between these modes there is transition with hysteresis. The behaviour of the system becomes increasingly stochastic with an increasing number of modes.

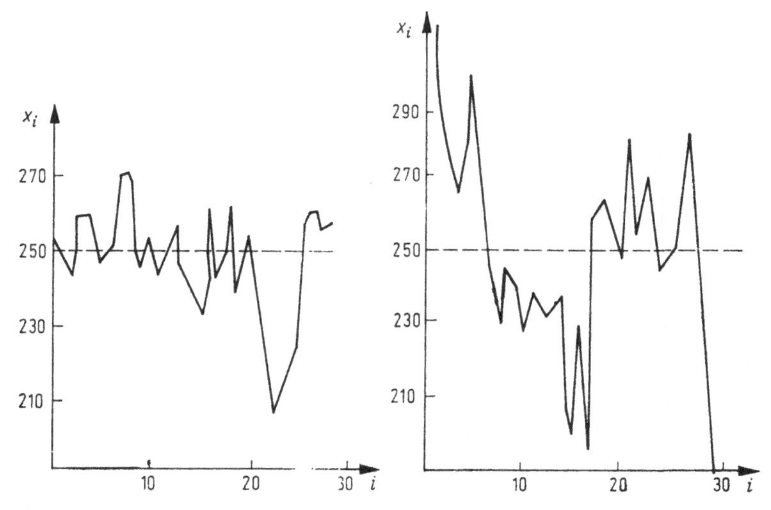

Fig. 8.2
Co-existence of a) homogeneous state and b) DS. During $\approx 1/15$ of the total time the system is in states near b), during the rest of the time the system is in the state near a). The computer process consisted of 70000 elementary acts: $A = 250$, $B = 10000 < B_{cr}$, $D_x = 11.2$, $D_y = 5D_x$, $N = 30$, $\Delta t = 10^{-7}$.

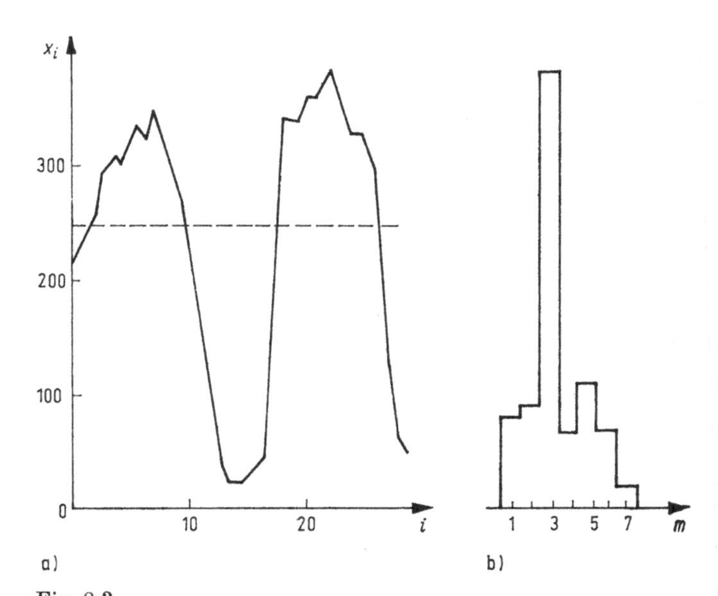

Fig. 8.3
a) DS shape in the presence of Poissonian fluctuations.
b) The spatial spectrum of the DS: m is the mode number, $B = 15000 > B_{cr}$, $D_x = 11.2$, $D_y = D_x\, 5$, $N = 30$, $\Delta t = 10^{-7}$. The duration of calculations is about $4 \times 10^4\, \Delta t$.

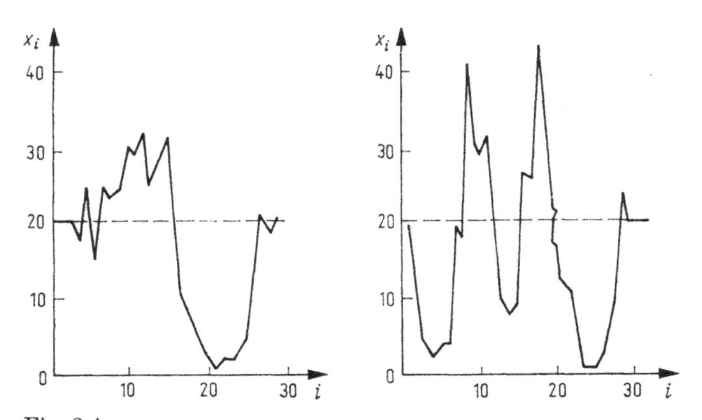

Fig. 8.4
Large-scale fluctuations near the critical point. One notes the co-existence of the first and third modes. Jump transitions occur between these states.

8.5 External noise and dissipative structures—linear theory

We consider the simplest case of δ-correlated fluctuations which are additive in space and in time. The system (7.15) serves as a mathematical model for the Brusselator, random functions $\xi_1(t, r)$ and $\xi_2(t, r)$ being added to the right-hand sides. Here

$$\langle \xi_i(t, r) \rangle = 0 ,$$
$$\langle \xi_i(t, r)\, \xi(t + \tau, r + \Delta r) \rangle = \varepsilon^2 \delta_{ij} \delta(\Delta r)\, \delta(\tau) \tag{8.30}$$
$$(i, j = 1, 2) .$$

The case of additive white noise is of course an extreme one. However considering it is of interest in itself and a few general rules may be obtained from it. Such studies were made most thoroughly in a series of papers by B. N. Belintsev, M. V. Volkenstein and M. A. Lifshits (see Belintsev et al., Belintsev 1979) which we follow here.

The correlation analysis of kinetic fluctuations which occur in systems with additive noise near the homogeneous stationary state as the bifurcation points are approached is provided by a linear approximation. The linear theory fails at and very near to the bifurcation points. The linearized two-component model in the vicinity of the homogeneous stationary states of the system (3.14) can be written as

$$\begin{aligned} \frac{\partial v_1}{\partial t} &= f_{11} v_1 + f_{12} v_2 + D_x \frac{\partial^2 v_1}{\partial r^2} + \xi_1(t, r) , \\ \frac{\partial v_2}{\partial t} &= f_{21} v_2 + f_{22} v_2 + D_y \frac{\partial^2 v_2}{\partial r^2} + \xi_2(t, r) . \end{aligned} \tag{8.31}$$

The coefficients f_{ij} are determined according to (3.4) and in the particular case of the Brusselator (7.14) we obtain

$$f_{11} = B - 1 , \quad f_{12} = -f_{21} = A^2 , \quad f_{22} = B .$$

The four space-time correlation functions

$$R_{ij}(\Delta r, \tau) = \langle v_i(r, t)\, v_j(r + \Delta r, t + \Delta\tau)\rangle$$

can in principle be evaluated because of the linearity of the system (8.31). However it is easier to find first the single-time correlation functions

$$\overline{R}_{ij}(\Delta r, t) = \langle v_i(r, t)\, v_j(r + \Delta r, t)\rangle.$$

The procedure for obtaining equations for these functions based on (8.31) is as follows. We multiply the first and second equation by v_1 and v_2, respectively, at $r + \Delta r$ and average over the ensemble of realizations of the random fields $\xi_i(t, r)$. We obtain the following equations

$$\frac{\partial \overline{R}_{ij}}{\partial t} = \sum_{k=1}^{2} (f_{ik}\overline{R}_{kj} + f_{jk}\overline{R}_{ik}) + (D_x + D_y)\frac{\partial^2 \overline{R}_{ij}}{\partial r^2} + \frac{\varepsilon^2}{2}\,\delta_{ij}\delta(\Delta r). \tag{8.32}$$

We are interested in stationary correlation functions with $\partial R_{ij}/\partial t = 0$ for different relations between the parameters. The most important kind of solutions of the Brussellator and any other two-component system are those that occur for small values of the bifurcation parameter

$$\alpha = \frac{D_1}{D_2} - \frac{\left(1 - \sqrt{B}\right)^2}{A^2}, \tag{8.33}$$

which indicates the distance from the point of Turing instability. We have

$$\overline{R}_{ij}(\Delta r) \sim \alpha^{-1/2} \exp\left\{-\frac{u_1\alpha^{1/2}}{2}\,\Delta r\right\}\cdot \cos(u_1\,\Delta r + \delta). \tag{8.34}$$

Here u_1 is the minimal wave number corresponding to the first permitted DS shape or mode. The parameter α describes the degree of coherence of the fluctuations of the kinetic variables and gives the scale of the region in which the fluctuations of the variables are spatially correlated:

$$\Lambda = 2/u_1\alpha^{1/2}. \tag{8.35}$$

Even if the system is disordered on average the space distribution of the variables is nearly periodic within the limits of the region $\Delta r_{\text{cor}} \sim \Lambda$. The value of Λ increases as the threshold of DS formation is approached (Fig. 8.5). The growth of Λ as shown

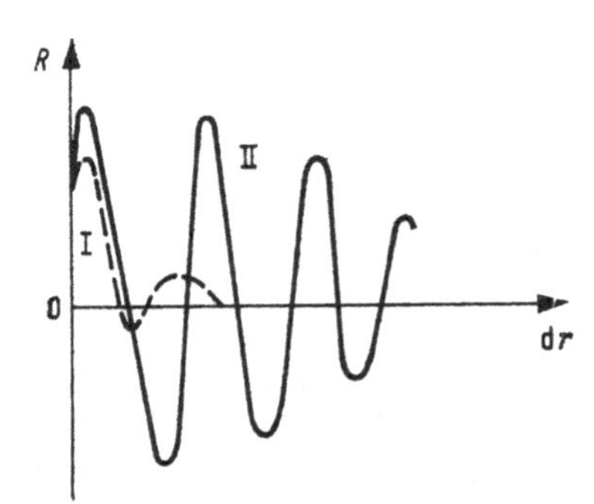

Fig. 8.5
A typical form of the stationary correlation function $(R_{ik}(\Delta r))$ for two values of the bifurcation parameter α. Line I corresponds to large distances from the bifurcation point $\alpha \approx 1$ and line II to the neighbourhood of the self-organization threshold $(0 < \alpha \ll 1)$.

corresponds to a real "long-range" order which has its origin in the parameter region below the threshold ($\alpha \leqq 0$). If $\alpha \leqq 0$ the macroscopic state of the system is ordered and the spatial period of the dissipative structure at $\alpha = 0$ coincides with the period of the oscillating factor in the correlation function (8.34). An immediate effect of the growth of the correlation radius $\Lambda \sim \Delta r_{\text{cor}}$ is the increase in the level of volume fluctuations as defined by the following integral:

$$\langle x_i^2 \rangle \sim \frac{1}{v} \int R(r) \, \mathrm{d}^2 r \sim \frac{1}{L} \int_0^\infty \bar{R}_{ij}(\Delta r) \, \mathrm{d}r. \tag{8.36}$$

Given the spatial function $\bar{R}_{ij}(\Delta r, t)$ it is easy to obtain an equation for the space-time function $R_{ij}(\Delta r, \tau, t)$:

$$\frac{\partial R_{ik}}{\partial \tau} = \sum_j f_{kj} R_{ij} + D_k \frac{\partial^2 R_{ik}}{\partial r^2}. \tag{8.37}$$

The initial conditions are expressed in terms of R_{ik}:

$$R_{ik}(\Delta r, \tau, t)|_{\tau=0} = \bar{R}_{ik}(\Delta r, t).$$

As in solving the linear equations for the excitations (3.8) we seek R_{ik} in the form

$$R_{ik}(\Delta r, \tau) = C_{ik} \exp(pt) \exp\left\{\mathrm{i} \frac{2\pi}{\lambda} r\right\} = \tilde{R}_{ik}(u) \, \mathrm{e}^{pt}. \tag{8.38}$$

Here p and $u = \dfrac{2\pi}{\lambda}$ have the same meaning as in (3.5) and are connected by a dispersion relation. The quantity

$$\tau_{\text{cor}} = [p_{\text{cr}}(\lambda)]^{-1} \tag{8.39}$$

plays the role of a correlation time. The correlation time of the mode λ_0 which is expected to occur increases as the threshold of instability of the homogeneous state is approached. Besides the properties of solutions in a system with additive noise B. N. Belintsev also studied the stability of systems in which the coefficients f_{ij} in (8.31) undergo small fluctuations. He showed that the threshold of Turing instability can be altered and in particular lowered by external multiplicative noise, i.e. parametric noise can cause an "untimely" transition from the homogeneous state to the dissipative structure.

8.6 Nonlinear effects—the two-box model

The results of studies made in the framework of the linear approximation emphasize the essential role played by fluctuations in distributed systems during the approach to the threshold of self-organization. However as noted above the linear approximation becomes inapplicable in the immediate vicinity of the self-organization threshold where, in particular, the correlation functions diverge and the level of fluctuations of the kinetic variables also tends to infinity. One should bear in mind that the branching off from the homogeneous state accompanied by symmetry breaking is an essentially nonlinear effect in the deterministic theory of DS. Calculating the contribution of the nonlinear terms in a two-component distributed system is in general a highly complicated problem.

However, in a small neighbourhood of the instability threshold it is possible to provide analytical considerations based on the method of separation of motions. A natural parameter occurs near the instability threshold: the reciprocal ratio of the relaxation time of the critical mode (τ_{cor}, in the simplest case the mode is a single mode and produces the shape of the bifurcating dissipative structure) to all other characteristic times of the linearized system. Thus in the time scale of the relaxation of the critical mode only this "slow" degree of freedom varies to any extent. The problem then reduces to the investigation of the corresponding stochastic equation of motion instead of a system of such equations at the starting point. This approach is based on Landau's idea of the separation of an order parameter in the theory of phase transitions of second order. It was realized for the case of self-organization in Turing systems in the paper by Belintsev (1979) for a two-box model. We emphasize that the two-box model as given in (7.1) also yields a number of principal results in the deterministic case.

As a "slow" mode we take the antisymmetric co-ordinate (cf. Section 3.6)

$$v(t) = a(x_1 - x_2) + b(y_1 - y_2). \tag{8.40}$$

During a time of order $\tau_{\text{cor}} \sim p_{\text{cr}}^{-1}$ and in the absence of noise this co-ordinate is determined by the equation

$$\frac{\mathrm{d}v}{\mathrm{d}t} = p_{\text{cr}} v - k^2 v^3. \tag{8.41}$$

We assume that the parameters A and B in the Brussellator model fluctuate around their mean values and that these fluctuations are small and have a Gaussian distribution

$$\begin{aligned} &A = \langle A \rangle + \varepsilon_a \xi^a(t), \quad B = \langle B \rangle + \varepsilon_b \xi^b(t), \\ &\varepsilon_a \ll \langle A \rangle, \quad \varepsilon_b \ll \langle B \rangle. \end{aligned} \tag{8.42}$$

Equation (8.41) is then extended by adding fluctuation terms

$$\frac{\mathrm{d}v}{\mathrm{d}t} = p_{\text{cr}} v - k^2 v^3 + \varepsilon^a \xi^a(t) + \varepsilon^b \xi_b(t)\, v. \tag{8.43}$$

The corresponding Kolmogorov-Fokker-Planck equation for the probability density is then

$$\frac{\partial W}{\partial t} = -\frac{\partial}{\partial v} \{[p_{\text{cr}} v - k^2 v^3 + \varepsilon_b v]\, W\} + \frac{\partial^2}{\partial v^2} [(\varepsilon_a + \varepsilon_b{}^2 v^2)\, W]. \tag{8.44}$$

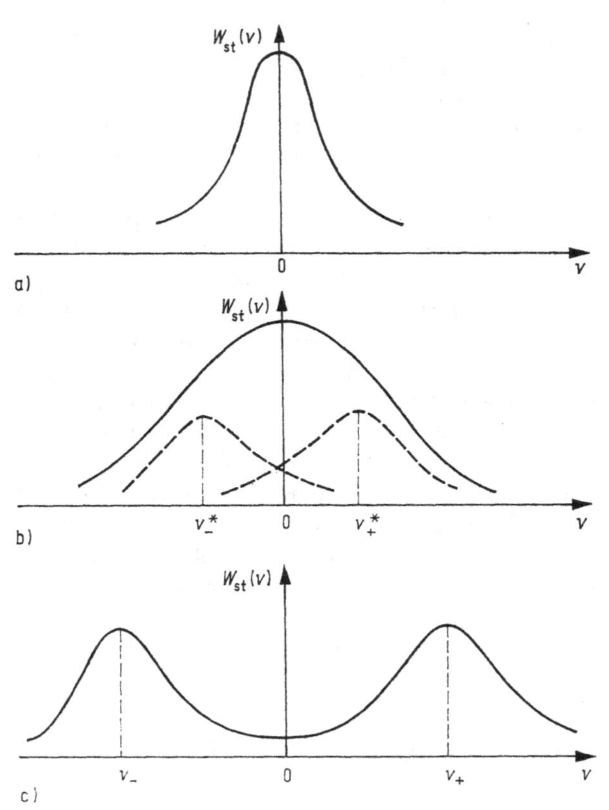

Fig. 8.6
The form of the stationary probability $W(v)$ for the antisymmetric mode:
a) $p_{cr} < -\varepsilon_b^2$,
b) $-\varepsilon_b^2 < p_{cr} < 0$,
c) $p_{cr} \geqq \varepsilon_b^2$

If $\partial W/\partial t = 0$ in (8.44) we find the stationary probability distribution whose shape is shown qualitatively in Fig. 8.6 for different parameter values. In the probabilistic treatment a macroscopic stratification effect occurs in that the maximum of $W_{st}(v)$ at $v = 0$ splits into two maxima symmetrically located at $v^*_{+-} = \pm |(p_{cr} - \varepsilon_b^2)/k^2|^{1/2}$. This is realized for $p_{cr} = \varepsilon_b^2$ (cf. Fig. 8.6) although in the deterministic description stratification occurs for $p_{cr} = 0$. However the system still has "knowledge" of its subsequent stratification up to the instability threshold in the parameter region $-\varepsilon_b^2 < \lambda < 0$ since it spends some time in one of the two stratified states given by

$$v^*_{+-} \simeq \pm \frac{(p_{cr} - \varepsilon_b^2)^{1/2}}{k}. \tag{8.45}$$

We note that ε_a, which determines the intensity of the additive fluctuations only affects the width of the stationary distribution $W_{st}(v)$ and does not modify the value of the stratification threshold.

The fact that the distribution remains unimodal in the region $-\varepsilon_b < p_{cr} < 0$ is the result of multiple jumps between the two neighbouring states. In conclusion we note that bimodal distributions are also observed in the case of internal distributed multiplicative noise (cf. Section 8.4) if the excitation condition for two modes is fulfilled. New states or maxima of the distribution do not however occur.

8.7 Wave propagation and phase transitions in media with distributed multiplicative noise

In Chapter 4 we considered the classical one-component model of Kolmogorov-Petrovskii-Piskunov. This model describes the propagation of an excitation front in the case that the corresponding homogeneous system has an unstable point and boundedness is saved by a cubic nonlinearity. In practice the following situation can occur in media which permit spontaneous autocatalytic replication of the matter or living species involved. The conditions for autocatalysis are not fulfilled in all elementary volumes of the system but only in isolated subsystems. The simplest model of such a process is given by an equation of the form

$$\frac{\partial x}{\partial t} = -\alpha x - \beta x^3 + D \frac{\partial^2 x}{\partial r^2} + \xi(r, t)\, x . \tag{8.46}$$

Here $\xi(r, t)$ is a random function of the co-ordinate and time or an external distributed parametric noise source. The properties of the model (8.46) were investigated by A. S. Mikhailov and I. V. Uporov with a sufficient degree of completeness for different statistical characteristics of the noise function $\xi(r, t)$ (see Mikhailov, Uporov 1980; Mikhailov 1981). Below we only consider the simplest model in the context of which the mean value is assumed to vanish, $\langle \xi(r, t) \rangle = 0$, and the correlations are given by exponential relations,

$$\langle \xi(r, t)\, \xi(r', t') \rangle = S \exp\left\{- \frac{|r - r'|}{r_0} - \frac{|t - t'|}{\tau_0}\right\}. \tag{8.47}$$

It is furthermore assumed that the function $\xi(r, t)$ has a Gaussian distribution. The shape of the correlation function implies that $\xi(r, t) > \alpha$ is possible in local regions of the space (sections of the straight line) and during certain intervals of time. The condition of autocatalysis is then fulfilled in these regions for certain times. Passive waves of the variable x will propagate outwards from these "active" regions because of diffusion. With increasing intensity S regions with autocatalysis will grow and unite. If S exceeds the critical value S_{cr} then on average the condition of parametric excitation is fulfilled for the whole space and the mean concentration $\langle x \rangle$ will increase as long as it is not bounded by the term $-\beta x^3$. Such an effect is called a noise-induced phase transition by Mikhailov and Uporov.

The aim of the theory is to determine the critical value S_{cr} in dependence on the parameters of the problem, α, r_0, τ_0. The following characteristic lengths can be obtained from the given parameters: the diffusion length $r_{\text{Diff}} = \sqrt{D/\alpha}$ which we have already used occasionally, the stationarity length $l = \sqrt{D\tau_0}$ which is equal to the distance covered by a particle during the lifetime of an autocatalysis or replication centre, and the correlation radius r_0.

The problem is solved by an approach which is analogous to the self-consistent field method in the theory of phase transitions of second order (Klimontovich 1980). The idea behind the solution is that the concentration can be separated into a fast-fluctuating component against a spatially smoothed background. The smoothed component is defined for a one-dimensional space by

$$\eta(r, t) = \frac{1}{L} \int\limits_L x(r + \varrho, t) \, \mathrm{d}\varrho = \langle x \rangle_L , \tag{8.48}$$

where L is larger than all characteristic lengths of the system (r_0, l and r_{Diff}). The fast-fluctuating component $\tilde{x}$ is given by

$$\tilde{x} = x - \eta . \tag{8.49}$$

Spatial smoothing of (8.46) gives

$$\frac{\partial \eta}{\partial t} = -\alpha\eta - \beta\eta^3 + D \frac{\partial^2 \eta}{\partial r^2} + \bar{\xi} \cdot \eta - 3\beta\eta\langle\tilde{x}^2\rangle_L + \langle\tilde{\xi}\tilde{x}\rangle_L - \langle\tilde{x}^3\rangle_L , \tag{8.50}$$

where $\tilde{\xi}$ is the fast-fluctuating component of the random function $\xi(r, t)$ and $\bar{\xi}$ is the slow component obtained by spatial smoothing

$$\tilde{\xi}(r, t) = \xi(r, t) - \bar{\xi} . \tag{8.51}$$

Subtracting (8.50) from (8.46) gives

$$\begin{aligned} \frac{\partial \tilde{x}}{\partial t} = {} & -\alpha\tilde{x} - 3\beta[\eta^2\tilde{x} - \eta(\tilde{x}^2 - \langle\tilde{x}^2\rangle_L)] + D \frac{\partial^2 \tilde{x}}{\partial r^2} \\ & + \bar{\xi}\tilde{x} + \tilde{\xi}\eta + \tilde{\xi}\tilde{x} - \langle\tilde{\xi}\tilde{x}\rangle_L - \beta[\tilde{x}^3 - \langle\tilde{x}^3\rangle_L] . \end{aligned} \tag{8.52}$$

$\langle\tilde{x}\tilde{\eta}\rangle_L$, $\langle\tilde{x}^3\rangle_L$ and $\langle\tilde{x}^2\rangle$ follow readily from (8.52).

It is then possible to write down an approximative equation for the smoothed component of the concentration η. Here we only give the basic results which were obtained under the assumption

$$\sqrt{\langle\tilde{x}^2\rangle} \ll \eta , \tag{8.53}$$

(for details see Mikhailov, Uporov 1980). This equation is a linearization of (8.52) in $\tilde{x}$.

On the other hand the linearized equation is solved for the fast component $\tilde{\xi}$ of the noise by methods of perturbation theory. Finally we find the following equation

$$\frac{\partial \eta}{\partial t} = -\alpha \left(1 - \frac{S}{S_{\text{cr}}}\right) \eta - \beta\eta^3 + D \frac{\partial^2 \eta}{\partial r^2} + \bar{\xi}\eta , \tag{8.54}$$

where

$$\frac{S}{S_{\text{cr}}} = \alpha^{-1} \left(\frac{\langle \tilde{x} \tilde{\xi} \rangle}{\eta} \right)_{\tilde{\xi}=0}$$

and the renormalization is assumed to be small. This equation is valid near the threshold, i.e. for small η. We note that for $D \equiv 0$ (8.54) is analogous to the averaged equation for the amplitudes (8.18) in a nonlinear parametrically excited circuit. We also emphasize the similarity of this equation to the nonstationary Landau-Ginzburg equation in the theory of second order phase transitions (cf. Landau, Lifshits 1976).

Finally one must find out under which assumption (8.53) is valid in order to be able to linearize the initial equation. To do this we consider calculated values of the parameter

$$G = \left[\frac{\tilde{x}}{\eta^2} \right]_{S \cong S_{\text{cr}}} \tag{8.55}$$

which characterizes the inequality (8.53) and the quantity S_{cr}. Calculations show that the following cases can be considered for different combinations of the parameters $r_{\text{Diff}}, l, r_0, \tau_0$:

a) $$r_{\text{Diff}} \gg r_0 \gg l [\alpha \ll D/r_0^2 \ll \tau_0^{-1}],$$

$$S_{\text{cr}}^{1/2} = \alpha r_{\text{Diff}}/l, \quad G = \left(r_0/\sqrt{2}\, r_{\text{Diff}} \right) \left| \frac{S}{S_{\text{cr}}} - 1 \right|^{-1/2};$$

b) $$r_{\text{Diff}} \gg l \gg r_0 [\alpha \ll \tau_0^{-1} \ll D/r_0^2],$$

$$S_{\text{cr}}^{1/2} = \alpha \left(r_{\text{Diff}}/\sqrt{r_0 l} \right), \quad G = \left(l/\sqrt{2}\, r_{\text{Diff}} \right) \left| \frac{S}{S_{\text{cr}}} - 1 \right|^{-1/2}. \tag{8.56}$$

These conditions imply that the fluctuations are not small near the excitation threshold. The theory must therefore be improved for $G \gtrsim 1$. The fluctuations are strong even far from the transition point ($G \geqq 1$) for the remaining relations between the parameters. The following question arises: Does (8.54) describe wave propagation with velocity $v \sim \sqrt{\alpha(S/S_{\text{cr}} - 1)\, D}$ for $S > S_{\text{cr}}$, i.e. for strong noise pumping, as follows from the theory of Kolmogorov, Petrovskii and Piskunov? If an initial excitation is given at a certain section $\tilde{L} \ll L$ it will propagate as a front, this front however vanishes in the noise or will be destroyed by parametric fluctuations after some time. A similar situation arises in modelling excitation propagation in media with "negative" friction even in the case of additive noise (Romanovskii, Sidorova 1967; see also Schimansky-Geier et al. 1983).

We now assume $L \to \infty$. At the instant of time considered η is then the mean value $\langle x \rangle$. Eq. (8.54) thus has the form

$$\frac{\mathrm{d}\langle x \rangle}{\mathrm{d}t} = \alpha \left(\frac{S}{S_{\text{cr}}} - 1 \right) \langle x \rangle - \beta \langle x \rangle^3. \tag{8.57}$$

The following function is its stationary solution depending on the initial conditions

$$\langle x \rangle_{st} = \begin{cases} D & \text{for } S < S_{cr}, \\ \sqrt{\dfrac{\alpha}{\beta}\left(\dfrac{S}{S_{cr}} - 1\right)} & \text{for } S > S_{cr}. \end{cases} \tag{8.58}$$

It follows from this solution that the medium becomes transparent for the diffusing matter for sufficiently strong fluctuations, despite the decay of this matter it dominates on average ($\alpha > 0$).

By its shape (8.58) reminds one of the expression for the order parameter above and below the transition point in the Landau theory of second order phase transitions. Apart from this more substantial similarities known as "critical deceleration" are observed, i.e. the characteristic time of variation of the quantity η reaches infinity at the threshold $S = S_{cr}$. The penetration depth of the matter diffusing into the medium also becomes infinite which also applies to the correlation radius of order fluctuations in the Landau theory. However the described theory differs essentially from the theory of second order phase transitions in the fact that the mean-square of the fluctuations is directly proportional to the quantity η^2. As shown in the original papers of Mikhailov and Uporov this difference is connected with the positivity of the order parameter in the present problem. Moreover this property renders the method applicable up to the transition point ($S = S_{cr}$) in two and three-dimensional media.

Chapter 9
Autowave mechanisms of transport in living tubes

In this chapter we consider a class of AW mechanisms in biology which effect fluid transfer inside a tube and are completely or partially caused by changes in the shape of the tube. Such a fluid transfer is called a peristaltic current and its objective is the transport and/or mixing of the fluid. In the first case the change in the shape of the tube usually has the character of a travelling wave, while in the second more complex wave processes take place. Technical examples of peristaltic systems are for instance various roll pumps used in devices for artificial blood circulation or for precise dosage. Biological examples are the organs of the gastrointestinal tract (oesophagus, stomach, small intestine, colon), the ureter, myometer, some blood-vessels and other organs with smooth muscle tissue as well as small transport veins in plants or even unicellular organisms.

It is the behaviour of the tube wall that determines the organization of peristaltic motion. In a sufficiently general case it can be described by the equation

$$Q(s, p, P_e, \gamma) = 0, \tag{9.1}$$

where s is the cross-section area, p is the pressure in the current, P_e is the pressure outside the tube and γ is a measure of wall activity. Special cases of (9.1) are

(i) the wave of external loading $P_e = P_e(z, t)$,

(ii) the parametric wave $P_e = \text{const}$, $\gamma = \gamma(z, t)$ and

(iii) $P_e = \text{const}$, $\gamma = \gamma(p, s)$.

In the first two cases AW processes may act as factors organizing peristaltic currents, in the third case AWPs appear after local feedback between the hydrodynamic pressure and the contraction mechanism of the wall.

Both kinds of AW phenomena in peristaltic motion may be found in biology. Later we will discuss them using three examples of biological objects. It should be remarked that despite the fact that this field has gained the attention of researchers for a long time many questions remain unsolved. The concepts and models proposed in the following should therefore be regarded as a first attempt to review the accumulated experimental material from the point of view of AW theory.

9.1 Autowaves in organs of the gastrointestinal tract

In the last century it was reported that an isolated small intestine of a rabbit, if completely denerved and placed in a special solution continues to work as a peristaltic pump for a long time. This activity is effected by contraction waves running along the intestine and powered by accumulated stocks of energy. As the process takes place in an isolated organ it may be accepted that the autowave phenomenon does in fact occur in the natural state. In this section we shall present the basic experimental facts concerning the structure and function of smooth muscular walls and discuss some results of modelling the AW processes occuring in them.

It should be noted that all peristaltic organs have very similar structure and properties. Their walls consist of bundles of smooth muscle cells. Usually there are two layers, one longitudinal and the other circular (the latter being nearer to the inner surface). In the longitudinal layer the cells in the bundles are mainly oriented along the tube and in the circular layer around the tube. Smooth muscle cells do not form genuine synctium even in one bundle. However good contacts are possible between cells inside one bundle. The connection between cell bundles is weaker. Bundles are united into fibres etc., the smooth muscle wall thus has a hierarchical structure.

Essential differences between the smooth muscles of different organs exist with regard to their innervation, i.e. the organization of the transfer of control signals from the central and vegetative nervous system to the muscles. The number of synapses in the tissues and their distribution vary over wide ranges. This is such an essential feature that it is sometimes used to classify smooth muscle tissues (Bogach, Reshodko 1979). The principal difference to the innervation of other kinds of muscle is that control signals are not transmitted to a "point" but instead affect the whole organ. The study of the role of the nervous system in the organization of peristaltic AW processes is an essential part of physiological research. However many aspects of these processes may be understood without using explicit results of this research, and in the following we will make use of this circumstance.

The spontaneous electrical activity of smooth muscle tissue has two components, a relatively slow potential variation (slow waves, "SW") and short pulses (spikes or action potentials, "AP"). Slow waves have small amplitudes (5 to 20 mV) and their characteristic periods are of the order of seconds or longer. The period and shape of a SW is relatively stable despite considerable variations according to the organ or the species of animal involved. APs have amplitudes of 60 to 90 mV and characteristic periods of the order of milliseconds. They arise only on the crest of a SW at random instants of time. However the distribution function of the intervals between neighbouring spikes is relatively stable. It should be emphasized that SWs are not directly related to contractions, they can even be found in the absence of contraction processes.

A smooth muscle cell may only contract in the presence of an AP. Chemicals inhibiting the peak activity of tissues also inhibit contraction activity. The contraction force depends on the number of spikes per unit time, single spikes usually do not cause significant contractions. In this way a summation of APs occurs during the characteristic time of a SW.

APs as well as SWs propagate in cell ensembles (tissues). Taking the facts mentioned before into account one may say that the intensity of the fast component of the electrical activity determines the degree of contraction, whereas the SW is the factor that synchronizes and coordinates the contraction. Slow waves propagate over considerable distances, the propagation velocity varies along the intestine, for example it is 10 to 12 cm/s in the duodenum, 6 to 10 cm/s in the jejumum and 0.3 to 1.0 cm/s in the stomach. The period of the SW also differs, in the small intestine it attains its minimum in the highest section of the duodenum, whereas in the large intestine the minimum is located in the middle section. The correlation length is of the order of several wave lengths and is appreciably smaller than the length of the intestine.

APs propagate along the bundles of smooth muscle cells. In the organs of the gastrointestinal tract an AP pulse usually traverses a length of not more than 1 cm. In rare cases this distance can increase up to 10 cm. The mean velocity does not exceed the range of 5 to 10 cm/s. The relative independence of AP and SW propagation can lead to various different propagation patterns of peak activity along the intestine. As it determines contraction activity of smooth muscles the latter is characterized by a large variety of regimes of spatio-temporal organization.

From a functional viewpoint all types of contraction activity of the intestine can be divided into two main groups: the first provides the mixing and churning of bowel contents and the second effects the transport of the contents through the intestine. Accordingly one distinguishes four types of contraction organization, (i) rhythmic segmentation, (ii) pendulum-like contraction, (iii) peristaltic contraction (fast, slow and very slow) and (iv) antiperistaltic contraction (Bogach 1974).

Rhythmic segmentation is mainly effected by contractions of the circular muscle layer, which are characterized by the simultaneous appearance of strained portions of the intestine. The transport of the contents is insignificant. Rhythmic segmentation may be regular or irregular, in the latter case fast transfer of the contents over considerable distances occurs during a short time interval followed by the restauration of a clearly regular rhythmic segmentation. Peristaltic contraction is represented by a travelling wave propagating downwards from the duodenum. The contents is thrust forward synchroneously. Peristaltic waves may arise at arbitrary loci of the intestine. The types of contraction usually change in the following sequence: after intensive mixing of the contents in intestinal segments a pendulum-like motion occurs which later transforms into peristaltic motion.

Finding a quantitative relationship between the contractive and the electrical activity of the smooth muscle tissue is an extremely complicated problem which has not been solved. We will briefly discuss some aspects of the problem. The generation of APs, spontaneously or chemically induced, always leads to the appearance of contractions. The contractile system is very inert compared to the electrical one. Hence considerable contractions of the tissue begin 0.5 to 1.0 s after the appearance of the corresponding pulse packet. At high frequencies of SWs with a large number of APs on their crests either permanent contraction (complete tetanus) or oscillations with small amplitudes and a large constant component of contraction occur.

Further it should be noted that besides the contractions caused by APs, there exists another type of contraction, the so-called tonic contraction. It can be observed when APs are absent (Bogach, Reshodko 1979). The characteristic times of tonic contractions significantly exceed the periods of electrical processes in the corresponding tissues. All this suggests that in addition to the electrical activation of the contractile mechanisms there may exist another, probably purely chemical way of stimulating contraction. Many types of smooth muscle cells exhibit mechano-receptive properties (Orlov 1967). In response to the fast stretching of a strip of smooth muscle tissue an AP arises and induces an active contraction. This effect is not observed if the same specimen is slowly stretched. There is a strict mutual relationship between the relative stretching of the smooth muscle strip and the magnitude of the membrane potential on which the frequency of the AP generation depends in its turn. Apparently the increase of the AP pulse frequency due to the stretching plays an important role in the appearance of tension in the stretched tissue in addition to that of passive elasticity. Moreover the mechano-receptive properties are necessary for the existence of a feedback chain ensuring the self-regulation of the activity of smooth muscle organs (see also Sections 9.2, 9.3).

Another, passive, mechanism binding the motor and electrical activity of the organs probably plays an essential role in the propagation of excitations. Generally speaking, by analysing the properties of smooth muscle tissues and cells one may suggest several ways in which electrical excitations propagate through the tissue:

1. by means of nerve fibres;
2. reflectorically, through the intramural network;
3. by chemical interactions;
4. by electrical interactions between neighbouring cells;
5. by mechanical action of contracting cells on resting neighbouring cells.

The realization of the first two mechanisms is hardly probable. Propagation of electrical excitations in the other cases mentioned above is determined by the relative positions of the cells, by the number and the areas of contacts between them, by the membrane tensions and so on, i.e., it depends on the mechanical state of the tissue.

The transfer of the excitation due to the stretching of resting cells in the gastrointestinal tract organs can only be of restricted importance. Indeed very fast and strong stretching of a cell is necessary to produce a contraction reaction, which moreover has a latency period comparable to the oscillation period itself. Remember the fact that slow waves may propagate even if mechanical activity of the organ is absent. Nevertheless it would be wrong to deny the role of stretching in the propagation of excitations absolutely (see also Sections 9.2, 9.3).

One has to consider the electrical mechanism based on local ion currents as being fundamental for the propagation of an excitation in the tissues mentioned above. The main parameter determining the potential transfer is the density of intercellular contacts (Barret et al. 1968). At the same time the most likely propagation mechanism of SWs is the chemical one that explains the smallness of their amplitudes and the metabolic hypothesis of their nature. It is worth noting that in experiments both of these mechanisms lead to the same phenomenological pictures and in the theoretical description they lead to structurally similar mathematical models, in particular to the kinds of model considered in the foregoing chapters.

The essential difference in the propagation of SWs and APs is due to the particularities of the cell kinetics in their relation. It is supposed that smooth muscle cells have an auto-oscillating metabolism with a SW period of 3 to 10 s (Tomita 1975; Ohta et al. 1975). SWs modulate the spike activity of the cells and as a result the tissue displays the autonomous contractile activity. As was already mentioned completely different regimes of functioning are possible. As a rule one can usually point out the pieces of the tissue which are the pacemakers. In the organs of the gastrointestinal tract and in some others similar to it the muscle walls have in addition a smooth gradient of eigenfrequencies. Thus the frequency distribution in the small intestine has a maximum in the higher and a minimum in the lower part (the difference is nearly a factor of two). It was found that the pacemaker function goes over to the highest partition of the remaining intestine if a part of it has been ablated (Bogach 1974). But a complete synchronization of the intestine has practically never been observed. For different regimes essentially different lengths of synchronization are characteristic.

The present short review of physiological data of the organs of the gastrointestinal tract indicates that in these organs a variety of complicated autowave processes take place which determine their peristaltic function. The mathematical modelling of these objects constitutes the first steps in a qualitative analysis of selected processes in observable phenomena. According to the classification based on (9.1) the proposed models are of the second type—peristaltics due to parametric waves. Hydrodynamical considerations in the long-wave approximation were presented in a series of papers (Bertuzzi et al. 1978; Melvill, Deni 1979; Macagno 1975). There attention was focussed on accounting for rheological properties of the intestinal wall. Despite the restrictive assumption of small contraction amplitudes interesting results were obtained. In particular the possibility of reflux generation (reverse currents) was indicated.

Autowave processes that organize mobility are investigated using the basic models. Axiomatic models were employed in the study of the mechanisms leading to wave source generation, the conditions for their propagation in inhomogeneous tissues etc. (Bogach, Reshodko 1979). One of the problems concerns the synchronization of chains of coupled auto-oscillators with a smooth eigenfrequency gradient which model SW propagation (Linkens, Datardina 1977; Linkens, Nhone 1979; Linkens 1980). It was found that the results do not depend qualitatively on the type of local system but rather that they are defined by the type and magnitude of the coupling in the chain. The synchronization problem has been considered above in Chapter 6 in the case of diffusion coupling and a constant gradient of frequencies (see also Drendel, Khors, Vasilev 1984). The synchronization regime corresponds to propagation of waves through the entire chain. In this model the flow of the contents attains a maximum, i.e., we are dealing with the case of "ideal peristaltics". In reality however desynchronization regimes are more typical, in this case the chain breaks up into synchronously oscillating clusters in which propagation of travelling waves (mainly in the direction of the eigenfrequency gradient) takes place (see Chapter 6). The dependence of the propagation velocity on distance corresponds qualitatively to that observed in experiments. The existence of synchronous clusters has also been confirmed experimentally (Drendel, Khors, Vasilev 1984). Many other results of such modelling can easily be found to correspond to experimental data. In particular they indicate that organ control can be effected not from localized influence but as in some AW systems in the organ as an entity. Thus the number of clusters, distances traversed by travelling waves etc. (i.e. regimes of peristaltic pumping) will vary. However certain phenomena (in particular reverse peristaltics) cannot be explained in the context of such models without additional hydrodynamical considerations and (or) conditions for AP generation.

Only first steps have been made in the modelling of gastrointestinal peristaltics. In order to generate more realistic models it will become necessary to take into account many of the tissue properties mentioned above, in particular anisotropy, the dependence of wave propagation on the tissue's mechanical state, the delay of mechanical response to stimulation, the mechano-receptial characteristics of the cells etc. The importance of tissue properties for peristaltics will be demonstrated in the following sections of this chapter where specific models are considered in which AW processes are caused by the interrelation of the mechanical and kinetic systems of the objects.

9.2 Waves in small blood-vessels with muscular walls

The investigation of processes in the system of microcirculation is one of the possible fields of application of autowave concepts. In particular it is known that the activity of vessel muscles are not only controlled by external influences but also by the state of strain of the wall itself (Khayntin 1971; Mandel et al. 1981). Experimental results for an isolated vessel point to the possibility that a decreasing zone exists in the static pressure-radius characteristic (Johnson, P. S. 1968; Cow 1972; Konradi 1973). Examples of inhomogeneous and possibly wave-like contractions of small vessels are shown in Fig. 9.1 (Karaganov et al. 1982). The theoretical consideration of the behaviour of muscular vessels can be found in the work of several authors (Regierer, Rutkevich 1975; Rutkevich 1974; Skobeleva 1980; Klochkov 1981), but here we

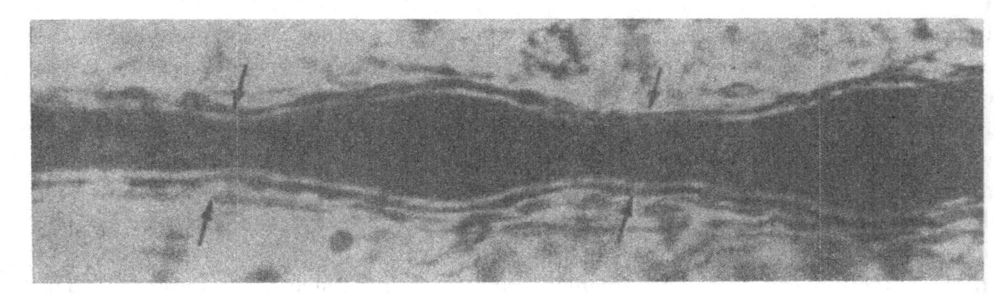

Fig. 9.1
A local contraction of arteriols (photograph reproduced from the atlas "Microangiography")

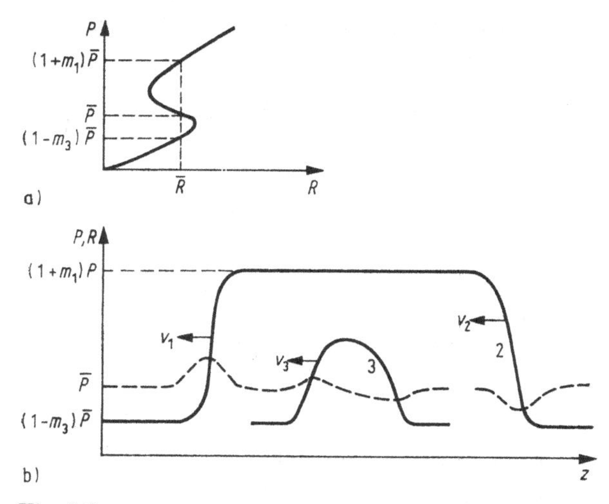

Fig. 9.2
a) Static pressure-radius characteristic for an S-type vessel;
b) qualitative representations of three stationary waves in system (9.1), (9.2). The continuous line represents the pressure changes, the dotted line represents those of the radius. v_1 v_3 v_2, $v_{1,2}$ are velocities of stable autowaves, and v_3 is the velocity of an unstable AW

restrict ourselves only to the results connected with active wave motions (Klochkov et al. 1985). The simplest model describing the combined motion of the liquid and the muscle wall of a vessel with circular cross-section was studied by S. A. Regierer and co-worker (1975). It can be written in the following form:

$$\frac{\partial P}{\partial t} - \beta[R - P_3(P)] = k\,\frac{\partial R}{\partial t}, \tag{9.2}$$

$$R\,\frac{\partial R}{\partial t} = \frac{1}{16\mu}\,\frac{\partial}{\partial Z}\left(R^4\,\frac{\partial R}{\partial Z}\right), \tag{9.3}$$

where P is the transmural pressure, R is the radius of the vessel, Z is the coordinate along the vessel axis, β and k are some positive fixed viscosity elasticity modules, μ is the viscosity of the fluid, $P_3(P_0)$ is a polynomial of third degree which approximates the static pressure-radius characteristic (see Fig. 9.2a) (Skobeleva 1980; Klochkov 1981). Equation (9.2) describes the rheological properties of the smooth muscle wall of an S-vessel phenomenologically. The continuity equation (9.3) follows from the assumption of a local Poiseuille flow in a pipe of variable cross-section. It may be noted that following the classification of "basis" models the system (9.2, 9.3) can be related to models of the propagation of a solitary excitation front.

For a qualitative understanding of possible processes in such system the study of stationary solutions is essential. Following B. N. Klochkov (1981) we consider a simplified case: small deviations of the variables $P - \bar{P} = \bar{P}_p{}'$; $R - \bar{R} = \bar{R}_x{}'$

from some chosen stationary values $(P, \bar{R})$. Then we get the following equations for dimensionless variables x'; $z' = z/\bar{z}$; $t' = t/\bar{t}$.

$$-v \frac{d}{d\xi} (mP' - nx') - cx' - p' \left(1 - \frac{P'}{m_1}\right) \left(1 + \frac{P'}{m_3}\right) = 0, \tag{9.4}$$

$$v \frac{dx'}{d\xi} + v_1 \frac{d^2p'}{d\xi^2} = 0, \tag{9.5}$$

where

$$\xi = z' - vt', \quad m = (\alpha \bar{t} \beta)^{-1}, \quad n = \frac{k\bar{P}}{\alpha\beta\bar{R}\bar{t}},$$

$$c = \frac{\bar{R}}{\alpha\bar{P}}, \quad v_1 = \frac{\bar{R}\bar{P}t}{16\mu\bar{Z}^2}.$$

Here m_1 and m_3 are the dimensionless roots of the polynomial $P_3(P)$, α is the coefficient of the linear term in expansion of $P_3(P)$ into series around $\bar{P}$.

There exist only three physically realizable stationary solutions of the system (9.4, 9.5) in the infinite interval $(-\infty < \xi < \infty)$. Two of them have similar shape, the pressure change corresponds to the solitary wave front of excitation and the change of the radius is connected either with a wave of local contraction or with a

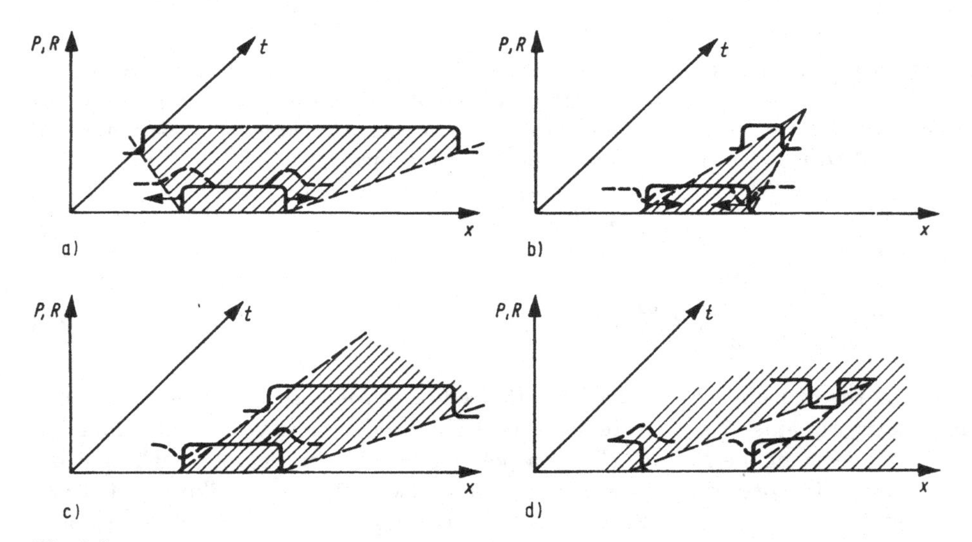

Fig. 9.3
Possible dynamic processes in the system permitting the existence of two stable fronts:

a) The scattering of boundaries with the local expansion of a vessel
b) self-liquidation of the higher pressure interval with initial conditions in the form of decreasing sections on vessel boundaries,
c) the propagation and increase of the field with higher pressure,
d) the propagation and self-liquidation of the field with higher pressure. Continuous lines represent-pressure autowaves, dotted lines represent deformation waves (radius changes)

wave of vessel expansion (see Fig. 9.3b):

$$P'_{1,2}(\xi) = \frac{m_1 - m_2}{2} \pm \frac{m_3 + m_1}{2} \tanh \frac{(\xi - \xi_0)(m_3 - m_1)}{2\sqrt{2n\nu_1 m_1 m_3}}, \tag{9.6}$$

$$x'_{1,2}(\xi) = \pm \frac{\nu_1(m_1 + m_2)^2}{2v_{1,2} 2\sqrt{2n\nu_1 m_1 m_3}} \cosh^{-2} \frac{(m_1 + m_3)(\xi - \xi_0)}{2\sqrt{2n\nu_1 m_1 m_3}}. \tag{9.7}$$

The propagation velocity of these waves corresponding to each sign ("$-$" or "$+$") in (9.6, 9.7) has two values

$$v_{1,2} = \frac{n\nu_1 v_0}{2m} \pm \sqrt{\left(\frac{n\nu_1 a}{2m}\right)^2 + \frac{c}{m}\nu_1}, \tag{9.8}$$

where

$$v_0 = \pm \sqrt{\frac{1}{2n\nu_1 m_1 m_3}}\,(m_1 - m_3).$$

Moreover, as follows from the conservation law (9.3) or (9.5), the velocity of wave propagation coincides with the velocity of the liquid in the vessel (caused by the wave of fall in pressure in the case of a local increase of the vessel radius. If the vessel radius decreases locally the velocity of the wave and the velocity of the liquid have opposite directions. The third stationary solution can be found using the equation (Rutkevich 1975)

$$\left(\frac{dP_3'}{d\xi}\right)^2 = \frac{1}{n\nu_1 m_1 m_3} \int_{m_i}^{p_3} (\tau - m_1)\,\tau(\tau + m_3)\,d\tau, \tag{9.9}$$

$$x' = -\frac{\nu_1}{a_3}\frac{dP_3'}{d\xi}, \tag{9.10}$$

where $m_i = m_1$ if $|m_1| \leqq |m_3|$ and $m_i = m_3$ if $|m_1| \geqq |m_3|$. It represents the pulse perturbation (see Fig. 9.3). The velocity of this wave is

$$v_3 = \pm\sqrt{\frac{c}{m}\nu_1}. \tag{9.11}$$

The analytical investigation of the stability of the stationary solution of the system (9.2), (9.3) has not been carried out. But using results of numerical calculations made by B. N. Klochkov and A. M. Reyman (1981) one may conclude that the first two solutions are stable while the third one is unstable with respect to small perturbations.

This conclusion is in agreement with the result obtained from the stability analysis of ordinary solitary excitation waves (waves of overfall type) (see 4.1). Consequently the third stationary solution $P_3'(\xi)$ and x_3' represents a certain kind of boundary separating the initial excitations leading to the formation of the stationary waves $P'_{1,2}$ and $x'_{1,2}$, from the excitations vanishing in the course of time. Such a relation between waves in the system (9.2, 9.3) is shown schematically in Fig. 9.2b.

Thus the qualitative difference between waves in the system (9.2, 9.3) just considered and ordinary solitary waves is connected with the possibility of the existence of two stable excitation waves propagating with different speeds in opposite directions.

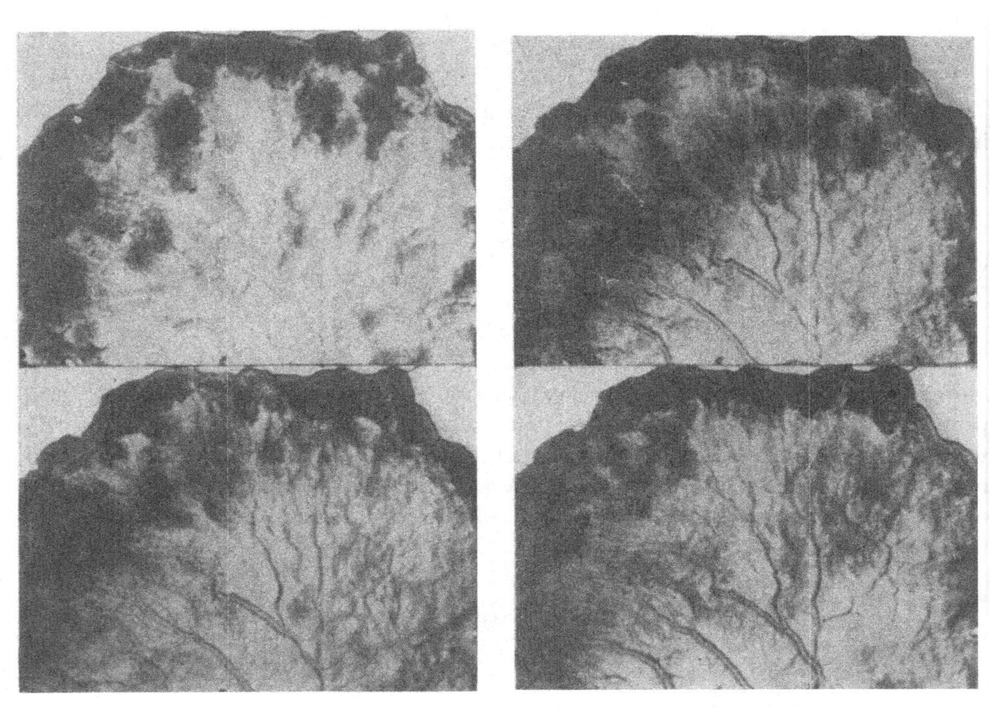

Fig. 9.4
Frames of the front zone for plasmodium Physarum taken every 30 seconds. One can see the autowave of a bulge propagating in the film (Baranovskii 1976).

Combining both these stable wave elements one may consider several variants of the dynamical behaviour of the system. Let for example the pressure in the interval Z_1, Z_2 be initially higher (see Fig. 9.3). Then the further motion of the boundary of this zone depends on whether the vessel contracts or expands at the boundaries Z_1 and Z_2. If there is a vessel expansion at the boundary the region size will increase. If the vessel contracts at the boundaries the region of higher pressure will disappear after some time (Fig. 9.3b). In the case that the vessel expands at one boundary and contracts at the other the region with higher pressure will propagate like a pulse perturbation and the length of the pulse will increase (Fig. 9.3b). We briefly note the possibility that autowave phenomena as described above may participate in the regulation of blood flow. Apparently at present great care is needed in ascertaining the role of such processes. Corresponding experimental results are scant. Theoretical studies are still far from being complete. It may be suggested for instance that the mechanism stabilizing blood consumption through a single arteriola for a given pressure interval at the entrance of the arteriola is connected with pressure autowaves.

Let us also remark that autowave concepts may turn out to be useful in studying blood circulation through a network of small vessels. Thus qualitative local "consumption-pressure" characteristics for a network of small blood-vessels have been constructed (Antonets et al. 1981). It was shown that these relations may have a decreasing section. Moreover the connection between individual parts of the network has diffusion character. Consequently the spatial structures of inhomogeneous blood distribution in the tissues may quite well be connected with active wave processes. Examples of such inhomogeneous blood distributions are known for some types of tissue. For example a spatial mosaic of blood flow in the brain cortex was discovered by Yu. N. Morgalev (1979) in experiments with cats. Microinfarcts of heart muscle testify the inhomogeneous blood flow distribution in the myocard. A clear example is the inhomogeneous blood supply of skin layers.

9.3 Autowave phenomena in plasmodia of Myxomycetes

The mould Physarum Polycephalum or a multi-nucleotic plasmodium of Myxomycetes is one of the most popular objects in studying intracellular mobility. The whole organism may occupy an area of up to 1 m^2. In nature it lives on the soil and the roots of trees in darker places in a thicket. At a certain stage of its development Physarum builds a strand (or tube) system with a frontal region which gives the direction in which Physarum can migrate (see Fig. 9.4). In the frontal region the strands fuse into a film pierced by a network of strands. The thickness of the tubes may reach 0.5 mm. Because of the contraction of active fibrils localized in the walls, there are periodical alterations of the radius that in turn generates oscillating (push-pull) currents of the protoplasma inside the tube and the frontal film. In Fig. 9.5

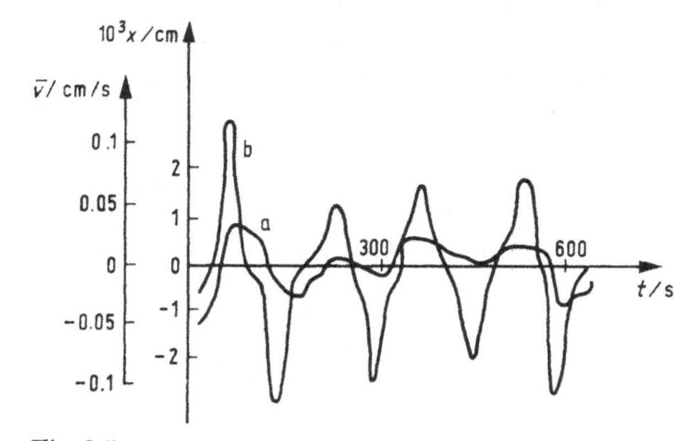

Fig. 9.5
Oscillograms of changes of a strand radius (a) and of mean velocity (b)

typical oscillograms of the changes of the tube radius and the mean current velocity of the protoplasma in the same section are shown. The oscillograms were obtained with a Doppler spectrometer by Evdokimov, Prieszhev, Romanovskii (1982) (see also Romanovskii, Khors 1982). The tube radius changes usually by about 5 to 10%, the maximum value of the velocity is 1—2 mm/s, and the oscillation period is 1—2 min. In every period more substance is transported to the frontal region than comes back, thus the frontal zone migrates and its velocity is 1—3 cm per hour.

One can distinguish the following kinds of autowave motions in a plasmodium:

(i) a wave-like propagation of bulges in the frontal layer with velocities from 5 to 30 μm/s (see Fig. 9.4). Sometimes a circulation of these waves can be observed;

(ii) quasistochastic oscillations of the film thickness;

(iii) nearly synphasic radial pulsations of the (tract) network including the tubes of small dimensions;

(iv) propagation of peristaltic contractions along tubes.

If one cuts out a tract fragment of 0.5 to 3 cm length and puts it on a nutritious medium then after 10 to 15 min push-pull motions of its protoplasma are restored.

The tube piece can be described by the following "conceptual" model. Let the tube cover be a cylindrical pipe with viscous-elastic walls. In the walls contractive elements, so-called fibrils, are circularly distributed; fibrils consist of bundles of active filaments. The pipe is filled with viscous protoplasma containing organelles, food pieces, conglomerates of actine filaments, myosine molecules in the form of small sized complexes, calcium ions (Ca^{++}), molecules of adenosine triphosphate acid (ATP) and a lot of other substances. In distinction to the striated muscles where actine and myosine filaments form well-organized contraction complexes these have not been observed in plasmodia. The structure of its contraction apparatus is similar to that of smooth muscle cells (Section 9.1, 9.2). Actine fibrils form a stochastic network in the matrix of walls where myosine oligomers are spotted. The simplest complex, myosine dimer, has two "heads". These heads can interact with active centres distributed along the actine filaments. The interaction act consists of a bending of the myosine molecule, if another end of a myosine dimer is fixed at the matrix or even at an active centre of another parallel fibril, then the bending of the dimer causes a force constraining the fibrils to slip relative to each other. If in the given tube section these motions proceed synchronously in the mean, e.g. on account of an increase of the "local" concentration of Ca^{++}, then it leads to a decrease of the tube radius.[1]) The protoplasma is pressed out of the contracting region, streams into neighbouring regions, and stretches the corresponding parts of the tube. These stretches can promote an increase of the local Ca^{++} concentration. In that way the conditions for

[1]) In agreement with the theory developed by A. S. Davydov (1980), the closure of contact or the bridge "myosin head-active centre of the actine filament" is a cooperative process. As one bridge act myosine molecules. This soliton propagates along the spiral and promotes the contacts between neighbouring heads. A closure of every bridge gives the pumping up of some energy into the soliton and it is that which compensates the losses stimulated by interactions between the spiral and the surroundings. The way this conception could be applied to stochastic fibril networks in the plasmodium is not yet clear.

a new active contraction are created, but already in the sections adjacent to the ones originally contracted. Let us note too that the action fibrils can slip relative to each other if there is a sufficient amount of ATP molecules which, by donating energy, take away the tension or open contacts made earlier (established terminology: "bridge") between the actine heads and active centres.

The given "conceptual" description can be expressed in terms of a mathematical model. Final models, more correctly speaking models of the same development and completeness as those of striated muscles (Deshcherevsky 1977) have not yet been constructed for plasmodia, but there exist sufficiently concise semi-phenomenological ones. We shall formulate such a model following Kamiya (1962), Wohlfart-Bottermann (1979, 1979), Teplov et al. (1981), Romanovsky et al. (1981), Kesler (1982), Allen and Allen (1978).

In the first approximation one can consider the protoplasma as a viscous Newtonian fluid. Let us denote the mean radius of the tube by $\bar{R}$, the kinetic viscosity of the protoplasma by ν, and a small deviation from the mean radius by x. As long as Reynold's number has a small value $\mathrm{Re} = V_z R/\nu \approx 10^{-3}$ the motion of the protoplasma may be described by the equation

$$\varrho \frac{\partial v_z}{\partial t} = -\frac{\partial P}{\partial z} + \mu \left[\frac{1}{r}\frac{\partial}{\partial r}\left[r\frac{\partial v_z}{\partial r}\right]\right]. \tag{9.12}$$

Here ϱ is the density, $\mu = \nu_\varrho$ is the viscosity of the protoplasma, $\partial P/\partial z$ is the longitudinal pressure gradient arising from the change of the radius, V_z is the longitudinal velocity component and v_r the radial one. As long as the wall deformation is small one may assume that

$$v_z|_{r=\bar{R}} = 0 \tag{9.13}$$

and that the profile of the longitudinal velocity in the canal is parabolic:

$$v_z(r, t) = v_0(t, z)\left(1 - \left(\frac{r}{\bar{R}}\right)^2\right). \tag{9.14}$$

The components v_z and v_r are connected by the continuity equation

$$\frac{\partial v_z}{\partial z} + \frac{1}{r}\frac{\partial (r v_2)}{\partial r} = 0. \tag{9.15}$$

Let us introduce the average velocity in a cross-section of the canal

$$\bar{v}(z, t) = \frac{2}{\bar{R}^2}\int_0^{\bar{R}} r v_z(r, z, t)\,\mathrm{d}r = \frac{v_0(z, t)}{2}. \tag{9.16}$$

Then one can integrate Eq. (9.12) over the cross-section area and gets

$$\varrho \frac{\partial \bar{v}}{\partial t} = -\frac{\partial P}{\partial z} + \frac{8\mu}{\bar{R}^2}\,\bar{v}. \tag{9.17}$$

The main problem to be solved is finding an expression for the pressure P depending on parameters of the envelope and properties of the contracting fibrils. The quantity P can be split into two parts, a passive pressure determined by the cover deformation and an active pressure depending on the degree of actine fibrils and myosine interaction

$$P = P_{\mathrm{P}} + P_{\mathrm{A}}. \tag{9.18}$$

At first we consider the passive component P_{p}. Let the relative lengthening of the tract element localized on the cover circumference be $\Delta l/l = x/R$. Then the passive part of the pressure resulting from the change of the radius is

$$P_{\mathrm{p}} = \frac{Eh}{\bar{R}^2}\, x + \frac{\varkappa h}{\bar{R}^2}\, \dot{x} + \mathscr{L}_z(x, \dot{x}). \tag{9.19}$$

Here h is the thickness of the wall, E and $\varkappa$ are the Young modulus and the viscosity coefficient, and $\mathscr{L}_z(x, \dot{x})$ is a linear differential operator depending on the Young modulus $E' \sim E$ and the coefficient $\varkappa' \sim \varkappa$ in the direction of the z-axis. For small radially symmetrical deformations of the tube one can use the approximate expression:

$$\mathscr{L}(x, \dot{x}) = \frac{E'I}{2\pi\bar{R}}\, \frac{\partial^4 x}{\partial z^4} + \frac{\varkappa' I}{2\pi\bar{R}}\, \frac{\partial^4 \dot{x}}{\partial z^4} + T\, \frac{\partial^2 x}{\partial z^2}\, \frac{1}{2\pi\bar{R}}. \tag{9.20}$$

Here I is the moment of tube section inertia ($I = \pi R h^3/6$), T is the longitudinal tension of the tube, which is small and may be described approximately by the formula $T = E' 2\pi R h \varepsilon$, where ε is the relative constant longitudinal deformation of the tube ($\varepsilon \leq 0.1$). Let there be a cosine-shaped standing wave along the strand:

$$x(t, z) = x_{0k} \cos \frac{\pi}{L}\, kz. \tag{9.21}$$

At the ends of the tract the loops appear and nearly always the strong inequalities

$$\lambda_k = \frac{L}{k} \gg \bar{R} \gg h \tag{9.22}$$

are fulfilled. Then it is easily seen that the quantity $\mathscr{L}$ can be neglected compared with the first terms of the expression (9.19).

Using (9.15) and (9.16) one can write down the condition for the radial velocity component at the wall of the pipe in the form

$$\dot{x} = v_r|_{r=\bar{R}} = -\frac{1}{\bar{R}} \int\limits_0^{\bar{R}} \frac{\partial v_z}{\partial z}\, r\, \mathrm{d}r = -\frac{\bar{R}}{2}\, \frac{\partial \bar{v}}{\partial z}. \tag{9.23}$$

After Eq. (9.17) has been differentiated with respect to z and the expressions (9.15) and (9.23) have been used, an equation describing an AW process of the contracting strand is obtained:

$$\frac{\partial^2 x}{\partial t^2} + \frac{8\mu}{\varrho \bar{R}^2} \frac{\partial x}{\partial t} = \frac{Eh}{2\varrho \bar{R}} \frac{\partial^2 x}{\partial z^2} + \frac{\varkappa h}{2\varrho \bar{R}} \frac{\partial^3 x}{\partial z^2 \, \partial t} - \frac{\bar{R}}{2\varrho} \frac{\partial^2 P_{\mathrm{A}}(x, \dot{x})}{\partial z^2}. \tag{9.24}$$

Finally $P_{\mathrm{a}}(x, \dot{x})$ must be given a concrete form. To do this we will use the following phenomenological requirements:

(i) the pressure must be given by a function which describes the existence of negative damping in the system;

(ii) an active force acts only during contraction;

(iii) $P_{\mathrm{a}}(x, \dot{x})$ must be bounded;

(iv) the active force is switched on with a delay τ that is small compared with the period of the auto-oscillation. This delay is caused by the finite characteristic time of Ca^{++}-ion diffusion from the "depots" to the fibrils, the rebuilding and polarization of the actine fibres, and for other reasons.[1]

The above requirements can be fulfilled by the function

$$P_{\mathrm{A}}(x, \dot{x}) = \begin{cases} \alpha \dot{x}(t-\tau)\,[1 - \beta x^2(t-\tau)] & (\dot{x} < 0), \\ 0 & (\dot{x} > 0). \end{cases} \tag{9.25}$$

Another equation as an alternative form representing the active pressure P_{A} is also relevant

$$\frac{\partial}{\partial t} P_{\mathrm{A}} = \frac{1}{\tau}\left[-P_{\mathrm{A}} + f\left(x, \frac{\partial x}{\partial t}\right)\right]. \tag{9.25a}$$

If time delay is small ($\tau \ll T$), then this equation is equivalent to introducing a certain constant delay (see Beylina et al. 1984; Archangelskaya et al. 1984).

The following is worthy of a note. The constant pressure (which is inside a strand) slightly exceeds the atmospheric value so that the tube retains its form. Thus, generally, the function $f(x, \partial x/\partial t)$ can not only be put in a form as in Eq. (9.25), but also in the form $f(x, \partial x/\partial t) = \alpha(\partial x/\partial t)\,(1 - \beta x^2)$. Hence the variations of x must be measured with respect to a certain level defined by the constant excessive pressure and not to zero. It is seen that Eqs. (9.24) to (9.25a) are equivalent to the set of equations for AW processes in blood-vessels (Section 9.2) if the following assumptions are valid: (i) The radius changes are small ($R = \bar{R}(1 + x)$ and $x \ll 1$); (ii) one may neglect the inertia term in Eq. (9.25).

We now introduce $t_1 = t - \tau$. Then, Eq. (9.24) can be written as

$$\frac{\partial^2 x(t_1+\tau)}{\partial t_1^2} + \frac{1}{\varrho \bar{R}}\left[\frac{8\mu}{\bar{R}} \frac{\partial x(t_1+\tau)}{\partial t_1} - \frac{\varkappa h}{2} \frac{\partial^3 x(t_1+\tau)}{\partial z^2 \partial t_1}\right] = \frac{Eh}{2\varrho \bar{R}} \frac{\partial^2 x(t_1+\tau)}{\partial z^2} + \frac{\bar{R}}{2\varrho} \frac{\partial^2 P_{\mathrm{A}}(t_1)}{\partial z^2}. \tag{9.26}$$

---

[1]) V. G. Kolinko (1983) first drew attention to these circumstances (see also Beylina et al. 1984).

Generally, one can solve this equation for different conditions at the strand butt-ends, for example taking into account the slow migration of the strand. Below we shall restrict ourselves to the simplest case in solving this equation, a so-called two-box model (Romanovskii, Khors 1982). A short piece of the tract of length $L \ll 1$ cm behaves like a system of two sequentially contracting chambers connected by a pipe. This can be illustrated by Fig. 9.6 showing contours of a microplasmodium. They

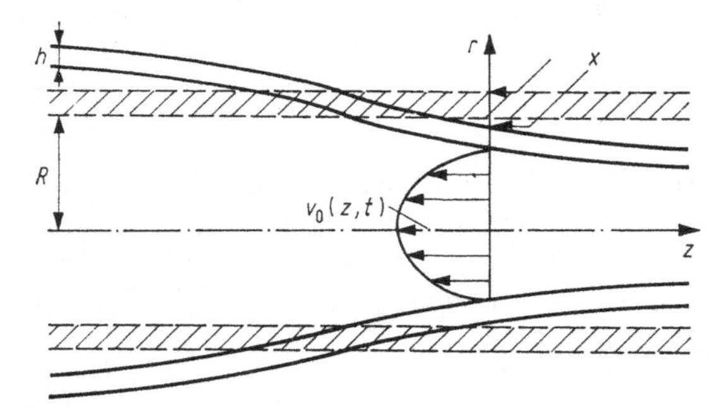

Fig. 9.6
The velocity profile of the flow in the contracting zone of strand Physarum

correspond to observations over 1/4 of a period. The solution of Eq. (9.21) with $k = 1$ corresponds to wave processes that occur in a distributed system. If τ is small compared with the auto-oscillation period, then the terms depending on $t_1 + \tau$ in Eq. (9.26) can be expanded into power series in τ, it is reasonable to take into account only the first two orders. We shall try solving Eq. (9.26) in the form $x(r, t) = x_0(t) \times \cos(\pi z/L)$ using the Galerkin procedure. First we multiply Eq. (9.26) by $\cos(\pi z/L)$ and integrate it from 0 to 1. Then we obtain an approximate representation of Eq. (9.26)

$$M \frac{\partial^2 x_0}{\partial t_1^2} + H \frac{\partial x_0}{\partial t_1} - A \frac{\partial x_0}{\partial t_1} \left(1 - \frac{1}{2} \beta x_0^2\right) + K x_0 = 0, \tag{9.27}$$

where

$$M = \frac{\tau \varkappa h}{\bar{R}}, \quad H = \frac{\varkappa h}{\bar{R}}, \quad A = \frac{\alpha \bar{R}}{4}, \quad K = \frac{Eh}{\bar{R}}.$$

Here, we have neglected the small terms

$$\tau \frac{\partial^3 x_0}{\partial t_1^3}, \quad \frac{\partial^2 x_0}{\partial t_1^2} \left(1 + \frac{8\mu\tau}{\varrho \bar{R}^2}\right), \quad \frac{\partial x_0}{\partial t_1} \left(\frac{8\mu}{\varrho \bar{R}^2} + \left(\frac{\pi}{L}\right)^2 \frac{Eh\tau}{2\varrho \bar{R}}\right)$$

taking into account particular parameter values: $\tau \ll T$, $L = 1$ cm, $\bar{R} = 0.1$ cm, $h = 10^{-2}$ cm, $\mu = 10$ Ps, $\varkappa = 10^6$ Ps, $E = 10^5$ dyn/cm^2. It is important to note that then Eq. (9.26) is fulfilled but the wall viscosity exceeds the protoplasma viscosity ($\varkappa \gg \mu$) by many orders of magnitude. Let us put Eq. (9.27) in a dimensionless form

$$y'' + \varepsilon y'(1 - y^2) + y = 0. \tag{9.28}$$

Here, the dimensionless time is $t' = t\Omega$; $y = \sqrt{\beta^*}\, x_{01}$;

$$\varepsilon = (A - H)/\sqrt{K - M}, \qquad \Omega = \sqrt{K/M} \approx \sqrt{E/\tau\varkappa}$$

and $\beta^* = \beta/\big(2\alpha(A - H)\big)$.

Let the amplitude of the active component of the pressure now be α and consequently A be such that $A \approx H$ and $\varepsilon = 3\cdots5$. This case corresponds to a relaxation auto-oscillation having a form much like that observed in experiments (Romanovskii, Khors 1982). The dimensionless auto-oscillation period equals $T' \simeq 1.6 \cdot \varepsilon$ and the dimensionless amplitudes of the radius and velocity are $y_{\max} \approx 2$ and $(\partial y/\partial t')_{\max} \approx \varepsilon$. The corresponding dimensioned quantities are $T = \varepsilon \cdot 1.6/\Omega$, $x_{\max} = 2/\sqrt{\beta^*}$ and $(\partial x/\partial t)_{\max} = 2\varepsilon\Omega/\sqrt{\beta^*}$. We now use the experimental value $x_{\max} \simeq 0.1 \cdot \bar{R} \simeq 10^{-2}$ cm to see if it is possible to choose values of $\tau \ll T$ such that the auto-oscillation period is approximately equal to the observed one ($T \approx 10$ s) and the protoplasma flow velocity is $v \approx 1$ mm/s. In accordance with Eq. (9.23), $v = (\partial x/\partial t)_{\max} \cdot 2L/R\pi$. Consequently we find that, e.g. $T \approx 10$ s and $v \simeq 5 \times 10^{-2}$ cm/s for $\tau = 0.1$ s.

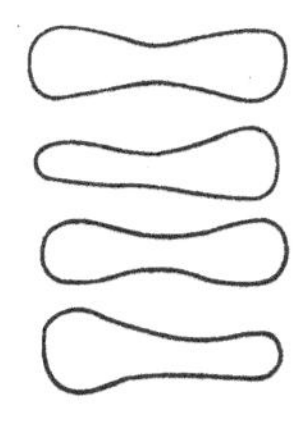

Fig. 9.7
Contours of a short strand Physarum in each quarter of an auto-oscillation period

For a quasiharmonic regime (if $\varepsilon \ll 1$) the auto-oscillation period is calculated without any difficulties when a standing wave forms in the strand (9.21). In this case, the period of the n-th mode is equal to

$$T_k = \frac{2\pi}{\Omega_k} = \sqrt{\frac{1 + 2\mu\tau/\varrho\bar{R} + (\pi k/L)^2\, \varkappa h\tau/2\varrho\bar{R}}{(\pi k/L)^2\,(Eh/2\varrho\bar{R})}}. \tag{9.29}$$

For the real values of the parameters the period is weakly dependent on n and on the tube size, and can be evaluated from the formula: $T_k \sim 2\pi\sqrt{\varkappa\tau/E}$. This circumstance and the presence of stopping regions in relatively long strands (this corresponds to the k-th mode of the protoplasma movements) are confirmed experimentally (see, e.g. Ermakov and Prieszhev 1984).

Thus, such a phenomenological model for autowave mobility in a plasmodium strand agrees well with the experimental data. This model can be still more refined using knowledge of the kinetics of actine-myosine complex interactions and of the concentration oscillations of Ca^{2+} ions and of other ingredients. Also, it is quite possible to construct a model for an active two-dimensional film of plasmodium in which complicated AW processes such as reverberators, quasistochastic waves and others are observed.

To conclude with our brief consideration of a peristaltic transport system of living objects it is not out of place to recall a thesis of Aristotle (Metaphysics, book IX, chap. 2), which reads: "And so it is clear that in any sort of thing, integrity is of a certain nature but this integrity is not the entire nature of each individual thing." The autowave character of the processes being considered is such an integrity. But in any such object autowave processes follow different patterns. This is caused not only by the intrinsic structure of each object but also by the variety of functions inherent in the system as a whole. Thus, in the brain, which is a marvellous creation of nature, an important role is also played by the active transport of neuron protoplasma. This transport associated with the activity of contractive proteins occurs on microscopic scales in axons, dendrites, synapses. Although autowave processes are without doubt essential to the explanation of superslow currents, particular models should be developed to substantiate ideas on their nature. The examples given in this chapter indicate that the problems of modelling for peristaltic phenomena are specific even at the initial stages. The authors hope that these thoughts are reflected in the content of this chapter.

Concluding Remarks

In conclusion we present a table in which the basic sufficient conditions for the existence of various types of AWPs are collected. The following abbreviations are used in this table.

(i) Local system of trigger type—T, of auto-oscillating type—A, of potential auto-oscillating type—PA.

(ii) The coefficients ($\varkappa_1$, $\varkappa_2$, $\varkappa$, σ) are completely determined by the local system parameters.

(iii) τ_x and τ_y are the characteristic times of the variables x and y.

(iv) x corresponds to an autocatalytic component and D_x is the corresponding diffusion coefficient.

There are only a few remarks to be made before concluding this book. First a question must be raised for discussion: Are the basic models considered here of sufficient generality, i.e. may we describe and investigate a whole domain of observed phenomena using only one model? It is possible to give an affirmative answer to this question. For example, the basic model of distributed auto-oscillating Belousov-Zhabotinsky reactions describes qualitatively all regimes observed experimentally. To achieve this the model can be switched from one regime to the other (LC to DS, SA to LC, etc.) by altering the coefficients of the corresponding equations. Such "pliability" of basic models can serve as a criterion of their completeness and perfection. Moreover, there are hopes of constructing a qualitative theory of AWPs (as discussed in this book) using such models as a basis.

Secondly, in addition to the main objective — the elaboration of a general qualitative theory of autowave processes — various new applied problems connected with the study of active kinetic systems arise in different fields of science. Their mathematical formulation, the identification of model coefficients in accordance with experimental data and the study of the simplest descriptions of the phenomena involved (albeit accurate enough to forecast model behaviour in time and space) are problems which can test the abilities of investigators from a variety of special fields.

Summary Table

Summary characteristics of autowave processes

Type of AWP	Minimal number of model variables	Minimal dimension of space	Type of local system	Diffusion coefficients	Velocity
1. Propagation of waves					
1.1. Solitary travelling front (TF)	1	1	T		$v \simeq \sqrt{\tau_x D_x}$
1.2. TF with pulsating velocity	2	1	T		
1.3. TF in regimes of spin burning	2	2	T		
1.4. Travelling pulses (TP)	2	1	PA, A, T	$D_x\tau_y \gg D_y\tau_y$	$v \simeq \sqrt{\tau_x D_x}$
2. Autonomous wave sources					
2.1. Mutual re-initiations of regions of an active medium — "Echo"	2	1	PA	$D_y \equiv 0$	
2.2. Fissioning stopping front	2	1	PA, A, T	$D_y \equiv 0$	
2.3. Stabilization of start region of TP/auto-pacemaker or leading centre (LC)	3	1	PA, A, T	$\begin{cases}\tau_y D_x \gg \tau_x D_y \\ \text{a) } D_z \ll D_x, \text{ b) } D_z \gg D_x\end{cases}$	
2.4. Spiral waves-reverberators	2	2	PA, A	$\tau_y D_x \gg \tau_x D_y$	
3. Auto-oscillations					
3.1. Spatially synchronous auto-oscillations (SA)	2	1	A	$D_x, D_y \gg \varkappa_1$, $\lvert D_x - D_y \rvert > \varkappa_2$	
3.2. Misphasing auto-oscillations	2	1	A	$D_y = 0$	
3.3. Standing waves	3	1	PA, A	$D_x \sim D_y < D_z$	
3.4. Migratory LC	3	1	PA, A, T	see Sect. 2,3, $D_z \sim D_x$	
3.5. Quasi-stochastic auto-oscillations	2	1	A	$D_y = 0$	
4. Stationary dissipative structures (DS)					$v = 0$
4.1. Quasi-harmonic DS	2	1	PA, A	$\tau_y D_x \lesssim \tau_x D_y$	
4.2. Contrast DS	2	1	PA, A, T	$\tau_y D_x \ll \tau_x D_y$	
4.3. Metastable DS	1	1	T	$D \sim \alpha x^\sigma$	
4.4. Pulsating DS	2	1	A	see Sect. 7.3	

References

Afraimovich, V. S., Rabinovich, M. I., Ugodchikov, A. D. (Афраймович, В. С., Рабинович М. И., Угодчиков, А. Д.): Критические точки и „фазовые переходы" в стохастическом поведении неавтономного ангармонического осциллятора. Письма в ЖЭТФ **38** (1983) 64.

Agladze, K. I. (Агладзе, К. И.): Исследование вращающихся спиральных волн в химической активной среде. Препринт ИБФ АН СССР, Пущино 1983.

Agladze, K. I., Krinsky, V. I.: Multi-Armed Vortices in an Active Chemical Medium. Nature **296** (1982) 424.

Aldushin, A. P., Kasparyan, S. G. (Алдушин, А. П., Каспарьян, С. Г.): Термодиффузионная неустойчивость стационарной волны горения. Препринт ОИХФ АН СССР, Черноголовка 1978.

Aldushin, A. P., Martyanova, T. M., Merzhanov, A. G., et al. (Алдушин, А. П., Мартьянова, Т. М., Мержанов, А. Г., Хайкин, Б. И., Шкадинский, К. Г.): Автоколебательное распространение фронта горения в гетерогенных конденсированных средах. Физика горения и взрыва № 5 (1973) 613.

Alekseev, V. V., Kostin, I. K. (Алексеев, В. В., Костин, И. К.): Устойчивость простейших биогеоценозов к случайным внешним воздействиям. В кн. Биологические системы в земледелии и лесоводстве. Наука, Москва **1974**, 105.

Allen, R. D., Allen, N. S.: Cytoplasmic streaming in amoeboid movement. Ann. Rev. Biophys. Bioengng. **7** (1978) 469.

Allessie, M. A., Bonke, F. I. M., Schopman, F. I. G.: Circus movement in rabbit arterial muscle as a mechanism of tachycardia. Circul Res. **33** (1973) 52; **39** (1976) 168; **41** (1977) 9.

Altov, V. A. et al. (Альтов, В. А., Зенкевич, В. Б., Кремлев, М. Г., Сычев, В. В.): Стабилизация магнитных сверхпроводящих систем. Энергия, Москва 1975.

Andronov, A. A., Vitt, A. A., Khaikin, S. E. (Андронов, А. А., Витт, А. А., Хайкин, С. Э.): Теория колебаний. Физматгиз, Москва 1959.

Antonets, V. A., Antonets, M. A., Kudryashov, A. V. (Антонец, В. А., Антонец, М. А., Кудряшов, А. В.): О возможности автоволновых явлений в сетях мелких кровеносных сосудов. В кн. Автоволновые процессы в системах с диффузией. ИПФ АН СССР, Горький 1981, 228–232.

Archangelskaya, T. A., Barch, G., Kolinko, V. G., Romanovskii, Yu. M. (Архангельская, Т. А., Барч, Г., Колинько, В. Г., Романовский, Ю. М.): Модели автоволновой подвижности плазмодия миксомицета. В кн. Динамика клеточных популяций. Изд. ГГУ, Горький 1984, 98–108.

Arshavskii, Yu. I., Berkinblit, M. B., Kovalev, S. A., Chailakhyan, F. M. (Аршавский, Ю. И., Беркинблит, М. Б., Ковалев, С. А., Чайлахян, F. М.): Периодическая трансформация ритма в нервном волокне с постепенно меняющимся свойствами. Биофизика **9** (1964) 365–371.

Arutyunyan, Kh. A., Davtyan, S. P., Rosenberg, B. A., Enikolopyan, N. S. (Арутюнян, Х. А., Давтян, С. П., Розенберг, Б. А., Ениколопян, Н. С.): Отвердение эпоксидианового олигомера ЭД-5 аминами в режиме распространения фронта реакции. ДАН СССР **223** (1975) 657–660.

Astashkina, E. V., Romanovskii, Yu. M. (Асташкина, Е. В., Романовский, Ю. М.): Спонтанное образование диссипативных структур в модели морфогенеза. В кн. Всесоюзная конференция по биологической и медицинской кибернетике. Москва 1978, **3**, 30–33; Флуктуации в процессе самоорганизации. В кн. Математические модели в экологии. Изд. ГГУ, Горький 1980, 74–82.

Autowave Processes in Systems with Diffusion (Автоволновые процессы в системах с диффузией) Под ред. М. Т. Греховой, ИПФ АН СССР, Горький 1981.

Babloyantz, A., Hiernaux, J.: Models for positional information and positional differentiation. – Proc. Nat. Acad. USA 71, N4, 1530–1533.

Babskii, V. G., Markman, G. S. (Бабский, В. Г., Маркман, Г. С.): О способности биологических систем к самоорганизации в свете новых результатов теории возникновения диссипативных структур. В кн. Математические модели в биологии. Наукова думка, Киев 1977, 169–178.

Babskii, V. G., Markman, G. S., Urintsev, A. L. (Бабский, В. Г., Маркман, Г. С., Уринцев, А. Л.): Значение брюсселятора как методологической модели теоретической биологии. Молекулярная биология (межвузовский сб.) **30**, Киев 1982, 82–93.

Balakhovskii, I. S. (Балаховский, И. С.): Некоторые режимы движения возмущения в идеальной возбудимой ткани. Биофизика **10** (1965) 1063.

Balescu, R.: Equilibrium and Nonequilibrium Statistical Mechanics, John Wiley & Sons, N.Y./London/Sydney/Toronto 1975.

Balkarei, Yu. I., Evtikhov, M. G., Elinson, M. I. (Балкарей, Ю. И., Евтихов, М. Г., Елинсон, М. И.): Гистерезис при точечном возбуждении диффузионной автоволновой среды. Микроэлектроника **9** (1980) 141–143; Формирование контрастных диссипативных структур при наличии шумов. Микроэлектроника **12** (1983) 65–69; Новые механизмы самодостройки пространственных диссипативных структур в активных средах диффузионного типа. Микроэлектроника **12** (1983) 171–176.

Balkarei, Yu. I., Nikulin, M. G., Elinson, M. I. (Балкарей, Ю. И., Никулин, М. Г., Елинсон М. И.): Твердотельные автоволновые системы. В кн. Автоволновые процессы в системах с диффузией. ИПФ АН СССР, Горький 1981, 117–134.

Baranovskii, Z. P. (Барановский, З. П.): Связь волновых явлений с миграцией плазмодия миксомицета. В кн. Немышечные формы подвижности. ОНТИ НЦБИ АН СССР, Пущино 1976, 47.

Barelko, V. V. (Барелко, В. В.): Процессы самопроизвольного распространения гетерогенно-каталитической реакции по поверхности катализатора. Препринт ОИХФ АН СССР, Черноголовка 1977.

Barelko, V. V., Barkalov, I. M., Vaganov, D. A., Zanin, A. M., Kiryukhin, D. P.(Барелко, В. В., Баркалов, И. М., Ваганов, Д. А., Занин, А. М., Кирюхин, Д. П.): К тепловой теории автоволновых процессов в низкотемпературных твердотельных радиационно-химических реакциях. ДАН СССР **264** (1982) 99–102.

Barelko, V. V., Beibutyan, V. M., Volodin, Yu. E., Zeldovitch, Ya. B. (Барелко, В. В., Бейбутян, В. М., Володин, Ю. Е., Зельдович, Я. Б.): Тепловые волны и неоднородные стационарные состояния в системе Fe-H_2. В кн. Автоволновые процессы в системах с диффузией. ИПФ АН СССР, Горький 1981, 135–148.

Barenblatt, G. I. (Баренблатт, Г. И.): О некоторых неустановившихся движениях жидкости и газа в пористой среде. Прикладная механ. и матем. **16** (1952) 67–78.

Barenblatt, G. I., Istratov, A. G., Zeldovich, Ya. B. (Баренблатт, Г. И., Истратов, А. Г., Зельдович, Я. Б.): О диффузионно-теплобой устоичибости ламинарного пламени. Ж. прикл. мех. и техн. физ. № 4, 1962, 21–26.

Barenblatt, G. I., Vishik, M. I. (Баренблатт, Г. И., Вишик, М. И.): О конечной скорости распространения в задачах нестационарной фильтрации жидкости и газа. Прикладная Механ. Матем. **20** (1956).

Barenblatt, G. I., Zeldovich, Ya. B. (Баренблатт, Г. И., Зельдович, Я. Б.): Промежуточные ассимптотики в математической физике. УМН **26** (1971) 115—151.
Barr, L., Berger, W., Dewey, M. M.: Electrical transmission at the nexus between smooth muscle cell. J. Gen. Physiol., **51** (1968) 347—369.
Baryakhtar, V. G., Ivanov, B. A., Sukstanskii, A. L. (Барьяхтар, В. Г., Иванов, Б. А., Сукстанский А. Л.): Динамика доменных границ в редкоземельных ортоферритах. Письма в ЖТФ **5** (1979) 853—859; К теории движения доменных границ в магнитоупорядоченных кристаллах. Письма в ЖЭТФ **27** (1978) 226—229.
Bazykin, A. D., Khibnik, A. I. (Базыкин, А. Д., Хибник, А. И.): Билокальная модель диссипативной структуры. Биофизика **27** (1982) 132—135.
Bazykin, A. D., Khibnik, A. I., Aponina, E. A., Neifeld, A. A. (Базыкин, А. Д., Хибник, А. И., Апонина, Е. А., Нейфельд, А. А.): Модель эволюционного возникновения диссипативной структуры в экологической системе. В кн. Факторы разнообразия в математической экологии и популяционной генетике, Ред. Молчанов А. М. и Базыкин, А. Д., НЦ АН СССР Пущино 1980, 33—47.
Belintsev, B. N., Livshits, M. A., Volkenstein, M. V.: On the Multi-Stationary State Transitions in the Spatial Kinetic Systems. Z. Phys. B **30** (1978) 211—218.
Belintsev, B. N. (Белинцев, Б. Н.): Динамические коллективные свойства развивающихся клеточных систем. Кандидатская диссертация, Москва **1979**; Диссипативные структуры и проблема биологического формообразования. УФН **141** (1983) 55—101.
Belousov, B. P. (Белоусов, Б. П.): Периодически действующая реакция и ее механизм. В кн. Автоволновые процессы в системах с диффузией. ИПФ АН СССР, Горький 1981, 176—189.
Belousov, L. V.: Synergetic and Biology Morphogenesis. In: Self-Organization, Ed. V. I. Krinsky, Springer Verlag, Berlin/Heidelberg/New York/Tokyo 1984.
Belousov, L. V., Petrov, K. V. (Белоусов, Л. В., Петров, К. В.): Роль межклеточных взаимодействий в дифференцировке индуцированных тканей зародышей амфибий. Онтогенез **14** (1983) 21—29.
Belousov, L. V., Luchinskaya, N. N. (Белоусов, Л. В., Лучинская, Н. Н.): Исследование эстафетных межклеточных взаимодействий в эксплантантах эмбриональных тканей амфибий. Цитология **25** (1983) 939—944.
Berman, V. S., Danilov, Yu. A. (Берман, В. С., Данилов, Ю. А.): О групповых свойствах обобщенного уравнения Ландау-Гинзбурга. ДАН СССР **258** (1981) 67—70.
Bertuzzi, A., Mancinelli, R., Rouroni, G., Salinari, S.: A mathematical model of intential motor activity. J. Biomech. II (1978) 41—47.
Beylina, S. I., Matveeva, N. B., Prieszev, A. V., Romanovskii, Yu. M., Sukhorukov, A. P., Teplov, V. A.: Mycsomiecate Plasmodium Physarum as Autowaves Self-organizing System. In: Self-Organization, Ed. V. I. Krinsky, Springer-Verlag, Berlin/Heidelberg/Tokyo/New York **1984.**
Bharuche-Reid, A. T.: Elements of the Theory of Markov Processes and Their Applications. McGraw-Hill Publ. Co., New York/Toronto/London 1960.
Blekhman, I. I. (Блехман, И. И.): Синхронизация в природе и технике. Наука, Москва 1981.
Bogach, P. G. (Богач, П. Г.): Моторная деятельность желудочно-кишечного тракта. В кн. Руководство по физиологии. Физиология пищеварения. Наука, Ленинград 1974.
Bogach, P. G., Reshodko, L. V. (Богач, П. Г., Решодько, Л. В.): Алгоритмические и автоматные модели деятельности гладких мышц. Наукова Думка, Киев 1979.
Bogolyubov, N. N., Mitropolskii, Yu. A. (Боголюбов, Н. Н., Митропольский, Ю. А.): Асимптотические методы в теории нелинейных колебаний. Наука, Москва 1974.
Bonch-Bruevich, V. L., Temchin, A. N. (Бонч-Бруевич, В. Л., Темчин, А. Н.): Нелинейная теория сверхрешетки электронной температуры в полупроводниках. ЖЭТФ **76** (1979) **1713—1726.**
Bunkin, F. V., Kirichenko, N. A., Lukyanchuk, B. S. (Бункин, Ф. Б., Кириченко, Н. А., Лукьянчук, Б. С.): Термохимическое действие лазерного излучения. УФН **138** (1982) 45—94 ;Термохимические и термокинетические процессы в поле непрерывного лазерного излучения. Изв. АН СССР, сер. физики, **47** (1983) 2000—2016.

Burns, B. D.: The uncertain nervous system, Edward Arnold Publ. LTD, London 1968.
Butenin, N. V., Neimark, Yu. I., Fufaev, N. A. (Бутенин, Н. В., Неймарк, Ю. И., Фуфаев, Н. А.): Введение в теорию нелинейных колебаний. Наука, Москва 1976.
Butuzov, V., Vasileva, A.: Singularly perturbed differential equations of parabolic type. In: Lecture Notes in Mathematics 985. Asymptotic Analysis II, Ed. F. B. Verhulst, Springer Verlag 1983, 38–75.

Cambell, R. D.: The Role of muscle processes in Hydra morphogenesis. In: Development of a Cellular Biology of Coelenterates, Ed. P. Tardent, R. Elsevier, North-Holland Publ. Co. (Amsterdam): Biomedical Press, 1980, 421–428.
Casten, R. C., Cohen, H., Lagerstrom, P. A.: Perturbation analysis of an approximation to the Hodgkin-Huxley theory. Quart. Appl. Math. **32** (1975) 365.
Chafee, N. A.: A stability analysis for a semilinear parabolic partial equations. J. Diff. Equat. **15** (1974) 522–540.
Chernavskaya, N. M., Chernavskii, D. S. (Чернавская, Н. М., Чернавский, Д. С.): Периодические явления в фотосинтезе. УФН **72** (1960) 627–652.
Chernavskii, D. S., Ruijgork, Th. W.: On the formation of unique dissipative structures. Biosystems **15** (1982) 75–81.
Chernavskii, D. S., Solyanik, G. I., Belousov, L. V.: Relation of the intensity of metabolism with processes of determination in embryonic cell. Biol. Cybern. **37** (1980) 9–18.
Coleman, A., Coleman, Y., Griffin, Y., Weltman, N., Chapmen, K. M.: Methylxathine Induced Escalation: A Propagated Wave Phenomenon Observed in Skeletal Muscle Developing in Culture. Proc. Nat. Acad. USA **69** (1972) N 63.
Conradi, G. P. (Конради, Г. П.): Регуляция сосудистого тонуса. Наука, Ленинград 1973.
Cow, B. S.: The influence of vascular smooth muscle on the viscoelastic properties of blood vessels. Cardiovasc. Fluid. Dyn. 8, **2** (1972) 65–100.
Czaikovsky, G., Ebeling, W.: Nonequilibrium phase transitions in enzyme reaction systems. J. Nonequil. Thermodyn. **2** (1977) I.

Danilov, Yu. A. (Данилов, Ю. А.): Групповой анализ системы Тьюринга и ее аналогов. Препринт ИАЭ 3287/1, Москва **1980**. Теоретико-групповые свойства математических моделей в биологии. В кн. Математическая биология развития. Ред. Зотин, А. И., Наука, Москва 1982.
Danilov, Yu. A., Kadomtsev, B. B. (Данилов, Ю. А., Кадомцев, Б. Б.): Что такое синергетика? В кн. Нелинейные волны. Самоорганизация. Ред. Гапонов-Грехов, А. В., Рабинович, М. И., Наука, Москва 1983.
Davydov, A. S. (Давыдов, А. С.): Биология и квантовая механика. Наукова Думка, Киев 1979.
Demidovich, B. D. (Демидович, Б. Д.): Лекции по математической теории устойчивости. Наука, Москва 1967.
Deshcherevskii, V. I. (Дещеревский, В. И.): Математические модели мышечного сокращения. Наука, Москва 1977.
Dombrovskii, Yu. A., Markman, G. S. (Домбровский, Ю. А., Маркман, Г. С.): Пространственная и временная упорядоченность в экологических и биохимических системах. РГУ, Ростов 1983.
Drendel, S. D., Khors, N. P., Vasilev, V. A. (Дрендель, С. Д., Хорс, Н. П., Васильев, В. А.): Режимы синхронизации клеток гладкомышечных тканей. В кн. Динамика клеточных популяций. ГГУ, Горький 1984, 108–117.
Dubinin, F. D. (Дубинин, Ф. Д.): Принцип построения оптоэлектронных моделей однородных биологических систем с Л. Т. и Р. В. Проблемы бионики вып. 11 ХГУ, Харьков 1973, 95. Гл. У11 Континуальные вычислительные системы в кн. Мясников, В. А., Игнатьев, М. Б., Покровский, А. М.: Программное управление оборудованием. Машиностроение, Ленинград 1974, 301–337.
Dudnik, E. N., Kuznetsov, Yu. I., Minakova, I. I. (Дудник, Е. Н., Кузнецов, Ю. И., Минакова, И. И.): Механизмы взаимной синхронизации в системах генераторов с однонаправленными связями. Изв. вузов Радиоэлектроника, **27** (1984), 37–42.

Dvornikov, A. A., Utkin, G. M. (Дворников, А. А., Уткин, Г. М.): Фазированные автогенераторы радиопередающих устройств. Энергия, Москва 1980.

Dvoryaninov, S. V. (Дворянинов, С. В.): О периодическом решении одной автономной сингулярно возмущенной параболической системы. Диффер. уравнен. **16** (1980) 1617—1622.

Ebeling, W.: Strukturbildung bei irreversiblen Prozessen. BSB B. G. Teubner Verlagsgesellschaft, Leipzig 1976.

Ebeling, W.: Statistisch-Mechanische Ableitung. Ableitung verallgemeinerter Diffusionsgleichungen. Ann. Physik **16** (1965) 147, Beitr. Plasmaphysik **7** (1967) 11.

Ebeling, W.: Zur statistischen Theorie der Diffusion geladener Teilchen. Beitr. Plasmaphysik **7** (1967) 11.

Ebeling, W., Czajkovski, G.: Phase separation in bistable enzyme reaction systems. Studia biophysica **60** (1976) 201.

Ebeling, W., Feudel, U.: Influence of Coulomb interactions on dissipative structures in reaction-diffusion systems. Ann. Physik **40** (1983) 68.

Ebeling, W., Klimontovich, Yu. L., Kovalenko, N. P., Kraeft, W. D., Krasny, Yu. P., Kremp D., Kilik, P. P., Riaby, V. A., Röpke, G., Rozanov, E. K., Schlanges, M.: Transport Properties of Dense Plasmas. Akademie-Verlag, Berlin 1983; Birkhäuser-Verlag, Basel 1984.

Ebeling, W., Klimontovich, Yu. L.: Self-organization and turbulence in liquids, Teubner-Texte für Physik, Bd. 2, Leipzig 1984.

Ebeling, W., Schimansky-Geier, L.: Stochastic dynamics of a bistable reaction system. Physica **98A** (1979) 587., Proc. 6th Internat. Conf. Thermodyn. Merseburg 1980 (p. 65).

Ebeling, W., Feistel, R.: Physik der Selbstorganisation und Evolution. Akademie-Verlag, Berlin 1982.

Eigen, M., Schuster, P.: The Hypercycle. Principle of Natural Self-Organization. Springer Verlag, Berlin/Heidelberg/New York 1979.

Elenin, G. G., Krilov, V. V., Polezhaev, A. V., Chernavskii, D. S. (Еленин, Г. Г., Крылов, В. В., Полежаев, А. А., Чернавский, Д. С.): Особенности формирования контрастных диссипативных структур. ДАН СССР **271** (1983) 84—87.

Elenin, G. G., Kurdyumov, S. P. (Еленин, Г. Г., Курдюмов, С. П.): Условия усложнения организации нелинейной диссипативной среды. Препринт ИПМ АН СССР № 106, Москва 1977.

Eleonskii, V. M., Kirova, N. N., Kulagin, N. E. (Елеонский, В. М., Кирова, Н. Н., Кулагин, Н. Е.): О предельных скоростях и типах волн магнитного домена. ЖЭТФ **74** (1978) 1814—1820.

Ermakov, V. B., Prieszhev, A. V. (Ермаков, В. Б., Приезжев, А. В.): Волновые режимы сократительной активности в плазмодии Physarum и их связь с переносом протоплазмы. Биофизика **25** (1984) 100—104.

Erneux, T., Herschkowitz-Kaufman, M.: The Bifurcation diagram of a model chemical reaction. II. Two dimensional time-periodic patterns. Bull. Math. Biol. **41** (1979) 767.

Evdokimov, M.V., Prieszhev, A. V., Romanovskii, Yu. M. (Евдокимов, М. В., Приезжев, А. В., Романовский, Ю. М.): Лазерный доплеровский анемометр на линии с ЭВМ для исследования медленных потоков протоплазмы в живых клетках. Автометрия № 3 (1982) 61—63.

Falkenhagen, H.: Theorie der Elektrolyte. Hirzel-Verlag, Leipzig 1971.

Falkenhagen, H., Ebeling, W.: Statistical derivation of diffusion equations. Physics Letters **15** (1965) 131.

Field, R. J., Noyes, R. M.: Oscillations in chemical systems. IY. Limit cycle behaviour in a model of real chemical reaction. J. chem. Phys. **60** (1974) 1877—1884.

Fisher, R.: The advance of advantageous genes. Ann. Eugenics **7** (1937) 335—369.

Fomin, S. V., Berkinblit, M. B. (Фомин, С. В., Беркинблит, М. Б.): Математические проблемы в биологии. Наука, Москва 1973.

Frank-Kamenetskii, D. A. (Франк-Каменецкий, Д. А.): Диффузия и теплопередача в химической кинетике. Наука, Москва 1967.

Galaktionov, V. A., Samarskii, A. A. (Галактионов, В. А., Самарский, А. А.): Метод построения приближенных автомодельных решений нелинейных уравнений теплопроводности. Математ. Сб. **118** (1982) 292—322; **160** (1982) 435—455.

Gaponov, A. V., Ostrovskii, L. A., Rabinovich, M. I. (Гапонов, А. В., Островский, Л. А., Рабинович, М. И.): Одномерные волны в нелинейных системах с дисперсией. Изв. вузов Радиофизика **13** (1970) 163.

Gaponov-Grekhov, A. V., Rabinovich, M. I. (Гапонов-Грехов, А. В., Рабинович, М. И.): Л. И. Мандельштам и современная теория нелинейных колебаний. УФН **128** (1979) 579—624.

Gaponov-Grekhov, A. V., Rabinovich, M. I.: Nonstationary Structures—Chaos and Order. Synergetics of Brain, Ed. H. Haken, Springer Verlag 1983.

Gelfand, I. M., Tsetlin, M. L. (Гельфанд, И. М., Цетлин, М. Л.): О континуальных моделях управляющих систем. ДАН СССР **131** (1960).

Gerish, G.: Cell aggregation and differentiation in Dictyostellium discoideum. Curr. Top. Deo. Biol. **3** (1968) 157. Cell interaction by Cyclic AMP in DIcytostelium. Biol. Cellulaire **32** (1978) 61.

Gierer, G.: The generation of biological pattern and form, some physical mathematical and logical aspects. Progr. biophysical molecular biol. **37** (1981) 1—41.

Glansdorff, P., Prigogine, I.: Thermodynamic theory of structure, stability and fluctuations. Wiley-Interscience, New York 1971.

Gorelova, N. A., Bureš, I.: Spiral waves of spreading depression in the isolated chicken retina. J. Neurobiology N 6 (1983).

Grigorov, L. I., Polyakova, M. S., Chernavskii, D. S. (Григоров, Л. И., Полякова, М. С., Чернавский, Д. С.): Модельное исследование триггерных схем и процесса дифференцировки. Молек. биол. **1** (1967) 410—418.

Gulyaev, Yu. V., Kalafati, Yu. D., Serbinov, I. A., Ryabova, L. A. (Гуляев, Ю. В., Калафати, Ю. Д., Сербинов, И. А., Рябова, Л. А.): Нелинейные волны при фазовом переходе полупроводник-металл и их применение для обработки информации. ДАН СССР **256** (1981) 357—360.

Gurija, G. T., Livshits, M. A.: Topology of Two-Dimensional Autowaves. Physics Letters **97 A** (1983) 175—177.

Haase, R.: Thermodynamik der irreversiblen Prozesse. Dietrich Steinkopff Verlag, 1963.

Haken, H.: Synergetics. Springer-Verlag, 1978.

Hodgkin, A. L., Huxley, A. F.: A quantitative description of membrans current and application to conduction and excitation in nerves. J. Physiol. **117** (1952) 500—544.

Hunter, P. I., Naughton, P. A., Noble, D.: Prog. Biophys. Molec. Biol. **30** (1975) 99.

Ibanes, I. L., Velarde, M. G.: Hydrochemical stability of an interface between two liquids; the role of Langmuir-Hinshelwood saturation law. J. Physique **38** (1977) 1479—1483.

Ilinova, T. M., Khokhlov, R. V. (Ильинова, Т. М., Хохлов, Р. В.): О волновых процессах в линиях с шунтирующим нелинейным сопротивлением. Радиотехника и электроника **8** (1963) 2006—2013.

Ivanitskii, G. R. (Иваницкий, Г. Р.): Борьба идей в биофизике. Знание, Москва 1982.

Ivanitskii, G. R., Krinskii, V. I., Selkov, E. E. (Иваницкий, Г. Р., Кринский, В. И., Сельков, Е. Е.): Математическая биофизика клетки. Наука, Москва 1978.

Ivanov, L. P., Logginov, A. S., Nepokoichitskii, G. N. (Иванов, Л. П., Логгинов, А. С., Непокойчицкий, Г. Н.): Экспериментальное обнаружение нового механизма движения доменных границ в сильных магнитных полях. ЖЭТФ **84** (1983) 1006—1022.

Ivanov, S. A., Romanovskii, Yu. M. (Иванов, С. А., Романовский, Ю. М.): Вынужденные колебания в контуре с флуктуирующей собственной частотой. Радиотехника и электроника **9** (1963) 551—560.

Ivleva, T. P., Merzhanov, A. G., Shkadinskii, K. G. (Ивлева, Т. П., Мержанов, А. Г., Шкадинский, К. Г.): О закономерностях спинового горения. Физика горения и взрыва **16** (1980) 3.

Jain, A. K., Likharev, K. K., Lukens, J. E., Sauvergeau, J. E.: Mutual Phase-Locking in Josephson Junction Arrays. Phys. Reports (A Review Section of Physics Letters), **109,** №. 6 1984, p. 310—426.

Johnson, P. S.: Autoregularity responses of cot mesentric arterioles measured in vivo. Circulation Research **22** (1968) 199—212.

Kabashima, S., Kawakubo, T.: Observation of noise-induced phase transition in parametric oscillator. Phys. Letter. **70A** (1981) 375.

Kahrig, E., Besserdich, H.: Dissipative Strukturen. VEB Georg Thieme, Leipzig 1977.

Kalafati, Yu. D., Serbinov, I. A., Ryabova, L. A. (Калафати, Ю. Д., Сербинов, И. А., Рябова, Л. А.): Нелинейные волны в среде с фазовым переходом полупроводник-металл. Письма в ЖЭТФ **29** (1979) 637. Об иерархии малых параметров в теории диссипативных структур. ДАН СССР **263** (1982) 862—864.

Kamiya, N.: Protoplasmic streaming. Springer-Verlag, Wien 1959.

Kanavets, V. I., Stabinis, A. Yu. (Канавец, В. И., Стабинис, А. Ю.): Сужение спектра генераторов с близкими частотами при взаимной синхронизации. Радиотехника и электроника **17** (1972) 2124—2130.

Kanel, Ya. I. (Канель, Я. И.): О стабилизации решений задачи Коши для уравнений, встречающихся в теории горения. Матем. сб. **59/101/** (1962) 245. О стабилизации решений теории горения при финитных начальных условиях. Матем. Сб. **65**, 107 (1964) **398.**

Karaganov, Ya. L., Kerdivarcnko, N. V., Levin, V. N. (Караганов, Я. Л., Кердиваренко, Н. В.. Левин В. Н.): Микроангоилогия. Атлас. Штиница, Кишенев 1982.

Kawczynski, A. L., Zaikin, A. N.: Spatial Effects in Active Chemical Systems. II. Model of Stationary Periodical Structure. J. non-equilibrium Thermodyn. **2** (1977) 139—152.

Kerner, B. S., Osipov, V. V., (Кернер Б. С., Осипов, В. В.): Нелинейная теория стационарных страт в диссипативных системах. ЖЭТФ **74** (1978) 1675—1696. Стохастически неоднородные структуры в неравно весных системах. ЖЭТФ **79** (1980) 2218. Пульсирующие „гетерофазные“ области в неравновесных системах. ЖЭТФ **83** (1982) 2201—2214. Свойства устойчивых диссипативных структур в математических моделях морфогенеза Биофизика **27** (1982) 137—142.

Kerner, B. S., Osipov, V. V. (Кернер, Б. С., Осипов, В. В.): Динамическая перестройка диссипативных структур. ДАН СССР **264** (1982) 1366—1370.

Kerner, B. S., Sinkevich, V. F. (Кернер, Б. С., Синкевич, В. Ф.): Многошнуровые многодоменные состояния в горячей электронно—дырочной плазме. Письма в ЖЭТФ **36** (1982) 359—361.

Kessler, D.: Plasmodial structure and motivity. In: Cell Biology of Physarum. Academic Press, New York 1982 (p. 145—208).

Khayutin, V. M. (Хаютин, В. М.): Механизм управления сосудами работающей скелетной мышцы. В кн. Проблемы современной физиологической науки. Наука, Ленинград 1971, 123—140.

Kholopov, A. V. (Холопов, А. В.): Распространение возбуждения в волокне, рефрактерность которого зависит от периода возбуждения. Биофизика **23** (1968) 670.

Khramov, R. N. (Храмов, Р. Н.): Циркуляция импульса в возбудимой среде. Критический размер замкнутого контура. Биофизика **23** (1978) 871.

Khramov, R. N., Krinskii, V. I. (Храмов, Р. Н., Кринский, В. И.): Стационарные скорости распространения устойчивого и неустойчивого импульса. Зависимость от ионных токов мембраны. Биофизика **22** (1977) 512.

Klimontovich, Yu. L. (Климонтович, Ю. Л.): Кинетическая теория электромагнитных процессов. Наука, Москва 1980; Springer-Verlag, 1983.
Статистическая физика. Наука, Москва 1982; Gordon and Breach, New York 1985 (in English).

Klochkov, B. N. (Клочков, Б. Н.): Автоволновые процессы в кровеносных сосудах мышечного типа. В кн. Автоволновые процессы в системах с диффузией. ИПФ АН СССР, Горький 1981, 233—242.

Koga, S., Kuramoto, Y.: Localized Patterns in Reaction-Diffusion Systems. Progr. theor. Phys. (Kyoto) **63** (1980) 106—121.

Kolinko, V. G. (Колинько, В. Г.): Модель автоколебательной подвижности многоядерной клетки — плазмодия Physarum с учетом структурной перестройки сократительного аппарата. В кн. Динамика биологических популяций. ГГУ, Горький 1983, 31—37.

Kolmogorov, A. N., Petrovskii, I. G., Piskunov, N. S. (Колмогоров, А. Н., Петровский, И. Г., Пискунов, Н. С.): Исследование уравнения диффузии, соединенного с возрастанием количества вещества и его применение к одной биологической проблеме. Бюлл. МГУ, секция А. 1, № **6** (1937) 1—26.

Koroleva, V. I., Kuznetsova, G. D. (Королева, В. И., Кузнецова, Г. Д.): Свойства распространяющейся депрессии в коре головного мозга крысы при создании двух калийных очагов. В кн. Электрическая активность головного мозга. Наука, Москва 1971, 130.

Kostin, I. K., Romanovskii, Yu. M. (Костин, И. К., Романовский, Ю. М.): Взаимная синхронизация релаксационных генераторов в присутствии шумов. Изв. вузов Радиофизика **18** (1975) 34.

Kowalenko, N. P., Ebeling, W.: Zur statistischen Theorie der Diffusionsprozesse in kondensierten Medien. phys. stat. sol. **30** (1968) 533.

Krinskii, V. I. (Кринский, В. И.): Распространение возбуждения в неоднородной среде (режимы, аналогичные фибрилляции сердца). Биофизика **11** (1966) 676. Фибрилляция в возбудимых средах. В кн. Проблемы кибернетики **20** (1968) 59—80.

Krinskii, V. I., Agladze, K. I. (Кринский, В. И., Агладзе, К. И): Вихри с топологическим зарядом 2,3 и 4 в активной химической среде. ДАН СССР **263** (1982) 335.

Krinskii, V. I., Kholopov, A. V. (Кринский, В. И., Холопов, А. В.): Проведение импульсов в возбудимой ткани с непрерывно распределенной рефрактерностью. Биофизика **12** (1967) 669.

Koga, S.: Rotating Spiral Waves in Reaction-Diffusion Systems. Progr. theor. Phys. (Kyoto) **67** (1982) 164—178.

Krinskii, V. I., Mikhailov, A. S. (Кринский, В. И., Михайлов, А. С.): Автоволны. Знание, Москва 1984.

Krinskii, V. I., Kokoz, Yu. M. (Кринский, В. И., Кокоз, Ю. М.): Анализ уравнений возбудимых мембран. 1. Сведение уравнения Ходжкина-Хаксли к системе второго порядка. Биофизика **18** (1973) 506—513. Мембрана волокна Пуркинье. Сведение уравнения Нобла к системе второго порядка и анализ автоматии по нуль-изоклинам. Биофизика **18** (1973) 1067—1074.

Krinskii, V. I., Reshetilov, A. N., Pertsov, A. M. (Кринский, В. И., Решетилов, А. Н., Перцов, А. М.): Исследование одного механизма возникновения эктопического очага возбуждения на модифицированных уравнениях Ходжкина-Хаксли. Биофизика **17** (1972) 271.

Krinskii, V. I., Zhabotinskii, A. M. (Кринский, В. И., Жаботинский, А. М.): Автоволновые структуры и перспективы их исследования. В кн. Автоволновые процессы в системах с диффузией. ИПФ АН СССР, Горький 1981, 6—32.

Krinskii, V. I., Yachno, V. G. (Кринский, В. И., Яхно, В. Г.): Спиральные волны возбуждения в сердечной мышце, В кн. Нелинейные волны. Стохастичность и турбулентность. ИПФ АН СССР, Горький 1980, 200.

Kurdyumov, S. P. (Курдюмов, С. П.): Собственные функции горения нелинейной среды и конструктивные законы построения ее организации. В кн. Современные проблемы математической физики и вычислительной математики. Ред. акад. Тихонов, А. Н., Наука, Москва 1982 217—243.

Kurdyumov, S. P., Kurkina, E. S., Potapov, A. B., Samarskii, A. A. (Курдюмов, С. П., Куркина, Е. С., Потапов, А. Б., Самарский, А. А.): Архитектура многомерных тепловых структур. ДАН СССР **274** (1984).

Kurdyumov, S. P., Malinetskii, G. G., Povestshenko, Yu. A., Popov, Yu. P., Samarskii, A. A. (Курдюмов, С. П., Малинецкий, Г. Г., Повещенко, Ю. А., Попов, Ю. П., Самарский, А. А.): Взаимодействие тепловых структур в нелинейных средах. ДАН СССР **251** (1980) 836—839.

Kuzmin, L. S., Likharev, K. K., Ovsyanikov, G. A. (Кузмин, Л. С., Лихарев, К. К., Овсянников, Г. А.): Взаимная синхронизация джозефсоновских контактов. Радиотехника и электроника (1981) 1067–1076.

Kuznetsov, Yu. A. (Кузнецов, Ю. А.): Существование и устойчивость бегущих волн в системе „реакция-диффузия" с одной пространственной переменной. Препринт НИВЦ АН СССР, Пущино 1982.

Labas, Yu. A., Belousov, L. V., Badenko, L. A., Letunov, V. N. (Лабас, Ю. А., Белоусов, Л. В., Баденко, Л. А., Летунов, В. Н.): О пульсирующем росте у многоклеточных организмов. ДАН СССР **257** (1981) 1247–1250.

Landa, P. S. (Ланда, П. С.): Автоколебания в системах с конечным числом степеней свободы. Наука, Москва 1980. Автоколебания в распределенных системах. Наука, Москва 1983.

Landau, L. D., Lifshits, E. M. (Ландау, Л. Д., Лифшиц, Е. М.): Статистическая физика. Наука, Москва 1976.

Lea, T. I. : Free calcium measurement in cells. Nature **269**, N 5624 (1977) 108.

Leão, A. A. P.: Spreading depression of activity in cerebral cortex. J. Neurophysiol. **7** (1944) 359. Spreading depression. In: Experimental Models of Epilepsy — A Manual for the Laboratory Worker, Eds. D. P. Purpura, I. K. Penly, D. B. Tower, D. M. Woodbury, R. D. Walter, Raven, New York 1972, 173.

Levandovsky, M., Childress, W. S., Spiegel, E. A., Hunter, S. H.: A mathematical Model of Pattern Formation by Swimming Microorganisms. J. Protozoology **22** (1975) 296–306.

Lifshits, E. M., Pitaevsky, L. P.: Physical Kinetics, Nauka, Moscow 1978 (in Russian), Pergamon Press 1981.

Lightfoot, E. N.: Transport phenomena and living systems. Biomedical aspects of momentum and mass transport. A Wiley-Interscience Publication. John Wiley & Sons, New York/London/Sydney/Toronto 1974.

Lindenmayer, A.: Developmental systems without cellular interactions, their languages and grammars. J. Theor. Phys. Biol. **30** (1971) 445.

Linkens, D. A.: Electronic Modeling of Slow-Waves and Spike-Activity in Intestinal Tissue. BME **27** (1980) 351–357.

Linkens, D. A., Datardina, S.: Frequency Entainment of coupled Hodgkin-Huxley Type Oscillator for Modeling. BME **24** (1977) 362.

Linkens, D. A., Mhone, P. G.: Frequency transients in coupled oscillators model of intestinal myoelectrical activity. Computers in Biology and Medicine **9** (1979) 295–303.

Lishutina, G. B., Romanovskii, Yu. M. (Лишутина, Г. Б., Романовский, Ю. М.): О колебаниях диффузионно связанных химических осцилляторов. В кн. Колебательные процессы в биологических и химических системах, Ред. Франк, Г. М. Наука, Москва 1967, 267–273.

Livshits, M. A., Gurija, G. T., Belintsev, B. N., Volkenstein, M. V.: Positional Differentiation as Pattern Formation in Reaction Diffusion Systems with Permeable Boundaries. Bifurcation Analysis. J. Math. Biology **11** (1981) 295–310.

Logginov, A. S., Nepokoichitskii, G. A. (Логгинов, А. С., Непокойчицкий, Г. А.): Свервысокие скорости волны опрокидывания магнитного момента в пленках ферритов-гранатов. Письма в ЖЭТФ **35** (1982) 22–25.

Lotka, A. J.: Elements of Physical Biology. Williams and Wilkins, Baltimore 1925.

Lvovsky, Yu. M., Lutsev, M. O.: Behaviour of normal zones in uniform superconductors. Cryogenics **19** (1979) 483–489.

Lykov, A. V. (Лыков, А. В.): Применение методов термодинамики необратимых процессов к исследованию тепло- и массообмена. Инженерно-физический жур. **9** (1965) 287–304. Тепломассообмен (справочник) Энергия, Москва 1978.

Malafeev, V. M., Polyakova, M. S., Romanovskii, Yu. M. (Малафеев, В. М., Полякова, М. С., Романовский, Ю. М.): О процессе синхронизации в цепочке автогенераторов, связанных через проводимость. Изв. вузов Радиофизика **13** (1970) 936.

Malakhov, A. N., Maltsev, A. A. (Малахов, А. Н., Мальцев, А. А.): О ширине спектральной линии системы взаимно связанных генераторов. ДАН СССР **196** (1971) 1065.

Malchow, H., Schimansky-Geier, L.: Noise in Diffusion in Bistabil-nonequilibrium Systems. Teubner-Texte für Physik, BSB B. G. Teubner Verlagsgesellschaft, Leipzig 1986.

Maltsev, A. A. (Мальцев, А. А.): Оптимальная самосинхронизация в автоколебательной системе с эквидистантным спектром. Изв. вузов Радиофизика **20** (1977) 1570–1572.

Mangel, A., Fahim, M., van Breemen, G.: Rhythmic contractile activity of the in vivo rabbit aorta. Nature **289** (1981) 692–694.

Markin, V. S., Pastushenko, V. F., Chizmadzhiev, Ya. A. (Маркин, В. С., Пастушенко, В. Ф., Чизмаджиев, Ю. А.): Теория возбудимых сред. Наука, Москва 1981.

Markman, G. S. (Маркман, Г. С.): О возникновении стационарных диссипативных структур, возникающих в модели Тьюринга-Пригожина. Биофизика. **25** (1980) 713–715.

Marridge, G. A.: The coordination of the protective retraction of coral polyps. Phil. Trans. Roy. Soc. (London) **B 240** (1957) 495.

Martinson, A. K., Pavlov, K. B. (Мартинсон, А. К., Павлов, К. Б.): К вопросу о пространственной локализации тепловых возмущений в теории нелинейной теплопроводности с поглощением. Ж. Вычис. Матем. и Матем. Физики **15** (1975) 891–905.

Meinhart, H., Gierer, A.: Generation and regeneration of sequences of structures during morphogenesis. J. theor. Biol. **85** (1980) 425–450.

Melville, J. G., Denli, N.: Fluid mechanics of longitudinal contractions of small intestine. Trans. ASME, J. Biomech. Eng. **101** (1979) 284–288.

Merzhanov, A. G. (Мержанов, А. Г.): Тепловые волны в химии. В кн. Тепло- и массообмен процессов горения. ОИХФ АН СССР, Черноголовка 1980, 36–58.

Meshcheryakov, V. N., Belousov, L. V. (Мещеряков, В. Н., Белоусов, Л. В.): Пространственная организация дробления. ВИНИТИ АН СССР, Москва 1978.

Mikhailov, A. S.: Effects of Diffusion in Fluctuating Media: A noise-induced phase transition. Z. Phys. **B 41** (1981) 277–282.

Mikhailov, A. S., Krinsky, V. I.: Rotating spiral waves in excitable media: The analytical results. Physica **9 D** (1983) 346–371.

Mikhailov, A. S., Uporov, I. V. (Михайлов, А. С., Упоров, И. В.): Спиральные волны и ведущие центры в модели Тьюринга. ДАН СССР **249** (1979) 733. Индуцированный шумом переход и переколяционная задача для флуктуирующих сред с диффузией. ЖЭТФ **79** (1980) 1958–1972. Кинетика гетерогенной цепной реакции со случайной во вариацией центров размножения. Жур. Физ. Хим. **56** (1982) 606–609.

Mishchenko, E. F., Rosov, N. Ch. (Мищенко, Е. Ф., Розов, Н. Х.): Дифференциальные уравнения с малым параметром и релаксационные колебания. Наука, Москва 1975.

Mitropolskii, Yu. A. (Митропольский, Ю. А.): Метод усреднения в нелинейной механике. Наукова Думка, Киев 1971.

Morgalev, Yu. N., Demchenko, I. T. (Моргалев, Ю. Н., Демченко, И. Т.): Пространственное распределение кровотока и PO_2 в коре головного мозга. Физиологический журн. **65** (1979) 985–990.

Mornev, O. A. (Морнев, О. А.): Об условиях возбуждения одномерных автоволновых сред. В кн. Автоволновые процессы в системах с диффузией. ИПФ АН СССР, Горький 1981, 92–98.

Murray, J. D.: Lectures on Nonlinear-Differential-Equations. Models in Biology. Clarendon Press, Oxford 1977.

Nagumo, I., Yoshizawa, S., Arimoto, S.: The active pulse transmission line simulating nerve axon. Proc. IEEE **ct – 12** (1965) 400.

Nicolis, G., Prigogine, I.: Self-Organization in Nonequilibrium Systems. A Wiley-Interscience Publ. John Wiley & Sons, New York/London/Sydney/Toronto 1977.

Noble, D.: A modification of the Hodgkin-Haxley equations applicable to Purkinje fibre action and pace-maker potential. J. Physiol. (Engl.) **160** (1962) 317–352.

Ohba, M., Sakamoto, Y., Tomita, T.: The slow wave in circular muscle of quinea-pig stomach. J. Physiol. No. 3 (1975) 505—516.

Orlov, R. S. (Орлов, Р. С.): Физиология гладкой мускулатуры. Медицина, Москва 1967.

Ortoleva, P., Ross, J.: On variety of wave phenomenon in chemical reaction. J. chem. Phys. **60** (1974) 5090—5107.

Ostrovskii, L. A., Yachno, V. G. (Островский, Л. А., Яхно, В. Г.): Формирование импульсов в возбудимой среде. Биофизика **20** (1975) 489.

Ovsyanikov, G. A., Kuzmin, L. S., Likharev, K. K. (Овсяников, Г. А., Кузьмин, Л. С., Лихарев, К. К): Взаимная синхронизация в много контактных джозефсоновских структурах. Радиотехника и электроника **27** (1982) 1613—1619.

Panfilov, A. V., Pertsov, A. M. (Панфилов, А. В., Перцов, А. М.): Спиральные волны в активных средах. Реверберaтор в модели Фитц Хью-Нагумо. В кн. Автоволновые процессы в системах с диффузией. ИПФ АН СССР, Горький 1981, 77—84.

Pertsov, A. M., Khramov, R. N., Panfilov, A. V. (Перцов, А. М., Храмов, Р. Н., Панфилов, А. В.): Резкий рост рефрактерности при подавлении возбудимости в модели Фитц-Хью. Новый механизм действия антиаритмиков. Биофизика. **26** (1981) 1077—1081.

Pikovskii, A. S., Sbitnev, V. I. (Пиковский, А. С., Сбитнев, В. И.): Стохастичность в системе связанных осцилляторов. Препринт ЛИЯФ-641, Ленинград 1981. Стохастичность колебаний в системе связанных осцилляторов. Дифференциальные уравнения **17** (1981) 2104—2105.

Polak, L. S., Mikhailov, A. S. (Полак, Л. С., Михайлов, А. С.): Процессы самоорганизации в физико-химических системах. Наука, Москва 1983.

Polyakova, M. S. (Полякова, М. С.): О диссипативных структурах в одной модели химических реакций с диффузией. Вестник Мос. Унив. сер. физики и астрономии. **15** (1974) 643—648.

Polyakova, M. S., Romanovskii, Yu. M. (Полякова, М. С., Романовский, Ю. М.): Математическая модель автоколебательной химической реакции с жестким режимом возбуждения в одномерном пространстве. Вестник Моск. Унив. сер. физики и астрон. **11** (1971) 441—447.

Polyakova, M. S., Romanovskii, Yu. M., Sidorova, G. A. (Полякова, М. С., Романовский, Ю. М., Сидорова Г. А.): О синхронизации автоколебательных химических реакций, протекающих в пространстве. Вестник Мос. Универ. сер. физики и астрономии, (1968) № 6, 86—89.

Prigogine, I.: Non-equilibrium Statistical Mechanics. John Wiley & Sons, N.Y. — London 1962.

Rabinovich, M. I. (Рабинович, М. И.): Автоколебания распределенных систем. Изв. вузов Радиофизика, **17** (1974) 477.

Regirer, S. A., Rutkevich, I. M. (Регирер, С. А., Руткевич, И. М.): Волновые движения жидкости в трубках из вязкоупругого материала. Волны малой амплитуды. Изв. АН СССР, Мех. жидкости и газа № **1** (1975) 45.

Reshodko, L. V., Bureš, J.: Computer Simulation of Reverberating Spreading Depression in a Network of Cell Automata. Biol. Cybernet **18** (1975) 181.

Reznichenko, P. N. (Резниченко, П. Н.): Преобразование и смена функций в онтогенезе низших позвоночных животных. Наука, Москва 1982.

Romanovsky, Yu. M., Stepanova, N. V., Chernavsky, D. S., Kinetische Modelle in der Biophysik. VEB Gustav Fischer Verlag, Jena 1974.

Romanovskii, Yu. M., Stepanova, N. V., Chernavskii, D. S. (Романовский, Ю. М., Степанова, Н. В., Чернавский, Д. С.): Математическое моделирование в биофизике. Наука Москва 1975. Математическая биофизика. Наука, Москва 1984.

Romanovskii, Yu. M. (Романовский, Ю. М.): Процессы самоорганизации в физике, химии и биологии. Знание, Москва 1981.

Romanovskii, Yu. M., Khors, N, P. (Романовский, Ю. М., Хорс, Н. П.): Модель механических автоколебаний в плазмодии миксомицета. Биофизика, **27** (1982) 707—710.

Romanovskii, Yu. M., Chernyaeva, E. B., Kolinko, V. G., Khors, N. P. (Романовский, Ю. М., Черняева, Е. Б., Колинько, В. Г., Хорс, Н. П.): Математические модели подвижности протоплазмы. В кн. Автоволновые процессы в системах с диффузией. ИПФ АН СССР, Горький 1981, 202—219.

Romanovskii, Yu. M., Sidorova, G. A. (Романовский, Ю. М., Сидорова, Г. А): О затухании колебательных химических реакций с учетом диффузии. В кн. Колебательные процессы в биологических и химических системах/ Ред. Франк, Г. М. Наука, Москва 1967, 258 до 266.

Rosenblum, L. A., Starobinets, I. M., Yakhno, V. G. (Розенблюм, Л. А., Старобинец, И. М., Яхно, В. Г.): О поведении неподвижного волнового фронта. В кн. Автоволновые процессы в системах с диффузией. ИПФ АН СССР, Горький 1981, 107—116.

Rössler, O. E.: Chaotic behavior in simple reaction systems. Z. Naturforsch. **31a** (1979), 1168.

Rozanov, N. N. (Розанов, Н. Н.): Гистерезисные и стохастические явления в нелинейных оптических средах. Изв. АН СССР, серия физическая, **46** (1982) 1886—1897.

Rutkevich, I. M. (Руткевич, И. М.): Волновые движения жидкости в трубках из вязкоупругого материала. Стационарные нелинейные волны. Изв. АН СССР, серия Механика жидкости и газа № 4, (1975) 86.

Samarskii, A. A. (Самарский, А. А.): Теория разностных схем. Наука, Москва 1977.

Samarskii, A. A., Elenin, G. G., Zmitrienko, N. V., Kurdyumov, S. P., Mikhailov, A. P. (Самарский, А. А., Еленин, Г. Г., Змитриенко, Н. В., Курдюмов, С. П., Михайлов, А. П.): Горение нелинейной среды в виде сложных структур. ДАН СССР **237** (1977) 1330—1337.

Samarskii, A. A., Zmitrienko, N. V., Kurdyumov, S. P., Mikhailov, A. P. (Самарский, А. А., Змитриенко, Н. В., Курдюмов, С. П., Михайлов, А. П.): Тепловые структуры и фундаментальная длина в среде с нелинейной теплопроводностью и объемными источниками тепла. ДАН СССР **227** (1976) 321—324.

Sbitnev, V. I. (Сбитнев, В. И.): Стохастичность в системе связанных вибраторов. В кн. Нелинейные волны. Стохастичность и турбулентность. ИПФ АН СССР, Горький 1980, 46—56.

Scott, A.: The electrophysics of a nerve fibre. Rev. mod. Phys. **47** (1975) N 2.

Scott, A. C.: Active and nonlinear wave propagation in electronics. Wiley-Interscience, John Wiley & Sons, New York/London/Sydney/Toronto 1970.

Scott, A. C., Chu, F. Y. E., McLaughlin, D. W.: Soliton: A New Concept in Applied Science. Proc. IEEE, **61** (1973).

Sedykh, E. I. (Седых, Е. И.): Модель непрерывной нейронной среды. Автоматика и телемеханика № **11** (1973) 62—69.

Selfridge, O.: Studies on flutter and fibrilation. Arch. Inst. Cardiology Mexico **18** (1948), 177 bis 187.

Sendov, B. Kh. (Сендов, Б. Х.): Математические модели процессов деления и дифференцировки клеток. Мос. Унив., Москва 1976.

Serbinov, I. A., Kalafaty, Yu. D., Ryabova, L. A. (Сербинов, И. А., Калафати, Ю. Д., Рябова, Л. А.): Диссипативные структуры при фазовом переходе полупроводник-металл. Письма в ЖТФ **6** (1980) 196—200.

Shchaban, V. M. (Щабан, В. И.): Электрофизиология гипокампа. Успехи физиологических наук **7** (1976) 57.

Shibata, M., Bureš, J.: Optimum topolographical conditions for reverberating cortical spreading depression in rats. J. Neurobiol. **5** (1974) 107—118.

Schimansky-Geier, L., Ebeling, W.: Stochastic theory of nucleation in nonequilibrium bistable reaction systems. Ann. Phys. **40** (1983) 10.

Schimansky-Geier, L., Mikhailov, A. S., Ebeling, W.: Effect of Fluctuation on Plane Front Propagation in Bistable Nonequilibrium Systems. Ann. Physik **40** (1983) 277.

Shkadinskii, K. G., Ivleva, T. P., Merzhanov, A. G. (Шкадинский, К. Г., Ивлева, Т. П., Мержанов, А. Г.): Математическая модель спинового горения. ДАН СССР **239** (1978) 1086—1088.

Shkadinskii, K. G., Khaikin, B. I., Merzhanov, A. G. (Шкадинский, К. Г., Хайкин, Б. И., Мержанов, А. Г.): Распространение пульсирующего фронта экзотермической реакции в конденсированной фазе. Физика горения и врыва № 1 (1971) 19–28.

Shnol, S. E. (Шноль, С. Е.): Физико-химические факторы эволюции. Наука, Москва 1979.

Skobeleva, I. M. (Скобелева, И. М.): Модель сосудистого тонуса (численный эксперимент) Механика композитных материалов, № 1 (1980) 107.

Smolyaninov, V. V. (Смолянинов, В. В.): Математические модели биологических тканей. Наука, Москва 1980.

Smolyaninov. V. V., Bliokh, Zh. L. (Смолянинов, В. В., Блиох, Ж. Л.): Механизмы движения фибробластов. В кн. Немышечные формы подвижности. ОНТИ НЦБИ АН СССР, Пущино 1976, 5.

Solitons in Action, Ed. *K. Lonngren* and *A. Scott*, Academic Press, New York/San Francisco/ London 1978.

Stazobinets, I. M., Yakhno, V. G.: One-dimensional Autowaves. Methods of Qualitative Description in Self-Organization. Autowaves and Structures for from Equilibrium (Ed. V. I. Krinsky). Springer-Verlag. Berlin/Heidelberg/New York/Tokyo 1984.

Stern, C. D., Goodvin, B. C.: Waves and periodic events during primitive streak formation in the chick. J. Embryol. Exper. Morphology **41** (1977) 15–22.

Stratonovich, R. L. (Стратонович, Р. Л.): Избранные вопросы теории флуктуаций в радиотехнике. Советское радио, Москва 1961 (Engl.: Pergamon Press 1963).

Stratonovich, R. L.: Nonlinear Nonequilibrium Thermodynamics Nauka, Moscow 1986 (in Russian); Springer-Verlag 1987.

Stratonovich, R. L., Romanovskii, Yu. M. (Стратонович, Р. Л., Романовский, Ю. М.): Параметрическое воздействие случайной силы на линейные и нелинейные колебательные системы. Научные докл. Высшей Школы № 4 (1958) 221–225.

Strelkov, S. P. (Стрелков, С. П.): Введение в теорию колебаний. Наука, Москва 1964.

Svirezhev, Yu. M., Gigauri, A. A., Razzhevaikin, V. N. (Свирежев, Ю. М., Гигаури, А. А., Разжевайкин, В. Н.): Волны в экологии. В кн. Нелинейные волны. Самоорганизация. Ред. Гапонов-Грехов, А. В. и Рабинович, М. И., Наука, Москва 1983, 32–46.

Talanov, V. I. (Таланов, В. И.): Стимулированная диффузия и кооперативные эффекты в распределенных кинетических системах. В кн. Нелинейные волны. Самоорганизация. Ред. Гапонов-Грехов, А. В. и Рабинович, М. И., Наука, Москва 1983, 47–56.

Teplov, V. A., Beilina, S. I., Evdokimov, M. V., Prieszhev, A. V., Romanovskii, Yu. M. (Теплов, В. А., Евдокимов, В. М., Бейлина, С. И., Приезжев, А. В., Романовский, Ю. М.): Автоволновые механизмы внутриклеточной подвижности. В кн. Автоволновые процессы в системах с диффузией. ИПФ АН СССР, Горгкий 1981, 190–201.

Tikhonov, A. N. (Тихонов, А. Н.): Системы дифференциальных уравнений, содержащих малые члены при производных. Математ. сб. **31** (1952) 575.

Tolubinskii, E. V. (Толубинский, Е. В.): Исследование по теплопроводности. Наука и техника, Минск 1967.

Tomita, T.: Electrophysiology of mammalian smooth muscle. Progr. Biophys. and Molec. Biol. **30** (1975) 185–203.

Tsanev, R., Sendov, Bl.: Possible molecular mechanisms for cell differentiation multicellular organisms. J. Theor. Biol. **30** (1971) 337–383.

Tuckwell, H. C.: Predictions and properties of a model of pottassium and calcium ion movements during spreading cortical depression. Internat. J. Neuroscience **10** (1980) 145–164.

Tuckwell, H. C., Miura, R. M.: Mathematical model for spreading cortical depression. Biophysical J. **23** (1978) 257.

Tuckwell, H. C., Hermansen, C. L.: Ion and transmitter movements during spreading cortical depression. Internat. J. Neuroscience **12** (1981) 109–135.

Turing, A. M.: The chemical basis of the morphogenesis. Proc. Roy. Soc. B **237** (1952) 37–71.

Vacquier, V. D.: Dynamic changes of the egg cortex. Developm. Biol. **84** (1981) 1—26.

Vasilev, V. A. (Васильев, В. А): Стационарные диссипативные структуры. и Динамические диссипативные структуры. В кн. Термодинамика биологических процессов, Ред. Зотин, А. И., Наука, Москва 1976, 186—197, 198—203.

Vasilev, V. A., Polyakova, M. S. (Васильев В. А., Полякова, М. С.): Некоторые математические модели ведущих центров. Вестник Моск. Универ. сер. физики и математики **16** (1975) 99—106.

Vasilev, V. A., Romanovskii, Yu. M. (Васильев, В. А., Романовский, Ю. М.): Связанные автоколебательные системы как модели ведущего центра. В кн. Теоретическая и экспериментальная биофизика. Вып. 5, Калининград 1975, 73—87. О роли диффузии в системах с автокатализом. Теоретическая и экспериментальная биофизика. Вып. 6, Калининград 1976, 73—87.

Vasilev, V. A., Romanovskii, Yu. M.: Stable Orbitals and Transitions in the Models of Biological Systems. In Book: Thermodynamics and Kinetics of Biological Processes. Ed. I. Lamprecht and A. I. Zotin, Walter de Gruyter and Co., Berlin/New York 1982, p. 473—483.

Vasilev, V. A., Romanovskii, Yu. M., Chernavskii, D. S. (Васильев, В. А., Романовский, Ю. М., Чернавский, Д. С.): Элементы теории диссипативных структур: связь с проблемами структурообразования. В кн. Математическая биология развития, Ред. Зотин, А. И. и Преснов, Е. В., Наука, Москва 1982.

Vasilev, V. A., Romanovskii, Yu. M., Yakhno, V. G. (Васильев, В. А., Романовский, Ю. М., Яхно, В. Г.): Автоволновые процессы в распределенных кинетических системах. УФН **128** (1979) 626—666.

Vasilev, V. A., Romanovskii, Yu. M. (Васильев, В. А., Романовский, Ю. М.): Два класса процессов самоорганизации и проблема их моделирования. В кн. Термодинамика и регуляция биологических процессов, Ред. Зотин, А. И. и Преснов, Е. В., Наука, Москва 1984.

Vasilev, V. A., Zaikin, A. N. (Васильев, В. А., Заикин, А. Н.): Волновые режимы в реакции окисления броммалоновой кислоты броматом, катализируемой комплексными ионами железа. Кинетика и катализ **17** (1976) 903.

Veinberg, M. M., Trenogin, V. A. (Вайнберг, М. М., Треногин, В. А.): Теория ветвления решений нелинейных уравнений. Наука, Москва 1969.

Velarde, M. G., Antorans, I. C.: Phase transition picture of the soret-driven convective instability in a two-component liquid layer heated from below. Phys. Letters **72A** (1979) 123—127.

Volkenstein, M. V. ((Волькенштейн, М. В.): Общая биофизика. Наука, Москва 1978.

Volodin, Yu. E., Beibutyan, V. M., Barelko, V. V., Merzhanov, A. G. (Володин, Ю. Е., Бейбутян, В. М., Барелко, В. В., Мержанов, А. Г.): „Кризис" теплообмена в системе платина-гелий. ДАН СССР **264** (1982) 604—607.

Volpert, A. I. (Вольперт, А. И.): Волновые решения параболических уравнений. Препринт ОИХФ АН СССР, Черноголовка 1983.

Volpert, V. A., Volpert, A. I., Merzhanov, A. G. (Вольперт, В. А., Вольперт, А. И., Мержанов, А. Г.): Применение теории бифуркаций к исследованию спиновых волн горения. ДАН СССР **262** (1982) 642—644. Анализ неодномерных режимов горения методами теории бифуркаций. ДАН СССР **263** (1982) 918—921.

Volpert, A. I., Ivanova, A. N. (Вольперт, А. И., Иванова, А. Н.): О диффузионной неустойчивости и диссипативных структурах в химической кинетике. В кн. Автоволновые процессы в системах с диффузией. ИПФ АН СССР, Горький 1981, 33—45.

Wiener, N., Rosenblueth, A.: The mathematical formulation of conduction of impulses in a network of connected excitable elements specifically in cardial muscle. Arch. Inst. Cardiologia de Mexico **16** (1946) N 3—4.

Winfree, A. T.: Spiral waves of chemical activity. Science **175** (1972) 634.

Winfree, A. T.: The geometry of biological time. Springer-Verlag, Berlin/Heidelberg/New York 1980.

Wohlfarth-Bottermann, K. E.: Contraction phenomena in Physarum: New results. Acta Protozool. **18** (1979) 59—73. Oscillatory. contraction activity in Physarum. J. exper. Biol. **81** (1979) 15—32.

Yakhno, V. G. (Яхно, В. Г.): Об одной модели ведущего центра. Биофизика **20** (1975) 669. О расчете скорости в возбудимой среде. Биофизика **21** (1976) 547.

Yakhno, V. G. (Яхно, В. Г.): Нестационарные процессы в одномерной возбудимой среде. 1У. Особенности распространения импульсов в среде с немонотонной зависимостью *a*(*v*). Биофизика **22** (1977) 8. Распространение импульсов через гладкую неоднородность возбудимой среды. Биофизика **23** (1978) 654.

Yakhno, V. G. (Яхно, В. Г.): Автоволновые процессы в одномерных релаксационных системах. В кн. Автоволновые процессы в системах с диффузией. ИПФ АН СССР, Горький 1981, 46–76.

Yakhno, V. G., Goltsova, Yu. K., Zhislin, G. M. (Яхно, В. Г., Гольцова, Ю. К., Жислин, Г. М.): Нестационарные процессы в одномерной возбудимой среде. Ш. Сложные режимы в автоколебательной возбудимой среде. Биофизика **21** (1976) 1067.

Yoneda, M., Kobeyakawa, K., Kubota, H. I., Sakai, M.: Surface contraction wave in amphibian eggs. J. Cell Sci. **54** (1982) 35–46.

Zaikin, A. N. (Заикин, А. Н.): Неустойчивость и распространение возбуждения в модели каталитической реакции. Биофизика **20** (1975) 616–620, 772–776.

Zaikin, A. N., Kokoz, Yu. M. (Заикин, А. Н., Кокоз, Ю. М.): Неустойчивость и распространение волн в модели каталитической реакции. Биофизика **22** (1977) 113–116.

Zaikin, A. N., Kawczynski, A. L.: Spatial effects in active chemical systems. I. Model of Leading Center. J. Non-Equilibrium Thermodyn. **2** (1977) 39.

Zaikin, A. N., Morozova, T. Ya. (Заикин, А. Н., Морозова, Т. Я.): Распространение возбуждения в активной одномерной среде с участком неоднородности триггерного характера. Биофизика **24** (1979) 124–128.

Zaikin, A. N., Zhabotinsky, A. M.: Concentration Wave Propagation in Two-dimensional Liquid-phase Self-oscillating Systems. Nature **225** (1970) 535.

Zaikin, A. N., Zhabotinskii, A. M. (Заикин, А. Н., Жаботинский, А. М.): Распространение концентрационных волн в двумерной жидкофазной автоколебательной системе. Журн. Физ. Химии **45** (1971) 265.

Zanin, A. M., Kiryukhin, D. P., Barkalov, I. M., Goldanskii, V. I. (Занин, А. М., Кирюхин, Д. П., Баркалов, И. М., Гольданский, В. И.): Твердотельные низкотемпературные превращения инициированные механическим разрушением. Письма в ЖЭТФ **33** (1981) 336.

Zanin, A. M., Kiryukhin, D. P., Barelko, V. V., Barkalov, I. M., Goldanskii, V. I. (Занин, А. М., Кирахин, Д. П., Барелко, В. В., Баркалов, И. М., Гольданский, В. И.): Инициирование и самоускорение низкотемпературных химических реакций при механическом разрушении облученных твердых образцов. ДАН СССР **260** (1981) 1397.

Zaraiskii, A. G., Belousov, L. V., Labas, Yu. A., Badenko, L. A. (Зарайский, А. Г., Белоусов, Л. В., Лабас, Ю. А., Баденко, Л. А.): Исследование клеточных механизмов ростовых пульсаций у гидроидных полипов. Онтогенез **15** (1984) 163–170.

Zaytsev, A. A., Dzherpetov, Kh. A. (Зайцев, А. А., Джерпетов, Х. А.): Слоистый высокочастотный разряд в инертных газах. ЖЭТФ **23** (1953) 516.

Zeldovich, Ya. B., Barenblatt, G. I., Librovich, V. B., Mikhviladze, G. M. (Зельдович, Я. Б., Баренблатт, Г. И., Либрович, В. Б., Михвиладзе, Г. М.): Математическая теория горения и взрыва. Наука, Москва 1980.

Zeldovich, Ya. B., Kompaneets, A. S. (Зельдович, Я. Б., Компанеец, А. С.): К терии распространения тепла при теплопроводности, зависящей от температуры. В кн. К 70-летиа А. Ф. Иоффе. Изд. АН СССР, Москва 1950, 61–71.

Zeldovich, Ya. B., Frank-Kamenetskii, D. A. (Зельдович, Я. Б., Франк-Каменецкий, Д. А.): Теория равномерно распространяющегося пламени. Журн. Физ. Химии **12** (1938) 100–105. К теории равномерно распространяющегося пламени. ДАН СССР **19** (1938) 693–698.

Zeldovich, Ya. B., Malomed, B. A. (Зельдович, Я. Б., Маломед, Б. А.): Сложные волновые режимы в распределенных динамических системах. Изв. вузов Радиофизика **25** (1982) 591–618.

Zhabotinskii, A. M. (Жаботинский, А. М.): Концентрационные автоколебания. Наука, Москва 1974.

Zhabotinskii, A. M., Zaikin, A. N.: Autowave processes in a distributed chemical system. J. theor. Biol. **40** (1973) 45.

Zhukov, S. A., Barelko, V. V., Merzhanov, A. G. (Жуков, С. А., Барелко, В. В., Мержанов, А. Г.): К теории волновых процессов на тепловыделяющей поверхности при кипении жидкости. ДАН СССР **242** (1978) 1064—1066. Динамика перехода между пузырьковым и пленочным кипением в режиме бегущей волны. ДАН СССР **245** (1979) 94—97.

Zmitrienko, N. V., Kurdyumov, S. P., Mikhailov, A. P., Samarskii, A. A. (Змитриенко, Н. В., Курдюмов, С. П., Михайлов, А. П., Самарский, А. А.): Локализация термоядерного горения в плазме с электронной теплопроводностью. Письма в ЖЭТФ **26** (1977) 620—623.

Zubov, V. I. (Зубов, В. И.): Методы Ляпунова и их применение. Атомиздат, Москва 1961.

Zykov, V. S. (Зыков, В. С.): Моделирование волновых процессов в возбудимых средах. Наука, Москва 1984.

Zhislin, G. M., Yakhno, B. G., Goltsova, Yu. K. (Жислин, Г. М., Яхно, В. Г., Гольцова, Ю. К.): Нестационарные процессы в одномерной возбудимой среде. 1. Деление остановившегося фронта. Биофизика **21** (1976) 792.

Zubarev, D. N.: Nonequilibrium Statistical Thermodynamics, Plenum Press New York 1974 (Engl.); Nauka, Moscow 1971 (in Russian).

Index